Arc Welding Processes Handbook

Scrivener Publishing
100 Cummings Center, Suite 541J
Beverly, MA 01915-6106

Publishers at Scrivener
Martin Scrivener (martin@scrivenerpublishing.com)
Phillip Carmical (pcarmical@scrivenerpublishing.com)

Arc Welding Processes Handbook

Ramesh Singh

Scrivener
Publishing

WILEY

This edition first published 2021 by John Wiley & Sons, Inc., 111 River Street, Hoboken, NJ 07030, USA and Scrivener Publishing LLC, 100 Cummings Center, Suite 541J, Beverly, MA 01915, USA

For more information about Scrivener publications please visit www.scrivenerpublishing.com.

Wiley Global Headquarters
111 River Street, Hoboken, NJ 07030, USA

For details of our global editorial offices, customer services, and more information about Wiley products visit us at www.wiley.com.

Library of Congress Cataloging-in-Publication Data

ISBN 978-1-119-81905-9

Cover image: Double headed GMAW system provided by the author
Cover design by Russell Richardson

Set in size of 11pt and Minion Pro by Manila Typesetting Company, Makati, Philippines

Printed in the USA

Contents

List of Figures

List of Tables

Foreword

The book, "Arc Welding Processes Handbook", brings together salient knowledge of arc welding methods used primarily in the industry and especially in the oil patch. The information presented about the welding process is usable and emulates the presence of your own welding engineer. Covering such welding methods as SMAW, GTAW, GMAW, FCAW and SAW with details in materials and techniques. This book is useful to both new welders as well as experienced welders. In the book, Ramesh covers these welding processes, how they work, and dives into the electrical side of welding. Welding machines, Transformers, Generators, Invertors, AC, DC, Sq. wave, Sine wave currents, Rectifiers, SCRs, Diodes, etc., as current control methods, all these are presented in a way that is easy to understand the functions of various welding machines. Most common weldable materials are discussed with welding guidance given that includes Aluminum, Nickel, Carbon steels, Stainless steels, Precipitation Hardened steels, Duplex Stainless steels, and others. The book is super comprehensive, easy to follow, and a welcome addition to any welding engineer's bookcase. It is a truly great guide for any budding engineer or welder to help them master their skills.

David Ammerman
Project Director at Gulf Interstate Engineering, MME, Texas PE
30+ years past-member of ASME, and member of API Committee:
Pipeline Construction Voting Group

Preface

The book **Arc Welding Processes Handbook** has been developed to address the need of a vast majority of people who want to know about welding, some of them also want to weld as hobbyists, or carry their passion for welding to be a professional welder. The book can also be used as a reference by field engineers and managers responsible for welding and fabrication activities. The book uses several figures and illustration that are available in the public domain, yet wherever it could be identified, the credit has been assigned to the source.

The book will provide readers and practitioners of the profession with an understanding of nearly all aspects of arc welding. The book covers the theory, the principles of the processes, the equipment, and the techniques that would improve the competency in welding, for each welding process. A good number of tables and illustrations are included to accentuate the points as well as to give readers familiarity with things that may or may not be available in their work or school trade workshops.

Chapter 1 introduces the reader to all possible welding process, including arc welding, electric resistance welding etc.

The practice welding procedure (WPS) given especially in Chapter 2 on SMAW process should prove a good basis to start welding and develop into an experienced welder. From here, one can move forward with other processes using the practice welding procedures included in Chapter 3 on GTAW, as well as GMAW processes in Chapter 4. For those who want to start welding, they can start with settings in these procedures and preparations and make changes to develop their skills around them. But it is not necessary to strictly follow this sequence, if someone has already developed the skills in any other process and wants to move to any other process.

The skills required to master FCAW process in Chapter 5 almost mimics the basic skills of GMAW in Chapter 4, and once this process is mastered, moving forward with the FCAW process should not be difficult at all.

The process of SAW in Chapter 6 is very different and very few welding schools will have this process in house. For students to practice on it, in most cases it will have to be learned and mastered on the job. But the chapter on SAW process gives the reader abundant information and familiarity with the process that they can step up to the opportunity when it becomes available.

Included in Chapter 7 is the welding symbols and how to use them, to read those symbols on fabrication drawings and weld accordingly. Reading and understanding the language of welding is an important step in becoming a successful welding professional. The chapter also includes other miscellaneous but important information that would come handy to any welding professional. The most important information is the detailed description of welding symbols and how to use and read them.

This book is best used in a workshop where the reader can pick up the welding torch or holder and try to convert the words from the book to an actual weld.

Ramesh Singh
Katy, TX
June 2021

1
Introduction to Welding Processes

1.1 Synopsis

The chapter introduces the most common welding and joining processes, by discussing the acceptable definition of welding, and the elementary understanding of skill development steps required to be a welder or a welding machine operator.

1.2 Keywords

Joining processes, definitions, welding, arc welding, arc efficiency, heat, heat affected zone (HAZ), solidification.

1.3 Welding

When we speak of welding, various images comes to our minds. Depending on persons' knowledge and experience with the process that can be various, simple or complex. But one thing that can be common to all those images and pictures is that the process of joining two pieces of metal to create a useful object.

This establishes one aspect of the term welding, that is, that the welding is a metal joining process. Let us explore a little more about what is welding, and how it is different from other Joining processes?

There have been discussions and sometimes arguments on describing if welding is an art or a science. Mundane as it might appear the question is pertinent and, in my experience, some well-meaning experts often miss the point as to which part about the term "welding" they are referring about to support their arguments. Welding as the physical and practical part of joining two materials in most part is an art, it requires dexterity in hand, and hand-eye coordination to do a good job. However, the study of the heat and melt flow solidifications prediction, prediction of material behavior under heating and cooling cycles associated with the term welding is a science, an essential pat under the science of physics. Hence it is both an art, and a science of joining metals by use of adhesive and cohesive forces between metals by welding, brazing, and soldering some of these joining processes produce metallurgical bonds. Person with the balanced knowledge of both science, and art parts of welding is expected to do much better work on either side of the argumentative divide. Further we get into the depth of the study, the line of separation from art to physics starts to become more evident.

Ramesh Singh. Arc Welding Processes Handbook (1–12) © 2021 Scrivener Publishing LLC

Both process metallurgy and physical metallurgy is involved in welding. Welding is a unique metallurgical activity as it involves a series of metallurgical operations similar to metal production, like steelmaking and casting but in a rapid succession and on a very small scale. In science side of welding the thrust of the study is on the materials behavior during application of localized heat, and cooling and solidification physics.

1.4 Defining Welding

The AWS definition for welding is "a materials joining process which produces coalescence of materials by heating them to suitable temperatures with or without the application of pressure or by the application of pressure alone and with or without the use of filler material".

Welding is often compared in a very rudimentary way with casting. The comparison with casting involves the fact that in welding a volume of molten metal is solidified (cast) within the confines of a solid base metal (mold). The base metal may have been preheated to retard the cooling rate of the weld joint just as in casting molds are preheated to slow down cooling and reduce "Chilling" of the casting. Upon solidification, the weld deposit or casting can be directly put into service, as the welds are often used in as-welded condition or may be heat-treated or worked further on as required. However, such comparison is not an accurate depiction of welding process, nor it is a fair comparison. For example, in welding the base metal "mold" is part of the weld, unlike the mold of a casting, which is removed after solidification, so unlike casting process, what happens to the "mold" is of significant importance in welding. Unlike casting, in welding the solidification and the nucleation of weld metal takes effect on the basis of the base metal grain structure that is just adjacent to the molten metal of welding and a unique set of metallurgy is created in the base metal that is heated to above austenitic temperature range, this small band of base metal is called heat affected zone (HAZ).

Welding involves small area relative to the full size of the structure, the base material. Thus, a weld is a very small mass of metal, mostly two metals that are heated very rapidly by intense heat and cooled rapidly, this rapidly heated and cooled small area often overlap each other in succession to create yet another complex metallurgical condition. The dissipation of heat is by all three modes; Conduction, Radiation and Convection. Often the large surrounding mass of colder base metal is heated by conduction process, which is the major source of heat transfer from weld. The heating and after welding the cooling process are dynamic, equilibrium conditions are seldom seen in conventional welding operations, in fact welding conditions represent a great departure from equilibrium. That is the reason weld zones often display unusual and verity of structures and properties, all this within the confines of a very small area affected by welding process.

It is thus important that a welding personnel have a very good understanding of "Heat" in welding. The understanding of the heat generation and physics of welding are important steps in making of a good welding engineer, and it helps being a good welder as well.

Welding is carried out based on a well thought out and specific plan in order to attain the required material properties. Many regulatory and industrial specifications have well developed process to get the plan in activation. Such plans are called Welding Procedures, and a well laid out sequence of operation is established for the welding qualifications, of both the procedure's ability to meet required metallurgical and mechanical properties and also a welder' ability to repeatedly produce the quality of weld desired through that welding procedure. Following is a brief discussion on welding procedures and their role in welding application.

1.5 Welding and Joining Processes

There are number of different approaches to welding, some of them are near universal in their application to most common materials, and are capable of adjusting to number of variables to be used on different positions, and conditions, while others are very specific and are no so universal in their application. With the welding we have included some other material joining processes that are in fact not a welding process. These are very often encountered in the industrial environment, and are often demanded that an accomplished welder knows how to use these processes. The Figure 1.1 below shows various welding and joining process.

The Table 1.1 below list s various welding and joining processes grouped as per the mode of energy used for that specific welding process. The table also includes other joining process that do not use Electric as the source of energy for joining. And there is other that are distinguished by the way they transfer the molten metal in to the metals being joined.

1.6 Arc Welding

The arc welding group includes eight specific processes, each separate and different from the others but in many respects similar. An introduction to those basic arc welding processes is presented here for some of those most common first-generation arc welding processes. Note that further variations have been made in some of these processes, some of them are discussed further in the book, but there are others that are proprietary developments, the information is covered under copyright laws, hence details on these developments are not included in the book.

1.6.1 Carbon Arc Welding

The carbon arc welding (CAW) process is the oldest of all the arc welding processes and is considered to be the beginning of arc welding. The Welding Society defines carbon arc welding as "an arc welding process which produces coalescence of metals by heating them with an arc between a carbon electrode and the work-piece. No shielding is used. Pressure and filler metal may or may not be added. It has limited applications today, but a variation or twin carbon arc welding is more popular. Another variation uses compressed air to force molten metal out to effect cutting.

1.6.2 Shielded Metal Arc Welding (SMAW)

The development of the metal arc welding process soon followed the carbon arc. This developed into the currently popular shielded metal arc welding (SMAW) process defined as, an arc welding process which produces coalescence of metals by heating them with an arc between

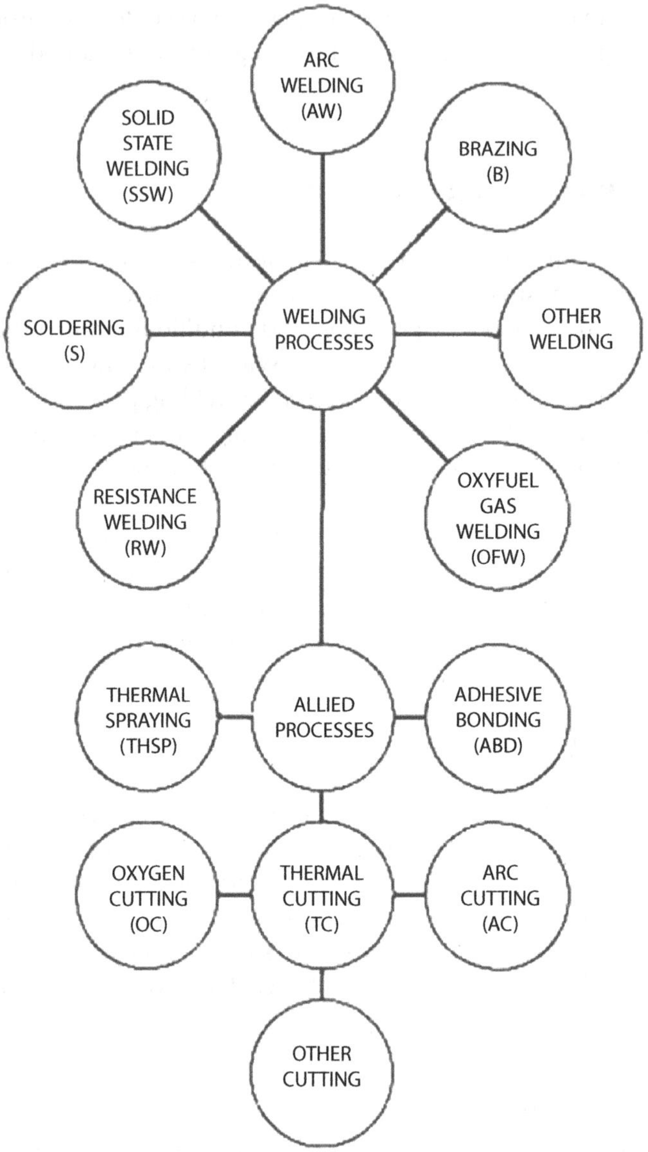

Figure 1.1 General lay out of welding and joining processes.

a covered metal electrode and the work-piece. Shielding is obtained from decomposition of the electrode covering. Pressure is not used and filler metal is obtained from the electrode.

1.6.3 Gas Tungsten Arc Welding (GTAW)

The need to weld nonferrous metals, particularly magnesium and aluminum, challenged the industry. A solution was found called gas tungsten arc welding (GTAW) and is defined as, an arc welding process which produces coalescence of metals by heating them with an arc between a non-consumable tungsten electrode, and the work piece. Shielding for the welding arc is obtained often from an inert-gas, or mixture gases that may not always be inert.

Table 1.1 Welding and joining processes, type of energy used, and their abbreviations as defined by the American Welding Society.

Group	Welding process	AWS letter designation
Arc Welding		
Electric Arc Welding	Carbon Arc	CAW
	Flux Cored Arc	FCAW*
	Gas Metal Arc	GMAW*
	Gas Tungsten Arc	GTAW*
	Plasma Arc	PAW**
	Shielded Metal Arc	SMAW*
	Stud Arc	SW
	Submerged Arc	SAW*
Electrical Resistance Welding	Flash Welding	FW
	High Frequency Resistance	HFRW
	Percussion Welding	PEW
	Projection Welding	RPW
	Resistance-Seam Welding	RSEW
	Resistance-Spot Welding	RSW
	Upset Welding	UW
Oxy-fuel Gas Welding (OFW)		
	Oxyacetylene Welding	OAW
	Oxyhydrogen Welding	OHW
	Pressure Gas Welding	PGW
Solid State Welding		
	Cold Welding	CW
	Diffusion Welding	DFW
	Explosion Welding	EXW
	Forge Welding	FOW
	Friction Welding	FRW
	Hot Pressure Welding	HPW

(Continued)

Table 1.1 Welding and joining processes, type of energy used, and their abbreviations as defined by the American Welding Society. (*Continued*)

Group	Welding process	AWS letter designation
	Roll Welding	ROW
	Ultrasonic Welding	USW
Capillary Action Transfer and Distribution of Metal		
Brazing	Diffusion Brazing	DFB
	Dip Brazing	DB
	Furnace Brazing	FB
	Induction Brazing	IB
	Infrared Brazing	IRB
	Resistance Brazing	RB
	Torch Brazing	TB
Soldering	Dip Soldering	DS
	Furnace Soldering	FS
	Induction Soldering	IS
	Infrared Soldering	IRS
	Iron Soldering	INS
	Resistance Soldering	RS
	Torch Soldering	TS
	Wave Soldering	WS
Other Welding Processes		
	Electron Beam	EBW
	Electroslag	ESW
	Induction	IW
	Laser Beam	LBW
	Thermit	TW

*Processes discussed in this book.
**Included with GTAW process.

1.6.4 Gas Metal Arc Welding (GMAW)

In the desire to increase the production rate, and widen the types of material being welded by one process the GMAW process was invented. Since its early days the process has gone through a number of improvements, and currently it is one of the most versatile welding processes among the arc welding processes. It has number of variants by the way the weld metal is deposited, and shielding gases used for various types of metal welding.

1.6.5 Submerged Arc Welding (SAW)

Earlier attempt to increase welding production lead the development of Automatic welding utilizing bare electrode wires in the early nineteenth century, but was not much popular primarily due to the open arc and the resultant quality of weld, which was always an issue.

The dissatisfaction with bare-wire welding and some innovative ideas lead to the development of the submerged arc welding (SAW) process, this was much better automated process and it made the automatic welding popular. Submerged arc welding is defined as "an arc welding process which produces coalescence of metals by heating them with an arc or arcs between a bare metal electrode or electrodes and the work piece. Pressure is not used and filler metal is obtained from the electrode and sometimes from a supplementary welding rod." It is normally limited to the flat or horizontal position.

1.7 Efficiency of Energy Use

Processes use electrical energy to initiate arc, but not all the arc energy is fully used to melt the metals being welded. There is significant energy loss in the process.

The process efficiency of various arc welding process differs significantly, based on number of factors like material being welded, type of gas being used if gas is used in the process. The effect of these variables depends on one or a combination of more factors.

The use of energy generated by a process is an important factor in determining how much current is needed to generate required heat for welding. The generation of heat also determines the effective use of a process for welding different materials. The table below gives the glimpses of Arc efficiency of various Arc Welding processes.

As is obvious, not all welding and joining process are equal, this leads to the fact that some are more versatile in usability for welding number of materials, while some are more specific to certain type of materials. The table below gives a general usability of various arc welding processes.

For example, GTAW process is nearly all the material listed, but SMAW and SAW processes stops short, and not suitable for welding Copper and alloys, or Titanium and its alloys. Table 1.2 is the generic information, while the Table 1.3 is more specific with numbers and includes more process and varients. Similarly Brazing is possible for nearly all material listed but soldering is not. The Table 1.4 blow presents a matrix that shows the ability and limits of various arc welding processes, the last two columns of the table show the applicability scope of soldering and brazing processes.

The Table 1.5 below is borrowed from another book Applied Welding Engineering Process, Codes and Standards. The table lists the electric arc process by the arc energy efficiency of each process. Note the highest efficiency of SAW process and the lowest in that of GTW process.

Table 1.2 Arc efficiency by welding process.

	Process	Arc efficiency
1	SMAW	Intermediate
2	GTAW	Low
3	SAW	High

1.8 Welding Procedures

A welding procedure is a statement of execution, a specific plan prepared by the welding contractor. The procedure details with listing of various variables associated with the proposed welding process giving an assurance that the resulting weld would guarantee that the required mechanical and metallurgical properties will be met. Any format of form may be used to develop a welding procedure giving essential details. Some international specifications especially addressing the welding requirements have developed a format for the purpose, AWS D1.1 has E-1 form for pre-qualified procedures, similarly ASME Section IX of Boiler and Pressure vessels code has a set of such forms for welding specifications, welding qualification records (PQRs) and welders' qualification records, they are numbered as QW- 482, QW- 483 and QW 484 respectively. Other international standards for welding are EN ISO 15609-1, EN ISO 15609-2, EN ISO 15609-3, EN ISO 15609-4, EN ISO 15609-5, and EN ISO 15614. Till the last revision, the EN ISO 15614 had 12 parts dealing with specific topics on welding various materials like Steel, Aluminum, Cast Iron, Titanium, Copper etc.

The plan details all essential and non-essential variables that are important to achieve the quality of weld. These variables are welding process specific. Some of these variables are discussed in this book. In ASME section IX, these variables are listed specific to the particular welding process, they are subdivided into essential, supplementary essential, and

Table 1.3 Shows the arc efficiency factors for various commonly used arc welding processes.

Welding process	Arc efficiency factor η	
	Range	Mean
Submerged Arc Welding	0.91 - 0.99	0.95
Shielded Metal Arc Welding	0.66 - 0.85	0.80
Gas Metal Arc Welding (CO_2 Steel)	0.75 - 0.93	0.85
Gas Metal Arc Welding (Ar Steel)	0.66 - 0.70	0.70
Gas Tungsten Arc Welding (Ar Steel)	0.25 - 0.75	0.40
Gas Tungsten Arc Welding (Ar Aluminum)	0.22 - 0.46	0.40
Gas Tungsten Arc Welding (He Aluminum)	0.55 - 0.80	0.60

Table 1.4 Indicates general limits of joining/welding processes that apply to the material listed in left column.

Material	Welding processes														Other joining processes	
	SMAW	SAW	GMAW	FCAW	GTAW	PAW	ESW	EGW	RW	OFW	DFW	FRW	EBW	LBW	B	S
Carbon Steel	x	x	x	x	x		x	x	x	x		x	x	x	x	x
Low alloy steel	x	x	x	x	x		x		x	x		x	x	x	x	x
Stainless steel	x	x	x	x	x		x		x	x	x	x	x	x	x	x
Cast Iron	x	x	x	x						x					x	x
Nickel and alloys	x	x	x		x	x	x		x	x					x	x
Aluminum and alloys	x		x		x	x	x	x	x	x	x	x	x	x	x	x
Titanium and alloys			x		x	x			x		x	x	x	x	x	
Copper and alloys			x		x	x						x	x		x	x
Magnesium and alloys			x		x				x			x	x	x	x	
Refractory alloys			x		x	x			x				x		x	

Table 1.5 Arc efficiency factor.

Welding process	Arc efficiency factor η	
	Range	**Mean**
Submerged Arc Welding	0.91 - 0.99	0.95
Shielded Metal Arc Welding	0.66 - 0.85	0.80
Gas Metal Arc Welding (CO_2 Steel)	0.75 - 0.93	0.85
Gas Metal Arc Welding (Ar Steel)	0.66 - 0.70	0.70
Gas Tungsten Arc Welding (Ar Steel)	0.25 - 0.75	0.40
Gas Tungsten Arc Welding (Ar Aluminum)	0.22 - 0.46	0.40
Gas Tungsten Arc Welding (He Aluminum)	0.55 - 0.80	0.60

nonessential variables. However, these variables are not specific to ASME but are in general agreement with welding technology.

Essential variables are those in which a change, as described in the specific variables, is considered to affect the mechanical properties of the weldments, hence any change shall require requalification of the welding procedure. The Supplementary essential variables are required for metals for which other Sections specify notch-toughness tests and are in addition to the essential variables for each welding process.

The Nonessential variables on the other hand are those in which a change, as described in the specific variables, may be made in the WPS without requalification.

Some special process like corrosion-resistant and hard-surfacing weld metal overlays may have different additional essential variables. Only the variables specified for special processes shall apply. A change in the corrosion-resistant or hard-surfacing welding process requires requalification.

The correct electrode diameter is one on of the variables, when used with the proper amperage and travel speed, produces a weld of the required size in the least amount of time. Selection depends on the thickness of the material being welded, the position of welding in relation to the gravity of the earth, and the type of joint to be welded. The welder's experience is also important since more skill is required to control the weld puddle in out of position welds, the different types of electrode coverings and fluxes, are important too. The inexperience may lead to poor quality welds that may have defects such as inclusions, porosities in the final welds.

Welding current can be either direct or alternating, depending on the process, type of electrode and available power supply and material being welded. DC provides a steadier arc and smoother transfer as well as good wetting action, and out of position control. Reverse and straight current polarities are used for specific applications. Reverse polarity produces deeper penetration and straight polarity produces higher electrode melting rates.

These topics are discussed in much detail in the subsequent chapters and in relation to specific welding process.

American Welding Society has developed the chart to describe all joining and allied process the chart above indicates those processes. Various welding process use different energy transfer modes, the table below groups those welding processes based on that.

1.9 Qualification of Welders and Operators

The skill of welders and welding machine operators is an essential step to ensure that only professionally skilled personnel are on the job. The qualification process also ensures that the welding personnel are continually upgraded in their skill and ability.

Welders are qualified on the essentials data established through a qualified welding procedure often referred as welding procedure specification or WPS. This welding procedure may comply with any or multiple construction specifications or codes.

The inherent intent of welders' qualification is to ensure (i) their ability to use the machine and (ii) that they can produce the weld, to match the essential data collected through the qualified welding procedures.

As a basic those essential data include the following;

- The material or material group being welded,
- The welding process,
- Thickness, and diameter of the material being welded,
- Joint design,
- Position of the weld, the plane in which the weld lies, and in case of tubulars, if it can be rotated for welding, or is in fixed position.
- Consumable type used,
- Electrical parameters used,
- Heating and cooling practices prior to, during, and after welding.

Other variables may be added to meet any specific requirements of the code or the job.

There are several options for the welder to get qualified, often the Local Union Halls have the Journeyman welder training, education and qualification program. These programs allow the aspiring welder to join the workforce as an apprentice, and gradually work their way up to become a fully qualified welder that is called Journeyman welder, through this route, a dedicated and regular person takes about 4 years to become a Journeyman welder, while still in the workforce.

Other than the union halls there are several state run or private trade schools in nearly all industrially active parts of the world, which train and qualify an aspiring welder to be able gain enough knowledge to make them enter the workforce. From there they develop themselves further and grow up the ladder. Several construction companies also have their own welding training schools that train and develop welders from among their own workforce.

The third way is the hardest way in which the aspiring person somehow gets an opportunity to weld, often as an apprentice with some welder, and from there just practices the trade whenever opportunity presents itself, seek guidance from the welder, and develop their skills. When ready, they qualify on the job, and grows from there.

In any case there is a lot of ground to cover from just able to weld to grow into a leading welder with specific expertise. The good news is that the goal is very much achievable and many committed personnel have made their career as welders.

2
Shielded Metal Arc Welding (SMAW)

2.1 Synopsis

This chapter introduces the Shielded metal arc welding process with the aim of giving basic understanding of the process and how it works, the goal of the chapter is to make a new entrant to the field knowledgeable enough to try to weld if they so choose, and gradually come to the entry point to the profession of welding.

The chapter includes the types of equipment required and choices available for the SMAW process. The safety required to work safely and necessary personal protective clothing associated with welding.

At the end of the chapter a section of questions is included to test the knowledge acquired from the chapter.

2.2 Keywords

Shielded metal arc welding, SMAW, power sources, constant current, constant voltage, alternating current, direct current, transformers.

2.3 Introduction

In the previous chapter we learned about welding in general, as a metal joining process. And we were also introduced to a group of welding processes thar are collectively called Arc Welding processes, because the required heat for melting the joining metals comes from an electric arc struck between the two electrical poles. The submerged arc welding (SMAW) process is one of that group called arc welding.

This is the most common welding process that an individual will see and come across. The process is also called by its unofficial name as "stick welding". It is also called manual metal arc welding or MMA welding in UK and many other countries.

For the sake of universality and consistency, in this book we will use the AWS nomenclature, Shielded Metal Arc welding (SMAW).

The process uses either alternating current (AC) or direct current (DC) to initiate arc for the welding. The arc generates heat in the range of 6500°F to 7000°F (3600°C to 3900°C), that heat is capable of melting most of the weldable material. The weld is completed by adding the molten droplets from the electrode passing through the arc column.

Ramesh Singh. Arc Welding Processes Handbook (13–114) © 2021 Scrivener Publishing LLC

Figure 2.3 A SMAW welder welding on a pipeline project.

2.4 Process Fundamentals

The process derives its name from how it works, the electric arc is created between the metal core of the electrode and the bare metal to be welded, and that arc is shielded from atmospheric contaminations by the gas and the slag created by the burning of the covering on the electrode, called flux. This completes the shielded metal arc welding process. Thus, the name Shielded Metal Arc Welding (SMAW). Some of these illurtrations are an image making devices for new entrants to welding example is Figure 2.4. Others are describded in the text with their subject for example Transformers that appears in the text for several times. And they are very easy to spot while reading the book. It is not worth pinning them down to one specific paragraph. Title says alot for their link to the subject in the text.

Shielded Metal Arc Welding (SMAW) process is one of the earliest arc welding processes and a versatile one, for welding ferrous and several nonferrous metals. The process is also called Manual Metal Arc (MMA) welding process in UK and some European countries. The process uses covered electrodes. An electrode consists of core metallic wire covered with silicate binders and other material that may include a combination of several chemicals including fluorides, carbonates, oxides, metal alloys and cellulose. The paste like consistency of the mixture is then extruded over the metal wire as core, the covering is then dried in an oven, and called coating. The electrode coating has several roles to play.

1. It works as an arc stabilizer,
2. It provides shielding from atmospheric contamination during molten state by evolving gases and during solidification by covering the weld metal with slag.

3. It provides scavengers, deoxidizers, and fluxing agents to cleanse the weld and prevent excessive grain growth in the weld metal.
4. It provides a slag blanket to protect the hot weld metal from the air and enhance the mechanical properties, bead shape, and surface cleanliness of the weld metal.
5. It is also a source of alloying elements to produce compatible weld metal.

2.5 How the Process Works

A SMAW process setup is shown in the Figure 2.4 above. Note that the power unit output could be either AC or DC, power sources are discussed further in the book, the electrode and the work piece are part of the full welding current flow circuit. The bare metal end of the electrode which is about 1 to 1.5 inches long, is clamped in the electrode-holder, the other end of the holder cable is connected to the power source. The second lead coming from the power source is connected to the work terminal.

The activity within the arc is depicted in the Figure 2.5 below, where the position of the electrode molten metal and its addition to the weld pool, solidified weld metal and slag are shown in the progressive sequence.

The arc is struck by bringing the electrode in contact with the work surface and then immediately pulling them apart about two to three mm (about 0.08 to 0.12 inch), thus

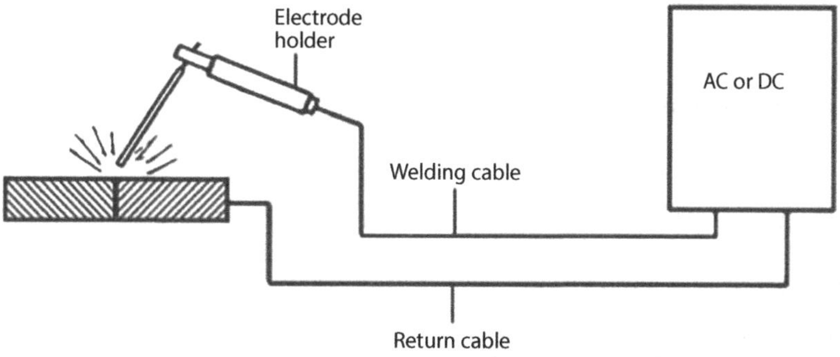

Figure 2.4 Typical SMAW setup.

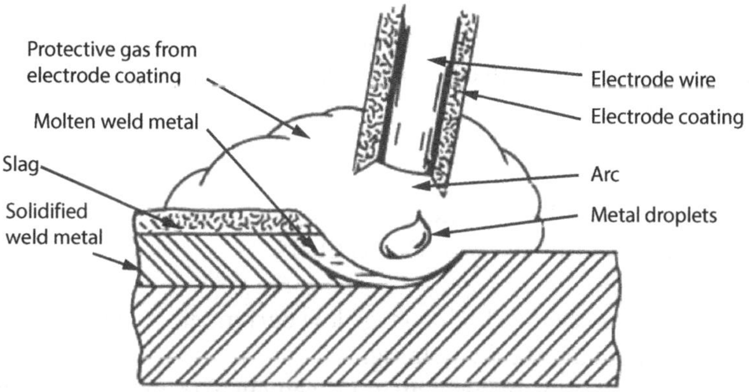

Figure 2.5 Welding arc action and various components of welding.

ionizing the gas between the two electrical ends. The resulting arc generates heat to simultaneously melt the work metal and the metal electrode. The coalescence of metals is produced by heat from an electric arc that is maintained between the tip of the covered electrode and the surface of the base metal in the joint being welded.

2.6 Power Sources

The welding machine for SMAW process are various and can be classified at the basic level on the basis of the type of current used. As alternating current (AC) power source or direct current (DC) power source.

However, more detailed description is required to properly identify power sources, as a minimum basic, the description of a power source must give at least the following three group of information.

- Type of power source,
- Whether it has a constant current or constant voltage,
- Whether it is AC or DC or both AC and DC.

Power sources are designed to produce either constant current or constant voltage. Current is measured in Amperes, and potential that is the voltage, is measured in Volts.

Alternating current power sources are further divided into following, by the type of equipment,

i. Transformers
ii. Motor or engine driven alternators,
iii. Invertors

The Direct Current power sources are following types of equipment.

i. DC rectifier transformers,
ii. Invertors,
iii. Power driven alternators with DC rectifiers,
iv. Motor or engine driven generators.

A third group of power source are also used, these are combined power sources that are capable of changing from alternative current (AC) to direct current (DC) output. These welding equipment are,

i. Invertors,
ii. Transformers with DC rectifier,
iii. Power driven alternators with DC rectifiers.

A transformer is used to change high voltage low current electricity into the low voltage but higher current output suitable for welding.

Addition of invertor to a transformer is an improvement to transformer method of getting welding current. The function of an invertor, in a transformer power source is to change the

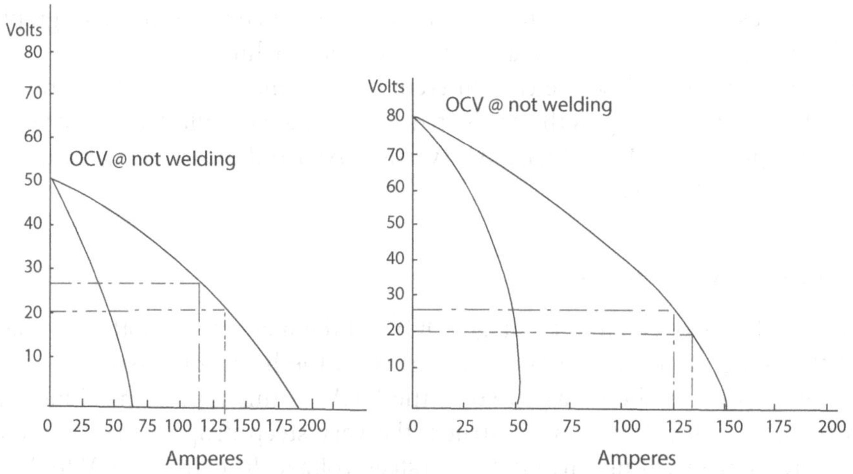

Figure 2.6 Above (2 graphs), graph 1 above, shows the volt-ampere curve, (output curve or slope) at lower stings. Graph 2 below, shows the volt-ampere curve, (output curve or slope) the steep slope of a "Drooper" type of constant current arc welder.

input AC to DC output, and then raise the current frequency through a very small but very efficient transformer built within, and produce (output) at very high frequency AC current.

Such machines are distinguished from Transformer type power source, as an independent power sources and are called just Invertors.

Another unique attribute of an invertor is that these are across the constant current or constant potential (voltage) classification of welding machines. They can be both.

Welding machines are manufactured to cover as wide attributes as possible to address the variety of job demands of a fabricator. For example, a machine may be capable of producing both AC and DC output. Another important way the machines are classified is the way the output current (welding heat) and voltage (length of welding arc) is manipulated, called the output slope, the way the open circuit volt and ammeter react as the arc is struck and welding progresses.

The following Figure 2.6 Graph 1, shows the Volt-Ampere curve, (Output curve or slope) at lower stings, while the graph 2 shows the constant current slope, which is also known as the drooper type slope.

Another factor that is part of the way a welding machine is properly described is the whether it is a constant current power source or a constant voltage power source. Now we know that the current varies thru the welding process, so the question arises as to what does the constant current mean?

This question is addressed in the subsequent description.

2.6.1 Constant Current and Constant Voltage Power Source

Power sources are not of much help if their output is not in control of the welder, or their output varies significantly with minor changes in the welders' dexterity, and the output is not reliable. In the two graphs in Figure 2.6 above, we see the characteristics of constant current machine's volt amperage curve from a constant current machine in graph 1. The graph 2 describes the flatter curve of a constant voltage machine.

We have two plots that describe two different types of machines. In both graphs we see the volt and current changing as the arc is struck, this change is plotted in the volt-ampere curves. Volts is plotted on the ordinate axis (vertical axis) and the amperage (Current) is on the abscissa (horizontal axis).

2.6.2 Constant Current Curve

In the first graph, the open circuit voltage (OCV) of the machine is plotted on the vertical axis, and the amperage drawn (while welding) is on the horizontal axis. We see that the OCV of the machine is shown as 80 volts, the OCV of this type of machines are often between 60 to 80 volts. As the arc is struck, the very steep drop in the voltage curve is observed to reach the welding current. Due to steep voltage drop, from OCV to the welding voltage of 20 volts to 25 volts, these types of machine and this curve is called the "drooper" or drooper-type machine. At the welding current, even when the voltage is varied by the welder from 20 volts to 25 volts the change in current is marginal from 136 amp, to 125 amps. It is about 6 to 7 % change this marginal change in current output, is what is considered as constant current output.

In the constant current machines, the welder is relatively secure from current increase, if he varies the arc length, his weld quality may not suffer significantly due to varying arc length.

This type of machine is preferred where manual welding of not so stringent quality is specified.

2.6.3 Constant Voltage Curve

The second graph, depicts a constant voltage type volt-amperes curve. This is another machine that has its OCV set to 50 volts, here at the same welding voltage of 20 volts to 25 volts the drop in current is significant, from 143 to 125 amps, which is about 13.3%. The curve is flatter, this slower sloping curve causes a lager change in the welding current (amperage), and this large change in current is associated with small change in the voltage. In this case the machine is able to hold the voltage constant, while allowing the current to change.

This type of machine would give the welder much better control over the weld pool, welder can control the electrode melt rate, an attribute much desired for out of position welding.

2.7 AC Power Sources

Alternating current welding machines are either the transformer and alternator type machines.

The transformers reduce the high voltage and low amperage supply electricity to high current to low voltage welding power. Figure 2.5 shows how the primary and secondary coils of a typical alternative current transformer are positioned. Transformers are constructed on a soft iron core, which is laminated. The iron core is used to build magnetic field, these iron-core are wound with a copper wire that carries the current. There are two such coils, called primary and secondary coils. Primary coils receive the primary (supply) current and the secondary coil outputs the changed (transformed) current for the welding. Primary coils have more copper wire winding turns, and the wire is thinner compared to the secondary coil, this is because the primary carries lower amperage - less current. The significance of more turns than secondary coil, is that the transformer is reducing the voltage and conversely increasing the amperage – the current. These transformers are aptly called the stepdown transformers. Number of coil turns indicate the force of the magnetic field created. In an alternating current this magnetic field collapse as the alternating cycles reverses its cycle, at this point the magnetic field collapses. Magnetic field is directly proportioned to the number of the turns the coil has.

The secondary coil has fewer turns of relatively thicker diameter copper wire.

These two coils are not connected by any means, they are set apparat from each other in one container, such that the magnetic field of the primary coil transfers to the secondary coil, as the current cycle reverses, and magnetic field collapses. The precise spacing of two coils plays very important role, if they are distanced too far apart, then not enough field from the primary coil will cross to the secondary core, reducing the current output.

The action of reversal and rebuilding of magnetic field induces the current in the secondary coil. This successive action occurs at 120 times per second, causes the current to flow from primary to the secondary coil.

2.7.1 The Alternator Type AC Welding Machines

These machines are different from transformer type welding machines discussed above, in that sense these machines generate their won electricity through a rotor, which may be driven by any hydrocarbon fuel, diesel or gasoline for example, or even by an electric motor driven by supply power source. The rotor is wrapped with multiple coils of copper wire and housed in a casing that has magnetic field. Thus, the rotating coils on the rotor create alternating current for welding.

The electrical parameters are very important variables for welding, hence the necessity to have better control on the current output cannot be over emphasized. The control in this context means the accuracy of current available for welding in various positions

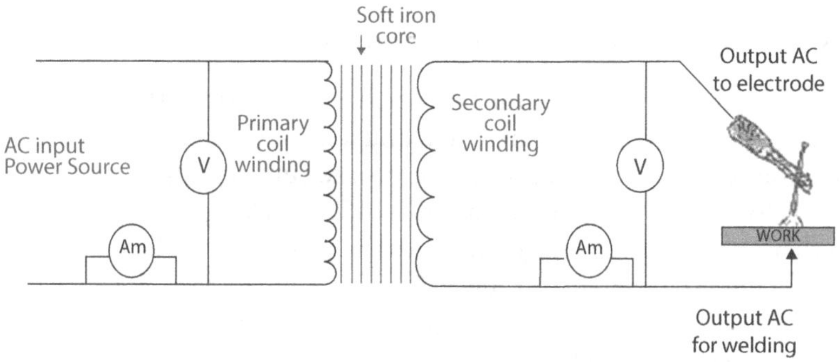

Figure 2.7 The schematic above shows the key components of an AC transformer.

and continuity of the available current for welding. The control for transformer type AC power sources is done either through the moving some parts of the system or through an electronic device fitted in the circuit. We can discuss them here for good understanding of the function of AC power system. The first three methods of control discussed below use the manual/mechanical means to vary and control output. The AC output is not limited by the frequency, it can be at any frequency.

Alternating current can also be obtained by a method called **Dual Source with inverter switching**. This machine uses inverter to deliver AC as the main welding current, a second power supply unit is included in the machine, by switching it, the machine supplies the direct current for welding. The second power supply adds its current to the main current flow to the DCEP part of the cycle. (The system is mainly used for the GTAW process).

Ac current output is generally very stable, for maintaining the line of the welding arc.

2.7.2 Movable Coil Type Control

This method is possibly the most used device used to control output from the transformer. The method uses the very principle of the transformer technology to vary the current output. In this approach the primary coil is moved farther away or brought closer to vary the output of the transformer. With the power on to the transformer, but machine is not open for welding the primary coil is moved farthest away from the secondary coil. Machine is set at the OCV, and minimum welding current flowing through the setting. As the adjusting liver is rotated to increase the current the primary coil is moved closer toward the secondary coil, this action increase the amperes flowing through the welding cable.

2.7.3 Movable Shunt Type Control

This method like the moving coil method, manipulates the basic principle of electricity and transformer to achieve required control. In this type of transformer both primary and the secondary cores are made stationary, the transformer has a movable shunt, this shunt is made out of same iron core as the main iron core. The shunt is moved in or out to control the power available for welding. When the shunt is fully withdrawn from the path of the magnetic flux, all flux lines are available for the flow of current, as the shunt is inserted in the path of the flow of magnetic flux it blocks certain flux lines and, free flow of current is restricted. The extent of magnetic flux blockage determines how much current is reduced.

2.7.4 Movable Core (Reactor) Type of Control

In this type of control system, the transformer has three coils. The usual two coils, the primary coil and the secondary coil are fixed, and a movable third coil is introduced. This movable third coil referred as the reactor, controls the current output.

The wound is to create a counter voltage, also called bucking voltage. If the movable coil is turned on then its inductive voltage increases the resistance of the secondary coil of the transformer, reducing the current output. When the inductive reactor (coil) is turned off, the output from the secondary coil of the transformer is in its full capacity.

Apart from the above some other transformers use the tapping devices to the secondary coil to mark stepped stages for the control of current output.

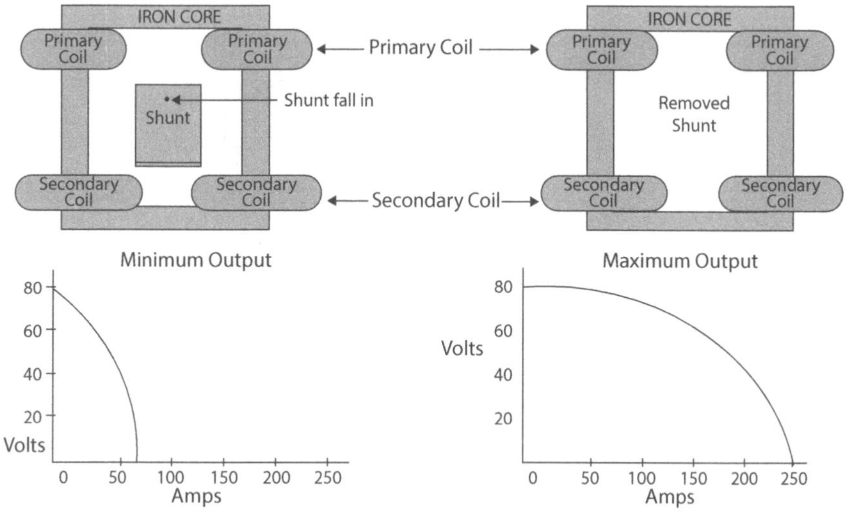

Figure 2.7.3 Schematic of a movable shunt type transformer control.

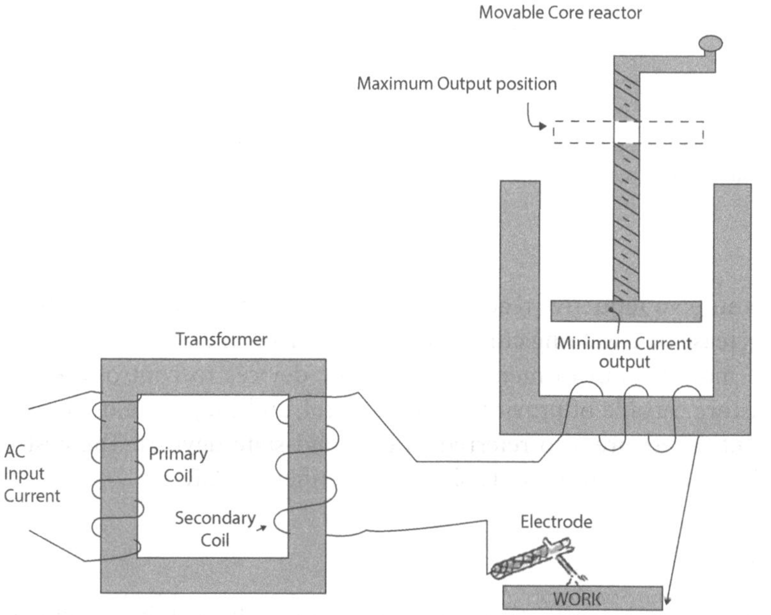

Figure 2.7.4 A schematic of a movable coil reactor, the position of the reactor coil causes the inductive reactance of the secondary output coil resulting in the variance in current output.

2.7.5 Magnetic Amplifier Method of Current Control

Yet another method of current control is the magnetic amplifier approach. This method uses the welding current coils and diodes in series with control coils. The load coil is used to assist the control coil to increase the magnetic field of the cores. The high magnetic field

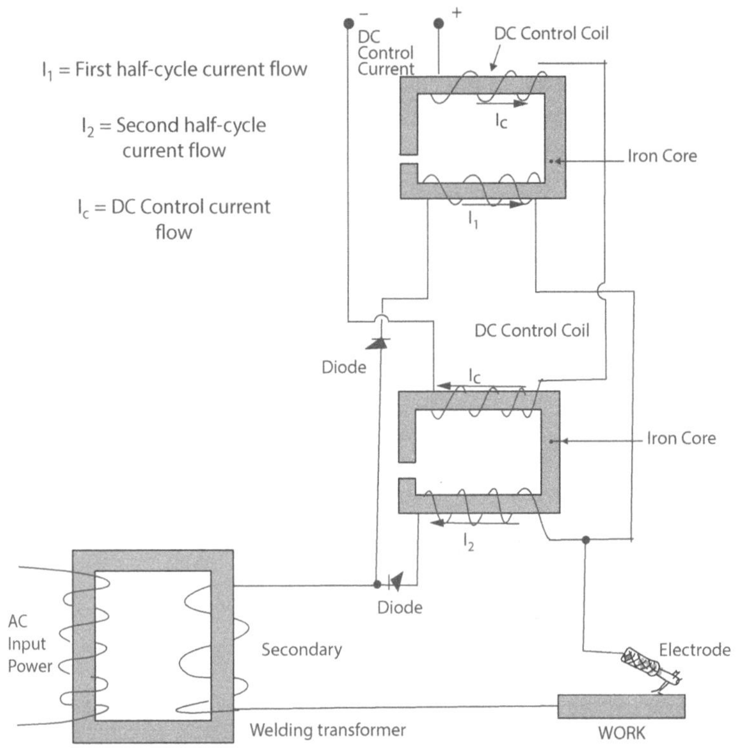

Figure 2.7.5 A magnetic amplifier transformer output control, the diode allows the current to flow in one direction, and this allows a remote control operation possible.

in the cores cause an inductive reactance in the secondary welding current. The increased reactance decreases the welding current from the transformer.

The more modern transformers use electronic devices to control the current output. These devices are capable of providing finer control, and they are more prevalent in modern equipment. These are also referred as the solid-state devices. The best example of a solid-state devices is the diode, and solid-state rectifier or SCR.

2.7.6 Diode

Diode is an electronic device that controls the directional flow of the current, it is like a one-way valve, a one-way electrical current flow device. The current can flow through the diode in one direction only, it does not allow the current to flow in the opposite direction. Diodes are used to convert alternating current to direct current.

Since the flow of current is controlled in one direction, the diodes are useful in converting AC to DC. As the current leaves the transformer, through the diode in the circuit. The alternative current entering into the rectifier is changing direction 120 times per second, the current exiting the rectifier through the diode is in one direction only, and it has changed to direct current.

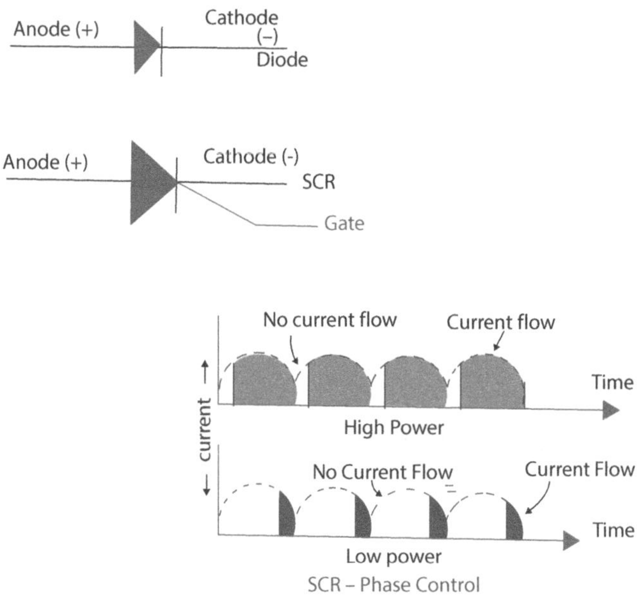

Figure 2.7.6 The top portion of the figure shows the use of diodes - shown in Red color, and it compare it with Silicon controlled rectifiers (SCRs).

2.7.7 Silicon-Controlled Rectifiers (SCRs)

Silicon-controlled rectifiers or SCRs, are also the devices that control the directional flow of current similar to the diodes, with one very important difference. SCRs get in to action only when they are "switched on". If they are not activated, they are not working as a direction control device. The SCR has a switch called Gate, when it is turned on it allows the current flow in the proper direction. Once the SCR is turned on the only way it can be turned off is by reversing the current direction or by stopping the current flow. As the reversing action is activated the SCR stops current flow from the wrong direction, as a result of this current flow from each direction is stopped. The current flow is restored only when the current starts to flow in the correct direction, and the Gate receives a signal to allow the current to flow.

SCRs are placed in the secondary circuit of arc welding power source. By turning on the SCR early in each half-cycle, more current can be obtained, conversely if the turn-on is programmed at the latter in each half-cycle, then less current is derived. The method of turning the SCR at different times allows the control of secondary current. This approach eliminates the need to move primary or the secondary coils, as in manual and mechanical approaches to current control. The method is very effectively used for controlling the transfer output. The way the electrical circuit is built it allows precise control over SCRs, which in turn allows accuracy in the welding current control. Use of SCRs allows the control of current accuracy required for the (i) pulse welding current, (ii) high initial current to start arc, followed by a study and lower current for the continued welding, (iii) programmed increase and decrease of current in out of position welding, where control over weld pool is extremely important. SCRs are used as current control device in constant current, or constant voltage devices and they are also used in inverter machines.

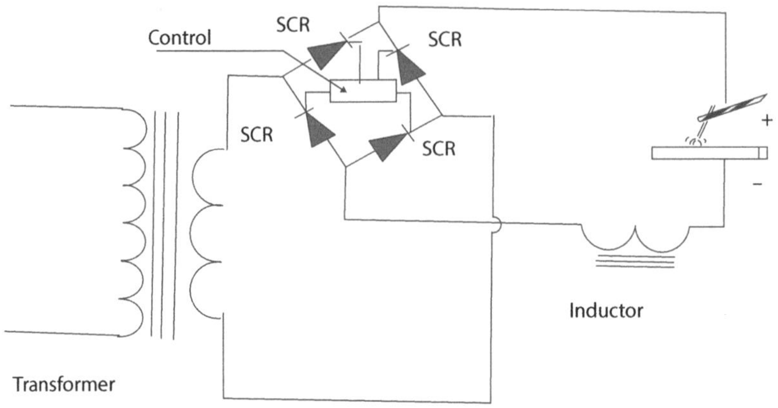

Figure 2.7.7 A schematic drawing of single-phase DC power source with SCR bridge control.

2.7.8 Transistors

Another solid-state device used for current control is a transistor. The application of transistors is mainly in very high-frequency-controlled over 10,000 Hz, inverters and other solid-state controlled power sources.

A transistor is similar to the SCR since it also controls current flow in one direction only, similar to SCRs it can turn the current on at different times. Then what is the difference? The difference is that a transistor can turn the current off without the need to reverse the current direction as is required for SCR. It can also allow different amounts of current to flow. The current amount flow is related to how much signal is applied to the transistor. More current would flow if more signal is applied, low current will allow low current to flow, and no signal to transistor will result in no current flow.

2.8 Direct Current Power Sources

The transformer-rectifiers and the generator types are the direct current power source machines. In the transformer rectifier type machine, there are two distinct sections, one is the transformer and the second is the rectifier, together these two make the DC power source.

The transformer section receives the supplied line voltage and current in either 220 volts, 440 volts, and at 60 Hz cycles. The transformer converts the AC line voltage and current to the welding current and voltage of 60 volts to 80 volts, also called open circuit voltage, and the welding current varies to the design of the equipment, and it could be several hundred amperes.

As this low-voltage and high-amperes current exits the transformer it enters in the rectifier section of the machine.

In the rectifier the AC changes to DC. A direct current constant-current output transformer rectifier may be single phase or a three-phase power source.

The rectifiers use devices that are called diodes to convert alternative current to direct current. In diodes the current can flow in one direction only, it does not allow the current to flow in the opposite direction.

Since the flow of current is controlled in one direction, through the diode in the circuit. The alternative current entering into the rectifier is changing direction 120 times per second, the current exiting the rectifier through the diode is in one direction only, and it has changed to direct current.

Figure 2.8.1 above shows the schematic of single-phase bridge type rectifier - diodes allow the current flow in one direction shown with arrows, this allows the change of AC current to DC current.

Figure 2.8.2 below shows the schematic of a three-phase bridge type rectifier - diodes allow the current flow in one direction shown with arrows, this allows the change of AC current to DC current.

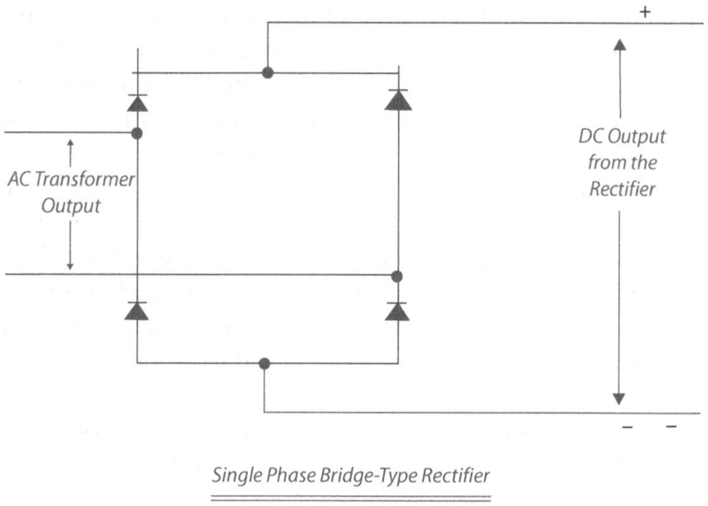

Figure 2.8.1 Shows the schematic of single phase bridge type rectifier.

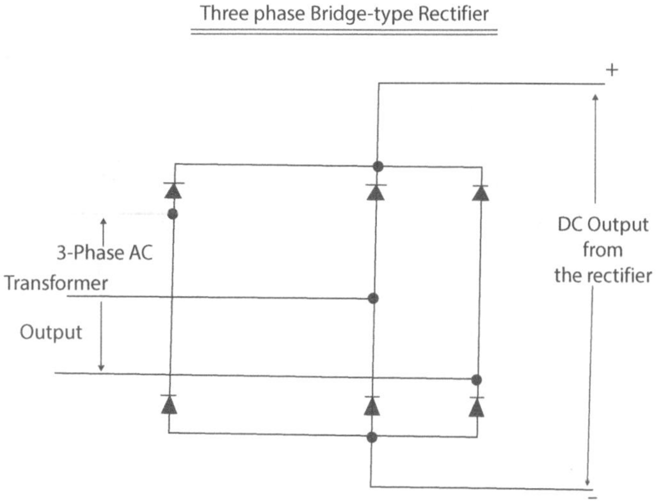

Figure 2.8.2 Three phase bridge-type rectifier.

2.8.1 Generator

The direct current is produced through a generator. Another option is to use an AC alternator with a rectifier. The construction of a generator is relatively simple as compared to an AC alternator with a rectifier.

The generator consists of a rotating armature shaft, that is wound with several independently wound coils. With ends of each armature coil soldered to a copper terminal called a commutator, the commutators use kind of half shells wrap on to the armature shaft, the externals of the half-shell commutator are in contacts with carbon or copper piece, that are called Brushes. The two brushes are positive (+) and negative (-) terminals. These brushes are in touch contact with commutator. Encasing the armature windings are stationary (not rotating) wire wound magnets, these are called field windings. The field windings are wires wrapped around an iron core. In this encloser of magnets, the armature is rotated by means of external power, either an electric motor, or a gas or diesel driven engine. As the armature winding rotates it cuts through the magnetic field of the field winding, this cutting of the magnetic field induces current in the armature coil. As the armature wire comes in the horizontal position is enters a position that it does not cut the magnetic field, creating a holiday of current induction, this occurs twice in one full rotation. This induction of current only when the armature is in vertical position in its rotation, induces current in only one direction, making it the direct current. The induced current is picked up from the commutator by the brushes. More the number of armature coils, field windings and the brushes, better is the flow of direct current. The voltage of a generator is varied by making changes in the current and the field windings. Welding generators are designed to produce low voltage and higher current output.

The study flow of current is essential not only for the efficiency of a welding generator but it is the very essence of the welding system. This consistency of output is ensured by

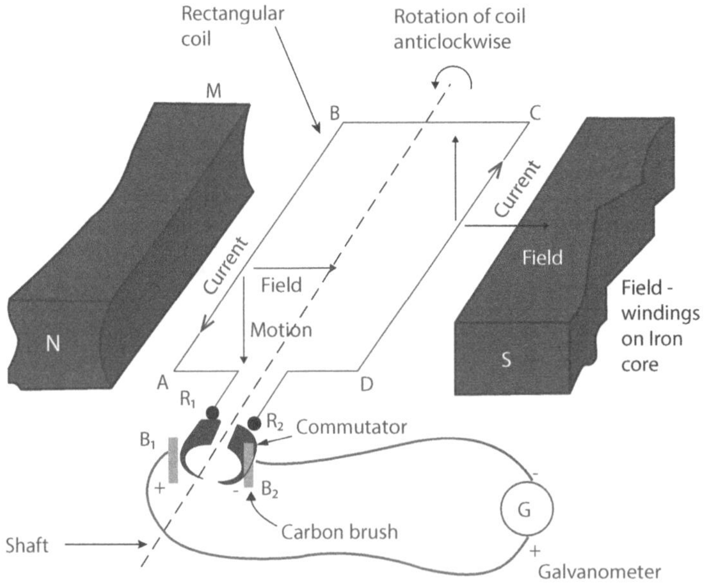

Figure 2.8.1.1 Schematic diagram of a DC generator.

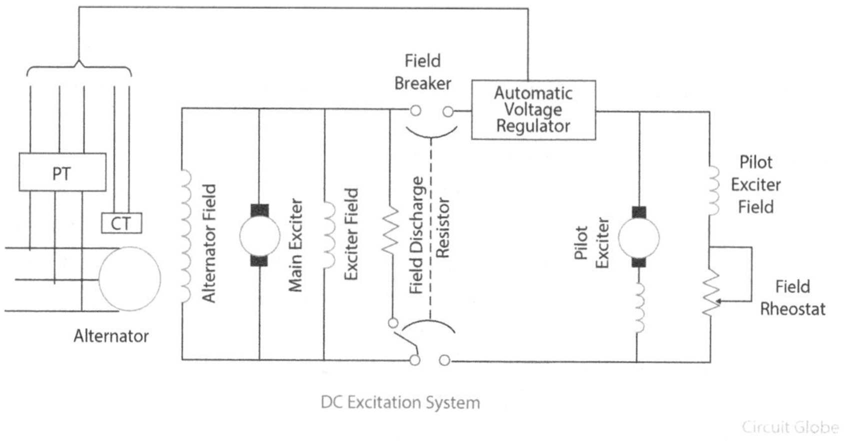

Figure 2.8.1.2 Circuitry of an exciter system.

introducing a small motor connected to the field winding, to ensure constant voltage on the main fields and prevent reversal of polarity. These independently powered motors are called exciter, see the Figure 2.8.1.2 for the circuitry of an excitation system.

2.8.2 Alternator

As we know that the alternator produces alternative current, so a rectifier is used to change the AC to direct current. The direct current will flow in one of the two direction. It may flow from the welding machine to the electrode, across the arc gap and return back to the machine through the workpiece lead, or it may flow from machine to workpiece and through the electrode back to machine. The direction of flow of direct current from machine to electrode through workpiece and back machine is called polarity, and in that the electrode negative or direct current electrode negative abbreviated as DCEN. Some people also call it direct current straight polarity and abbreviate it as (DCSP), however DCEN universally is more acceptable term.

Alternatively, the direction where the position of electrode and the work place is reversed, will flow the current flow direction. This is direct current electrode positive or DCEP. Alternative name is direct current reverse polarity or DCRP. Here too, the DCEP is universally accepted and understood term.

As stated above, the polarity can be changed by changing the work lead and electrode positions, however some welding machines have switch, to change the polarity as desired for specific welding task.

Further developments in welding machines have given options of selecting either an alternative or a direct current output from the same source. An AC transformer fitted with a rectifier in series in the circuit allows the selection of the desired current type for the specific welding task.

There are other machines that are called invertor arc welding power sources, these alternative current output machines are an improvement over the transformer-rectifier machines. These are more current in technology and can produce high frequency current in the rage of 1 kHz to 50 kHz across the main transformer. Whereas normal output frequency is 50Hz to 60 Hz alternative current. This improvement in frequency increase allows the welding machines to be reduced in size by about 60% to 70 %. Figure 2.8.3 below describes the stages of conversion and resulting wave forms that take place in an inverter type welding machine.

The step-by-step description of how an inverter works.

The inverter process starts with first step where input bridge rectifier is installed its function is to convert alternative current to direct current. The input 50Hz or 60 Hz AC may be one or three-phase. This direct current is passed through the inverter switcher that consists of series of SCRs or transistors that are turned on quickly. They chop the DC in to very high high-frequency square-wave alternating current. Single or multiple number of SCRs may be used to direct the one direction flow of the DC current. The chopping of DC current very rapidly by successive SCRs produces high frequency AC current. The SCRs can develop frequencies up to 10m kHz, when even higher frequencies are desired the SCRs are replaced with transistors, they can produce frequencies up to 50 kHz.

This higher frequency wave AC is now used as input to the step-down transformer. This input is high-voltage, low-current AC power. The output of this transformer is low-voltage high current AC power. At this point an output bridge rectifier is used to change the AC into DC for welding. In the final the device called inductor is used to smooth out the DC output to make is more adoptable for high quality welding processes.

One of the highlights of this type of welding machine is the feed-back control, this eliminates the need for two welding machines one with constant voltage output, and another for constant current output. Feed-back control allows the one power source to perform multiple welding and cutting functions, like SMAW, GMAW, GTAW, or plasma arc cutting, etc.

Due to electronic circuit and use of SCRs and Transistors, and high frequency power output, these machines are small in size, and light weight, easy to move and mange on work sites. The reduction in the size of the machine is related to the reduction of the size of the transformer. The product of the number of the transformer coils and cross-sectional area

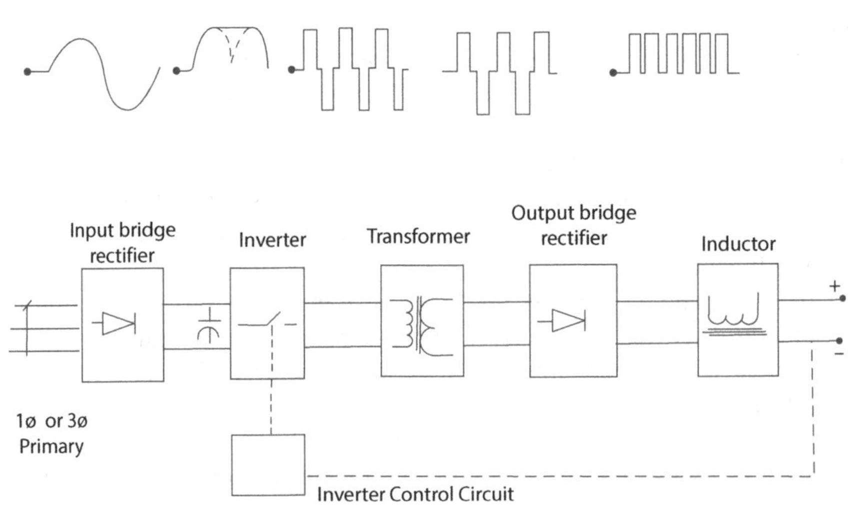

Figure 2.8.3 Current conversion and resulting wave forms.

of the transformer core is the product of invertor voltage divided by the product of the flux density of core material and the operating frequency of the invertor.

If the number of turns in the transformer coil (N), and invertor voltage (V), and the flux density of the core are held constant, then the cross-sectional area of the transformer (A) can be reduced, while the operating frequency (*f*) is increased. Transformer being the largest and the heaviest part of the system the reduced size of transformer results in machine being smaller and lighter. An advantage that is highly appreciated by the welders, especially more appreciated for site work. A dual source with inverter switching is described in the Alternating current power source.

Sometimes, the DC arc has tendency to wander the direction, if this happens and keeping a straight arc line becomes difficult. This phenomenon is unique to DC current, because it flows in one direction as compared to AC current that frequently changes the polarity, thus canceling the magnetic field created. The phenomenon of wandering DC arc called arc-blow. Arc blow affects the quality of weld as it pulls the molten metal away from the weld line.

There are several corrective measures that the welder can take to avoid the bad impact of arc blow in welding.

i. Connect the grounding cable as far away the weld line as possible.
ii. Move the ground connector at the end of the weld.
iii. Reduce the welding current.
iv. Wrap the welding cable 2 to 3 times around the weld line, and in the direction that would counter the magnetic field.
v. Use AC welding machine, with suitable electrode.

2.8.2.1 Power Source Remote Control

The diversity in terms of materials, and positions of welds are invariable, welders require more controls in their hands to control the arc, and weld pool for a good quality weld, in fact not just good quality welds but even just to weld. Arc behavior varies with number of factors that may include, type of electrode covering for SMAW welding, the current delivery, the position of the weld, and finally and most important factor is the distance from the machine to the location of the weld and the welder. All these challenges demand that the welder is provided with controls that they are able to adjust the current, the arc length, and subsequent weld pool for successful welding.

Most welding machines supplied these days include a small, portable remote-control panel that helps distance current adjustment. This remote control allows the welder to turn on or off the machine, adjust the current demand for specific type of welding conditions.

2.8.3 Installation of Welding Machines

The welding power sources are often connected to the one phase or three phase power supply points. The power factor that these welding machines develop, disturb the power supply if these machines are connected to the same circuit. This requires that other machines connected to the same supply circuit are provided with some power factor correction devices of their own. This is done by connecting a capacitor to give these machines a boost of power to improve their power factor. This arrangement requires the careful planning and consideration and should be done with the expert analysis and to comply with the local electrical code.

2.8.3.1 Cooling System for Welding Power Sources

Welding machines get hot in their operation, not due to the welding heat, but due to the operation within the welding machines itself. Most welding machines are naturally air cooled, smaller machines get the gravity feed air to flow through the machine to prevent them getting from overheated.

The lager machines need some help for the air to flow through all parts of the machine to keep them cool, they use forced air circulation. An electrical motor is used run a fan that provides the air to cool the machine.

For effective cooling, the air passageway should be designed for free flow of the air from entry to exit ends. Periodic maintenance of the duct and openings must be inspected, and cleaning of dust and any other dirt from the system, reduces the heating and overheating of the machine.

2.8.3.2 Welding Connections – Welding Cable and Electrode Holders

Welding cables used to carry current from the welding machine to the work and back are also called welding leads, these leads are super flexible large diameter electrical cables. The Figure 2.8.3.1 below shows various types of welding cables.

The lead that is connected to the electrode-holder and often carries current from the machine to the electrode are called electrode lead. The cable that is connected to the work place often by a clamp, is called ground cable or workpiece lead.

The need to be very flexible to meet the demands of welding activities, they also need to be very well protected since they are carriers of heavy current. The flexibility and insulation are provided by thick rubber covering which is often supported by a layer of reinforcement, by woven fabric layer to provide some rigidity and protection from damage.

Welding leads are produced to be flexible so as to reduce the strain on welders' wrist and hands. The flexibility is achieved by use of about 800 to 2500 strands of fine copper or aluminum wires, wrapped in one single bundle as a cable. Copper leads are more suitable for carrying higher currents, this attribute of copper cable also reduces the diameter cables. But copper cables are heavier that aluminum cables. Aluminum cables are lager in diameter as compared to the copper cable for carrying same current capacity, this is because aluminum can carry only up to 61% current. But the advantage of aluminum is that these cables are lighter in weight.

Figure 2.8.3.1 Copper and aluminum welding leads: note the number of fine wires that compose a cable, and the rubber sheathing that covers them.

The electrical capacity is the most key difference between application of copper and aluminum cable. The alloy mix is also determined by the intended use of the welding cables. Copper is considered a better conductor with a higher capacity per volume. However, aluminum has higher capability per weight. The weight difference also is determined by the specified material used. The Table 2.8.3.1 below gives the current capacities of various types of welding cables.

The changes caused by the metals thermal cycle is more prominent in aluminum than that is for copper. These changes are significant in aluminum due to its thermal growth coefficient, compared to copper.

The leads are produced in various sizes, and identified by universal numbering, the number indicates the diameter of the lead, which in turn indicates the current carrying capacity of the cable. Larger the number thicker the diameter and lower the current carrying capacity. This is clearly brought out in the table below. The table below also brings out another factor, which is the drop in current as the length of the cable is increased. The length shown in the table includes the length of electrode lead and the workpiece lead. So, the point to note here is that, the reduction in lead diameter, and the increase in the lead length reduces the current capacity of the welding lead.

The corresponding drop in voltage is very low, if all connections are tight and secured, the drop in a copper cable is about 4 volts.

2.8.4 Electrode Holders

Electrode holders are the end point of the welding cable, where a clamp like device is encased in an insulated handle for the welder to grip it, while the clamp holds the welding

Table 2.8.3.1 Welding lead and their capacity.

Significance of cable (welding lead) diameter and length and current carrying capacity					
	Lead diameter		**Cable length**	**Cable length**	**Cable length**
Lead no.	**Inch**	**MM**	**Amperes**	**Amperes**	**Amperes**
4/0	0.959	24.4	600	600	400
3/0	0.827	21.0	500	400	300
2/0	0.754	19.2	400	350	300
1/0	0.720	18.3	300	300	200
1	0.644	16.4	250	200	175
2	0.604	15.3	200	195	150
3	0.568	14.4	150	150	100
4	0.531	13.5	125	100	75

Note the drop in current as the length of the lead increases.

Figure 2.8.3.2 Different types of SMAW electrode holders.

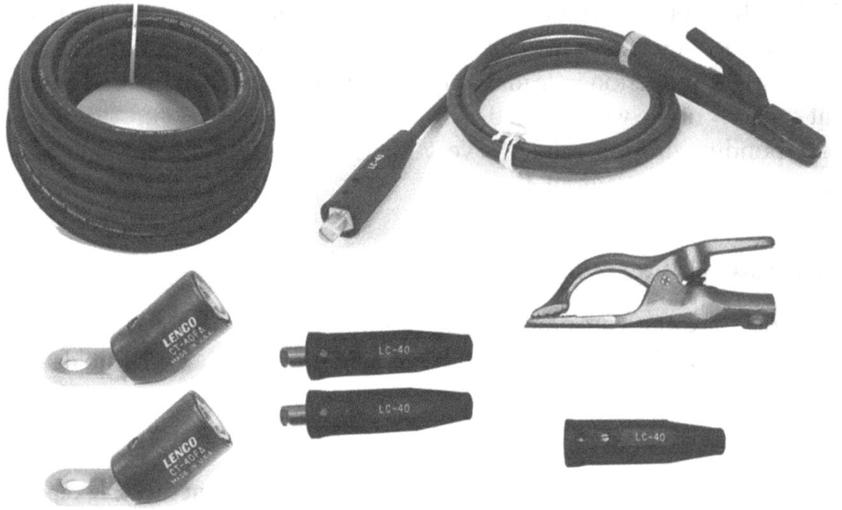

Figure 2.8.4.1 Various types of cable connectors, and ground clamp. Pictures courtesy of LENCO® catalogue.

electrode. Clean end of the covered electrode is held in the jaws of the electrode holder. The clean end establishes the electrical continuity, essential for striking arc for the welding.

Heavy duty electrodes are often the clamps with spring loaded grip handle that allows the jaw like end to open and electrode end is placed her for secured holding.

2.8.5 Arc Welding Power Source Classification by NEMA

Welding machines are electrical machines, so they covered under the guidelines of the National Electrical Manufacturers Association (NEMA). NEMA id a trade association of electrical machine manufacturers.

NEMA classifies the welding machines primarily on the basis of their rated duty cycle output. The NEMA classifications are given below.

1. NEMA Class I: Machines that deliver 60%, 80%, or 100% duty cycles are classified in this group.

2. NEMA Class II: Machines that deliver output at 30%, 40%, and 50% are classified in this group.
3. NEMA Class III: Machines that deliver output at 20% duty cycle are grouped in this class.

The arc welding machines are described by their following three attributes.

1. Rated current output,
   This is the amount of current measured in amperes that a welding machine is rated to supply at a given voltage. NEMA rated output current for different NEMA class rating described above is in the following table.

2.8.5.1 *Duty Cycle*

The term duty cycle is referred and used when speaking about a welding machine, and rating, its usability.

Duty cycle is defined as the following.

The length of time that a welding machine can be used continually at its rated output, in any 10-minute period.

Most welding machines are not in use 100% of the time, welding is stopped for loading or unloading of the weldment, or for inspection and cleaning of the weld etc. Normally machines are used at 60% duty cycles. However semi-automatic and automatic processes are required to operate at 100% duty cycle. And these machines are developed at 100% duty cycles.

Most of the hobby welding machines are rated for 20% duty cycles.

Rated output current		
Class I	**Class II**	**Class III**
200	150	18-230
250	175	235-295
300	200	
400	225	
500	250	
600	300	
800	350	
1000		
1200		
1500		

Figure 2.8.5.1 NEMA rating.

2.8.5.2 Power Requirement

The rated load or the rated voltage or welding voltage are the term often used for power requirement of a welding machine. For the Class I and Class II machines, the rated load voltage is the product of a constant and the rated amperage. The constant for machines rated up to 500 A is 20+0.04 x rated amperes. This formula if applied to a machine that is for 400 A machine will give the rated load voltage of 36 volts.

The rated load voltage for machines with current rating of 600 A, and higher is determined to be 44 volts.

2.9 Welding Safety and Personal Protecting Equipment

Welding activity is a rough environment; it is fraught with possibilities of accident and severe injuries to welders and people working around them. If not protected, the dangers include possibilities of cuts, from sharp metal edges, burn from hot metal surfaces, heat and radiation from welding arc, and dangerous visible and invisible rays that can cause serious damages to skin and eyes.

The welders and all those engaged in work in close proximity to welding activities must use suitable personal protective clothing to safeguard themselves from these hazards.

2.9.1 Shields and Helmets

The most important PPE for welder and operators can be easily said to be the Shield and Helmet. The arc welding helmet is used to protect the face and eyes from sparks, spatters and the heat, and most importantly from the ultraviolet rays emanating from the electric arc of the welding. While the purpose of both shield and helmet is same, both these terms are often interchangeably used. The difference in shield and helmet is that shields are often

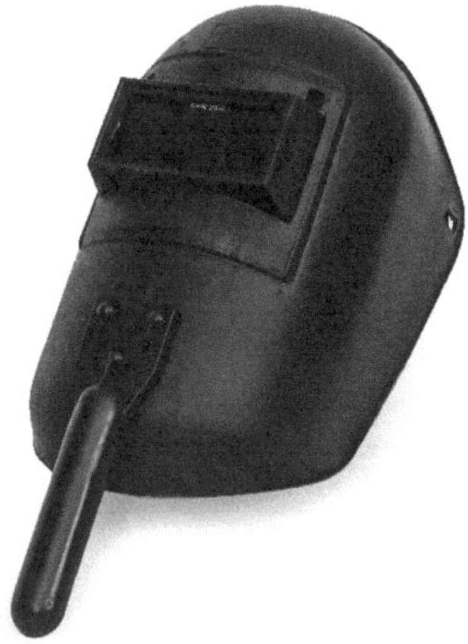

Figure 2.9.1 A typical hand-held welding shield.

hand held, while the welding helmet is worn on the head, leaving both hands of welder free to carry out their work. Obviously, the shields are suitable for smaller work, where the welders have to inspect their work much more frequently, and work that requires frequent viewing of the work for example, when fitting and aligning components prior to welding, or tacking as preparation for welding etc.

The helmet consists of a head mount that is supported by a head cover either full or partial, and a head band that wraps around the head on the forehead. The side knobs on the head band allow adjustment to fit different head sizes. The head and side coving protects the face in general, while the front of the helmet/shield has a window where a dark filter glass of suitable UV-Ray protection rating is fitted between two plain glass to filter UV rays, and heat to enter in the eye and protects the eye and face from any damage to the welders' eyes.

While welding activity is very safe if proper precaution is taken, the welding arc contains some very damaging rays, they can burn the skin, and damage the eye. The exposure to UV rays in the welding arc can cause eye pain, eye watering, and swelling with irritation as if sand is in the eye, and pain and discomfort can last for about 10 to 20 hours after arc exposure. However, exposure to infrared rays, can injure eyesight. Hence protection is extremely necessary.

The filter glass or lens is an important a part of the shield or helmet, it is required to protect eye of the welder from damaging UV rays, and heat of the welding, while the welder is able to see the arc and progress of his work. Due to its very critical role much study has been done and determined that various shades of lens are required to protect differing intensities of welding arcs. The density of filter shades is such that the welder cannot see through it until arc is struck.

Basic helmets have evolved with emerging technologies developed for other fields. Helmets have been developed with some "automatic" features, battery is sued to introduce photoelectric cells inbuilt in the helmet, and this allows the lens to be clear until the arc is struck. This is a great advantage to the welders; they can precisely locate their electrode tip on the weld location and the trike the welding arc. This helps produce cleaner welds, reduce damage to the parent metal near welds, less of arc strike outside the weld zone.

For further eye protection from back flashes some welders also wear a pair of ordinary welding glasses that has # 1 or # 2 lenses on them, this allows them to inspect the weld before or after welding, chip slag etc., from the weld.

Figure 2.9.2 shows a more modern welding helmet, laced with modern technology. This Millers™ welding helmet Technology optimizes contrast and clarity in welding and light states. 1/1/1/2 optical clarity rating allows for a lighter light state while not welding. This permits them to keep the helmet down thus maximizing safety and productivity. This type of helmet has four arc sensors and four modes: weld, cut, grind, and X-Mode.

While the first three modes are self-explanatory about their use, the X-Mode electromagnetically senses the weld to eliminate sunlight interference and continuously detects the arc even if sensors are blocked. Gen 3.5 headgear with comfort cushion has an ergonomic design that provides extensive adjustability, settings, and enhanced support. Digital controls easily allow welder to adjust shade, delay and sensitivity. AutoSense™ eliminates issues related to setting helmet sensitivity by allowing the welder to push and hold the AutoSense™ button to automatically set the helmet sensitivity for their environment. Auto-on/off power control triggers lens at the strike of an arc.

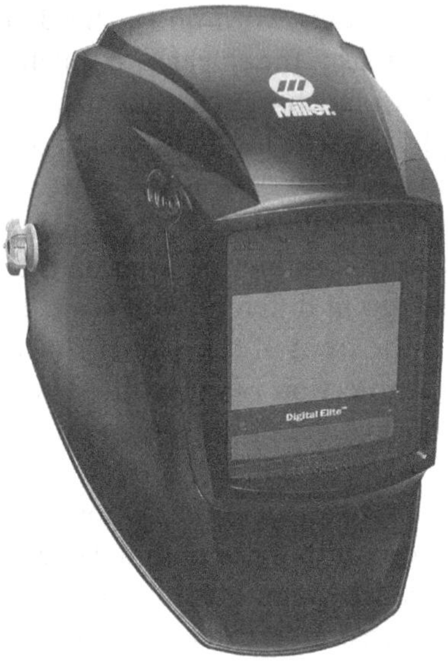

Figure 2.9.2 Miller Digital Elite helmet.

More complex jobs require more complex safety equipment and PPE. Helmets have been developed and are used where air quality around the welding work is improved by introducing fresh air into the helmet, and some are fitted with air filters that provides clean air for the welder to breath.

The following Table 2.9.1 gives the recommended safe shade numbers for various welding operations. The user must experiment with the most suitable and safe shade number that suits them. The following table gives an indicator and may work as the safe trials as start point in selecting most suitable shade for individual safety in welding.

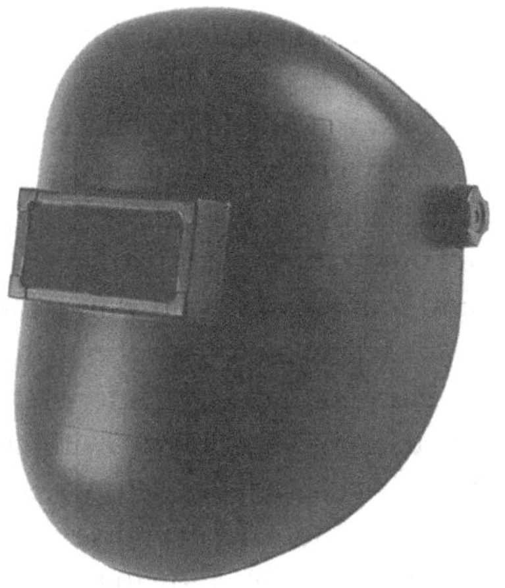

Figure 2.9.3 A typical welding helmet.

Table 2.9.1 Welding lens shades.

Welding process – intensity of arc	Lens shade number required to provide protection
SMAW	
Electrode diameter inch (mm)	Shade number
≤ 0.156 (4)	10
0.1875 to 0.25 (5 to 6)	12
0.3125 to 0.375 (8 to 10)	14
GTAW	
Any metal (Depending on the arc intensity)	10 to 14
GMAW (ferrous metal) wire diameter	
0.0625 to 0.15625 (1.6 mm to 4 mm)	12
GMAW (non-ferrous metal) wire diameter	
≥ 0.15625 (≥ 4 mm)	11

2.9.2 Optical Clarity for Welding

For a lot of welders, it's normal to see haze of smoke, spatter and other things through their welding helmet lens. But this vision can be significantly improved.

- Importance of the optical clarity while welding

The importance of clear vison for welders during welding, cannot be overemphasized, or ignored. What welder sees during welding has significant impact on the quality of the weld they are able to produce, and also overall health of the welders.

The modern helmet makers have used developing technologies of auto-darkening lenses, to improve upon the clarity of welding lenses and combined it with the helmets.

It is not possible to describe in writing and reading the difference these changes make, and it is best to use one of these auto darkening lenses and experience the difference. The differences in the lens ratings can have a marked impact on the productivity. Looking through a standard less for hours at a time, trying to see past haze of smoke, flying spatter and impurities, distortion or inconsistent shade can cause eye strain, making it more difficult to achieve a quality weld. The quality of visibility, contrast, and other ingredients of good vision have been studied and reacted for better understanding, comparison, and selection.

A rating of 1/1/1/1 means the lens has a perfect view, presents best color contrast, and free of defects and distortion, and less strain on the welders' eyes.

As the arc is struck the auto darkening lenses immediately adjust to the light, and that makes the difference. It gives the welder better vision of the workpiece, and they are more comfortable, with the glare of the arc. These lenses provide Color contrast. Normally a welder sees green color through the standard helmet lenses. The improved lenses eliminate that green hazy view with more realistic color of the work and weld area.

Table 2.9.2 Helmets with auto adjusting lenses.

Welding helmet	Optical rating
Miller Digital Elite™	1/1/1/2
Miller Classic Series™	1/1/1/3
Tecmen TM820 and TM 1000™	1/1/1/1 True Color
Linkcoln Viking 1840, 2450, 3350™	1/1/1/1
Jackson WH70™	1/1/1/1
3M Speedglas 9100XX, 9100X, 9100V™	1/1/1/2

- Improved welding results:

Welder is able to see subtle variations in the ever-changing arc and weld puddle, line of the weld, and the surrounding with near perfect clarity.

The ability to clearly see what's going on in the weld puddle when welder is adjusting the angle and speed, the improved lenses provide more accurate end results that is the better-quality weld. Number of welding equipment manufacturers have produced helmets with these modern lenses fitted to their helmets. This required some quality control, and that lead to testing and assigning ratings to various products markets.

The European standard has developed EN379 Ratings for helmets with auto darkening lenses.

Auto-darkening helmet brands have been rated according to the abilities of the lenses. The Table 2.9.2 gives glimpses of some popular helmets with auto adjusting lenses and their ratings.

It is important to understand what the auto adjusting lenses actually do, to improve quality of weld-vision, and what the number rating means.

These welding lenses on helmets are sometimes also referred as welding cartridges, the number rating is given after rigorous testing. The European Standard Commission developed the EN379 detailing the rating system for auto-darkening welding cartridges as a way of measuring the quality of optical clarity in auto-darkening helmet lens. To qualify for an EN379 rating, the auto-darkening lenses get tested and rated in 4 categories:

1. Optical class,
2. Diffusion of light class,
3. Variations in luminous transmittance class, and
4. Angle dependence on luminous transmittance class.

Each category is rated on a scale of 1 to 3, with 1 being the best (perfect) and 3 the worst.

2.9.3 Other Essential Clothing for Welders

In welding operation, the molten weld metal often in the form of spatter fly all over and can easily land on the person of the welder. This can cause sever burn and injuries. Welders need protection from such injuries. Most of these accessories are made from leathers, hence they are often collectively referred as *leathers*.

Table 2.10.1 Electrode classification and A-numbers.

Material group	AWS A number
Carbon Steel	5.1
Low Alloy Steel	5.5
Corrosion Resistant Steels	5.4
Cast Iron	5.15
Aluminum and its alloys	5.3
Copper and its alloys	5.6
Nickel and Nickel alloys	5.11
Surfacing	5.13 and 5.21

- Welding gloves,
- Welding Gauntlet sleeves,
- Aprons,
- Leggings,
- A Jacket, especially if welding in overhead position
- A cape, especially if welding in overhead position
- Heat protected gloves, these are insulated gloves in addition to the leather welding mentioned above, and used when welding on hot surfaces and for longer time.

2.10 Covered Electrodes Used in SMAW Process

Shielded metal arc welding process uses electrode that have solid steel wire as the core, the steel core is covered by a coating that is called flux. The electrodes are identified by the steel wire at the core, its identification, diameter and by a series of letters and numbers. The significance and the meaning of these number and letters is explained further in the description.

American welding society (AWS) and several other agencies around the world have developed methods to designate identifier system for electrodes, by far the AWS system is most commonly used and understood. But understanding other systems like Canadian and European identification system is also recommended, that allows a wider selection poll for very demanding work requirements.

AWS classifies electrodes for the group of metals that these welding electrodes are developed. Some of these are listed below. They are given a specific AWS specification number as indicated with the material group.

2.10.1 Coating Types

There are several different groups of coated stick electrodes for arc welding of unalloyed steels and fine-grained steel. These groups may be classified on the basis of their coating. The four main types of coatings are named after their main components:

- C (Cellulose) cellulose electrodes
- A (Acid) acid electrodes*
- R (Rutile) rutile electrodes
- B (Basic) basic electrodes

*Acid electrodes can hardly be found anymore, and have been almost completely replaced by rutile electrodes. So, based on the type of coating, in effect only three types of electrodes are in the market.

These coating types are the key difference among the unalloyed stick electrodes. Stick electrodes for high strength, creep resistant or high-alloyed steels are mostly coated with rutile or basic coating elements.

The following is the brief description of coating types.

2.10.1.1 Cellulose-Coated Electrodes

This is the primary type of electrode for out of position welding, one best example of their high demand sector of industry for this type of electrode is the pipeline welding. The high level of cellulose in the coating, they are able to maintain the stability of arc in the out-of-position welding, but they are poor selection for horizontal welding position. They are therefore mainly used for vertical-down welding on large pipes.

2.10.1.2 Rutile-Coated Electrodes

The rutile coated electrodes present good weld metal properties. The rutile coating allows welding arc to be stable and calm and is easy to reignite, the seams are finely rippled, and most of the slag comes off by itself. Rutile-coated electrodes have good toughness properties, but are not very useful in out-of-position welding, but they are used to a limited extent.

2.10.1.3 Basic-Coated Electrodes

The main advantages of basic electrodes are the outstanding toughness properties of the weld metal and its resistance to hot and cold cracks. Basic-coated electrodes have a coarse droplet material transfer, can be used to weld in all positions and have somewhat coarsely

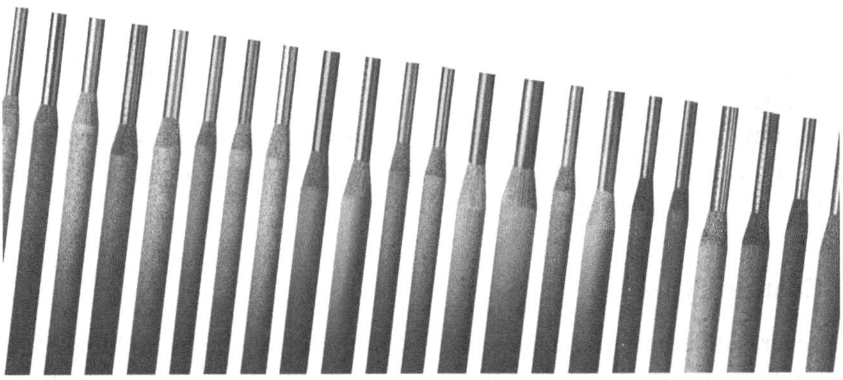

Figure 2.10.2 Portfolio of SMAW electrodes.

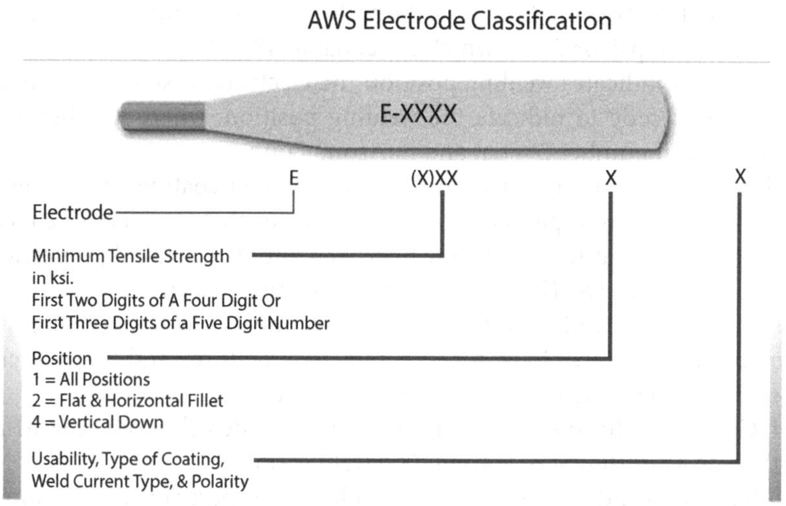

Figure 2.10.3 AWS electrode classification method.

rippled seams, not as fine as one can get from rutile coated electrodes. The slag can be relatively easily removed, but not as easily as with rutile-coated electrodes.

2.10.2 Portfolio of SMAW Electrode

Each type of steel, and type of work quality demand may need a different type of stick electrode. It can be differentiated between the following steel types:

- SMAW electrodes for unalloyed steels
- SMAW electrodes for high-strength steels
- SMAW electrodes for creep resistant steels
- SMAW electrodes for stainless steels
- SMAW electrodes for nickel and nickel alloys

2.10.3 Identification of Welding Electrode

Welding electrodes are identified by an alfa-numeric numbering code system. Both American (AWS) practice and European practices use similar basis, however their units and sometimes even the sequencing differ. Hence it is important to learn what system is being used to designate the electrode, but in most cases the American practice is prevalent, and offers very less chances of confusion. The description of AWS classification (identification) system is given in the following sketches.

Unalloyed electrodes are covered under AWS A5.1, also see the AWS classification figure given below. Example of how to read welding electrode classification is given below and also in the figure above.

Example: E7018-1H4R

E: Type of welding consumables. Here, E = Electrode

- **70:** These first two digits (70) is the weld metal's nominal tensile strength (70 x 1000) in psi, or 70 ksi which are equal to 480 MPa.
- **1:** This digit indicates welding position the electrode is suited for. This digit is a kind of code to indicate the welding position, where number 1 = All positions and number 2 = flat and horizontal only.
- **8:** This last of the four digits indicates, the Type of coating and current and polarity. In this example the number 8 indicates that the electrode has iron powder in the coating and its low hydrogen and its electrical properties are, usable in AC and DC (EP) with electrode positive polarity.
- **-1:** Next number indicates that the electrode meets low temperature impact requirements. (Note: look for the test temperature, and impact energy absorbed values when selecting proper electrode)
- **H4:** Code for diffusible hydrogen. And the electrode will produce weld-metal with hydrogen content of < 4mg/100g weld metal
- **R:** The letter R indicates that the electrode has property of reduced moisture absorption.

The American Welding Society (AWS) classifies electrodes on the basis of chemical composition of their undiluted weld metal or mechanical properties or both. Welding current and position are also indicated.

Carbon steel electrodes are included in AWS Specification A5.1. Under AWS 5.1 there are two strength levels: 60 and 70 ksi. Example of electrode designation system is E6010, which is explained below. Some of the other common use electrodes are E7011, E7015, E7018, E7024, for an example, in an electrode designated as E 6010, the letters and numbers have following explanations.

- Letter E designates an electrode.
- Number 60 signifies that the tensile strength of the deposited weld metal is minimum 60 000 psi.
- The second last digit (1) represents the welding position the electrode is suitable for use (1 = all positions).

Table 2.10.2 Shielded arc welding electrodes.

AWS classification	Type of covering	Suitable position of welding	Type of current
The following electrode are capable of depositing weld metal with tensile strength of 60,000 psi			
E 6010	High Cellulose Sodium	F, V, OH, H	DCEP
E 6011	High Cellulose Potassium	F, V, OH, H	AC or DCEP
E 6012	High Titania, Sodium	F, V, OH, H	AC or DCEN
E 6013	High Titania, Potassium	F, V, OH, H	AC or DC either polarity

(Continued)

Table 2.10.2 Shielded arc welding electrodes. (*Continued*)

AWS classification	Type of covering	Suitable position of welding	Type of current
E 6020	High iron Oxide	H-fillet	AC or DCEN
E 6022 *Used for single pass welds only*	High Iron oxide	F	AC or DC either polarity
E 6027	High Iron powder, iron oxide	H-fillet, F	AC or DCEN
Following electrodes are rated to deposit weld-metal with minimum 70,000 psi tensile strength			
E 7010	High cellulose sodium	F, V, OH, H	DCEP
E 7014	Iron powder titania	F, V, OH, H	AC or DC either polarity
E 7015	Low hydrogen sodium	F, V, OH, H	DCEP
E 7016	Low hydrogen potassium	F, V, OH, H	AC or DCEP
E 7018	Low hydrogen Potassium, iron powder	F, V, OH, H	AC or DCEP
E 7024	Iron powder titania	H-fillet, F	AC or DC either polarity
E 7027	High iron oxide, iron powder	H-fillet, F	AC or DCEN
E 7028	Low hydrogen potassium, iron powder	H-fillet, F	AC or DCEN
E 7048	Low hydrogen potassium, iron powder	F, OH, H, V-down	AC or DCEP

- The last digit, (0) refers to the covering type and current type, in this case 0 indicates the covering is of cellulose, and electrode is good for all positions of welding.

Low alloy steel electrodes are included in AWS Specification A5.5. Their numbering system is similar to that used for carbon steel electrodes. A letter or letter number combination suffix is added to indicate the alloy content (E7010-A1, E8016-C2). Weld metal strengths of alloy steel electrode range from 70 to 120 ksi minimum tensile strength. American Welding Society (AWS) specifications use suffixes at the end of the electrode designation to classify the alloying elements, and time to time reviews the new developments and adds or withdraws as required. The following Table 2.10.2 shows some of the most common carbon steel electrodes, with their coating types, suitability of welding position and type of electrical current they are suitable for use.

For alloy steel electrodes there are further additions of code letter and number identifications. These are used as suffixes to the primary identification system. The following are, some of the most common suffixes and their intended meanings,

Suffix	Meaning of the suffix in Carbon and alloy steel SMAW electrodes
A_1	Contains 0.5% Molybdenum (Mo)
B_1	Contains 0.5% Chromium (Cr) and 0.5% Mo
B_2	Contains 1.25% Cr and 0.5% Mo.
B_3	Contains 2.25% Cr and 1% Mo.
B_4	Contains 2% Cr and 0.5% Mo.
B_5	Contains 0.5% Cr and 1% Mo.
C_1	Contains 2.5% Nickel (Ni).
C_2	Contains 3.5% Mo.
C_3	Contains 1% Ni, 0.15% Cr and 0.35% Mo.
D_1	Contains 1.75% Manganese (Mn) and 0.25% Mo.
D_2	Contains 1.75% Manganese (Mn) and 0.45% Mo.
G	0.5%Ni, 0.3%Cr, 0.2%Mo, 0.1%V, 1%Mn (only one of these elements has to meet the requirement to qualify as "G" electrode).
L	Controlled elements (example low carbon)
M	Meets military requirements.
HZ	Meets weld metal diffusible hydrogen requirements. H1 up to 15 ml/100 gram of weld metal H2 up to 10ml/100 gram of weld metal H3 up to 5ml/100 gram of weld metal H4 ≤ 5ml/100 gram of weld metal
R	Meets absorbed moisture test requirements.
Numerical	Numerical suffixes following the above listed; indicate toughness properties of the weld metal.

Corrosion resistant steel electrodes are included in AWS Specification A5.4. Their classification is based on chemical composition, position of welding, and the type of welding current. For example; E310-15 is an electrode with nickel and chromium alloying suitable for use in all positions with DC current. The example of their designation is given below.

Example: E308L-16

- **E:** Code for type of welding consumables. Here E = Electrode
- **308:** Code for chemical composition of the weld metal,
- **L:** Code for low Carbon
- **16:** Usability Designation: where
  - 15 = DCEP only, basic type;
  - 16 (and 17) includes arc stabilizing elements (Rutile). AC possible.

AWS specifications exist for nickel alloy (A 5.11), aluminum alloy, and copper alloy electrodes as well as cast iron welding, hard-surfacing and overlaying. A list of these AWS specifications is given in table above.

2.10.4 Need for the Covered Electrode

The covering on a shielded metal arc welding electrode is called *flux*. The flux on the electrode performs many different functions. These include:

- Producing a protective gas around the weld area.
- Providing fluxing elements and deoxidizers.
- Creating a solid coating over the weld as it cools.
- Establishing electrical characteristics.
- Adding alloying elements.

During the arc process, some of the flux covering changes to neutral gases or reducing gases, such as carbon monoxide (CO) or hydrogen (H). These gases, as they surround the arc proper, prevent air from coming in contact with the molten metal. They prevent oxygen in the air from combining with the molten metal. However, the gases usually do not protect the hot metal after the arc leaves that area of the weld.

The covering also contains special fluxing ingredients that work to remove impurities from the molten metal. Impurities are floated to the top of the molten weld pool. As the electrode flux coating residue cools, it forms a coating of glass like and brittle material over the weld called slag. This slag sometimes easily peels off, but sometimes it requires an extra effort to de-slag the weld. This is primarily due to the type of flux used as covering of the electrode. This slag coating prevents the air from contacting the hot metal. The slag covering also allows the weld to cool more slowly and helps prevent a hard, brittle weld.When welding with AC, the current changes direction and actually stops 120 times per second. To maintain an arc as the current changes the direction, ingredients are added to the covering of the electrode to create an ionized gas. This ionized gas allows good arc stability when welding with alternating current.

The flux covering on a shielded metal arc electrode can also contain alloying elements. These alloying elements are added to the weld pool as the covering is melted. Iron powder and iron oxide also can be added to the coverings of steel electrodes. These electrodes deposit metal into the weld at a faster rate than standard electrodes. Electrodes are often referred to as drag electrodes because they can only be used in the flat and horizontal positions. Certain coverings on steel electrodes are designed to be low in hydrogen. Hydrogen causes low ductility and cracking in certain cases. The use of low-hydrogen electrodes helps to eliminate these problems. A good flux-covered electrode can produce a weld that has excellent physical and chemical properties.

2.10.5 Electrode Conditioning

Electrode conditioning refers to the storage and handling of covered electrodes to maintain their optimum moisture content. Low-hydrogen electrodes such as E7018 must be maintained in a holding oven set at 150°F to 300°F. Excessive moisture can cause porosity or lead to hydrogen cracking.

Holding ovens are electrically heated oven of various types as shown in the Figure 2.10.5.1 below. These ovens are often placed in a workshop or sometimes in lager fabrication yards as close to the work station as possible for bulk holding of electrodes. Welders transfer the required electrode type, and quantity to their portable quivers, as shown in the Figure 2.10.5.2 below, and take it to their place of work.

Figure 2.10.5.1 Shop use electrode drying oven.

Figure 2.10.5.2 Portable electrode holder also called quivers.

Cellulose electrodes (E6010, E6011) are not required to be conditioned, they do not operate properly if they are dried out and should be stored in a clean, dry place rather than in an oven.

2.11 Welding Training – Making of a Welder

As is pointed out in the introduction chapter of this book welding is both and art and a science. To be a good welder means to acquire the skills of an artist, and for that skill like any other art form, the importance of training cannot be over emphasized. The training is not only to become a welder but also to maintain the skill and dexterity in hand muscles and coordination training is very essential part of being a welder.

2.11.1 Joint Design and Preparation

There are number of types of joints made using welding technology. These are depicted in the following figure. Note the weld profile of a butt weld is shown as bead, fillet is a right-angle triangle, and groove welds are shown at the top of the figure, their various applications

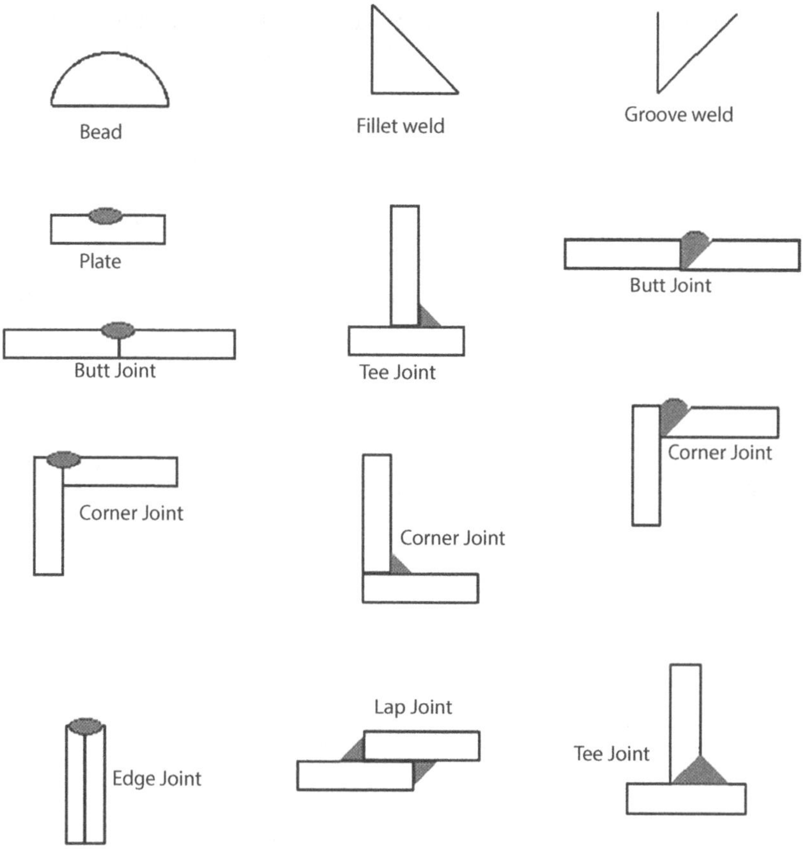

Figure 2.11.1 Different types of weld joints.

are shown in the subsequent rows under these profiles. There can be more complex joints, but for initial knowledge these types are way sufficient.

A would-be welder should start practicing using the plate on the bead method shown first figure titled Plate, and then graduate to next level with square butt weld shown as the Butt weld, and progress from there.

Welds are made on various materials, of thickness, and to meet number of structural design demands placed by engineering team of the project. Those demands are matched in combination with the material type, joint type and the thickness of the material.

These weld designs are shown in the figure below. Welds can be made as square groove butt welds, Single V-groove butt welds, and Single Bevel grove butt welds, J-groove butt welds, U-groove butt welds and they can be combined with fillet welds and used in combination with flare welds as shown in the figure below. The selection of what type of weld design is made is an engineering decision, and is often based on the available material, the stress loading and strength required of the welded member.

Although square groove joints are the most economical to prepare, the thickness is limited to about 6mm (¼-inch). For thicker members, the edges must be prepared to a contour that will permit the arc to be directed to the point where the weld metal must be deposited. Standard 30° bevel, J-groove and U-groove joints are desirable for intermediate thick sections since they allow access to the root with the least amount of filler metal required. Higher thicknesses may have composite and double bevel preparations for welding by SMAW process.

Fillet welds require little or no preparation except that they need to fit to the required work design, combined with groove welds. Minimum stress concentration at the toes is obtained with concave fillets.

Welder skill efficiency and quality is demanded to make full penetration welds, which are often required in load bearing structures, and pressure containing or even lethal and dangerous chemical storage equipment.

Where access to both sides of weld is present these types of welds may be made from one side, and the go back and deposit additional weld from the back side to complete one full sound weld. However, that may not be possible in the restricted spaces like welding

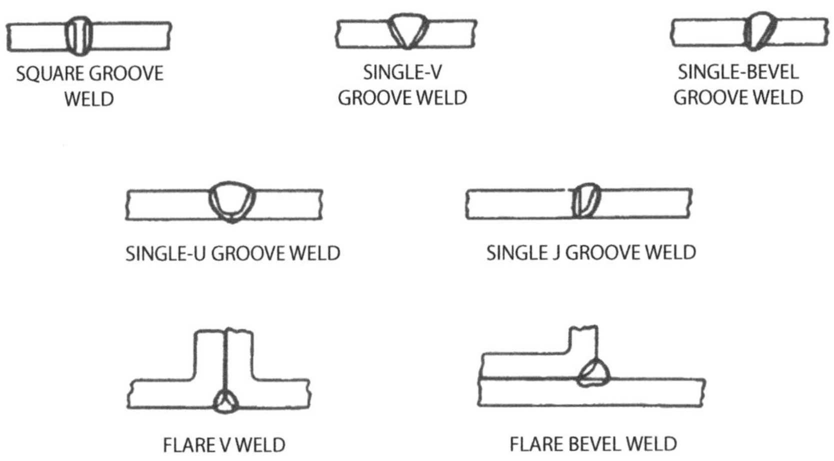

Figure 2.11.2 Different types of weld designs.

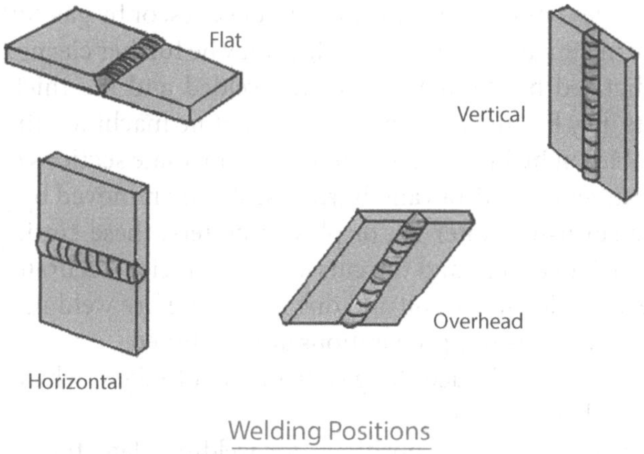

Figure 2.11.3 Welding positions for welding a plate, the positions are primarily designated in relation to the position of the weld to the horizontal surface of the earth.

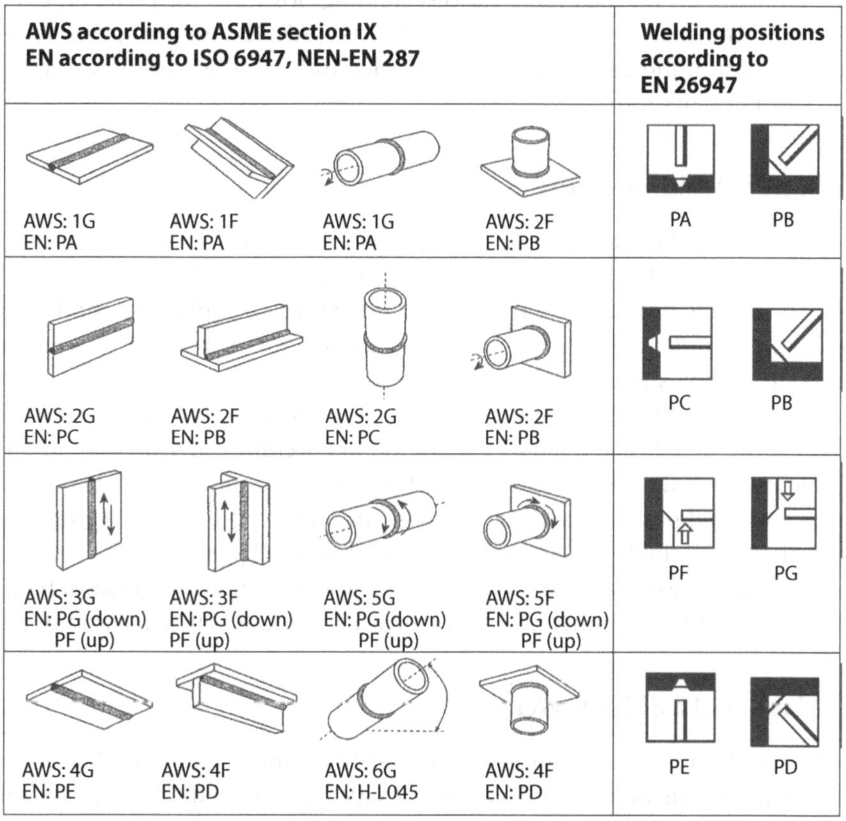

Figure 2.11.4 Positions of plate and pipe butt welds and fillet welds with both AWS and European designations.

pipeline, pipping, tight structural construction, or small sized storage containers. When full penetration joints are required welders" efficiency is in great demand.

Depending on the quality of the fabrication and therefore the welding, the methods of welding edge preparation differ extensively.

From rough cut edges either by machine press cut edges, or by gas cutters, to very precise machined edges are in demand. The gas cut edges may be further cleaned by grinders. Some of these are also dictated by the material being welded and the thickness. For example, stainless steel sheets may be cut using snips and guillotine machines, the thicker sections of stainless are invariable cut by Plasma arc cutters. Intermediate sections of CS steels are often cut by oxy-fuel gas cutters, and dross and burned scales are removed by grinding. Very high thick steels may be cut using water jet, or plasma cutters, these cutting process are often available in specialized locations, and the cuts are very precise, accurate and clean reducing any further work except cleaning of oil and dirt residue before welding.

Mostly J-groove, and U-groove preparations are machined.

In most case the welder will face the gas or press-cut edges, which may require some grinding and cleaning before welding.

Figure 2.11.3 depicts the welding positions, for welding plate, the positions indicate the working with earth's gravity and its impact on the flow of the molten metal. Which affect quality of weld and tests welder's ability to maintain arc length, flow and deposition of material.

The above Figure 2.11.4 illustrates various welding positions for plate, pipe butt welds and fillet welds. The chart shows corresponding AWS/ASME and European designations. The right had column indicates the position of electrode or filler-rod/wire corresponding to the European designations.

2.11.2 SMAW Welding of Plate

The SMAW process is the most versatile welding process for any welder to start their career in welding. The initial welding practices to familiarize with the welding activity, and feel the striking of arc and the heat and arc behavior, is done on a steel plates of modest thickness and sizes. As the confidence of striking arc and manipulation of arc, and movement of electrode grows the complexities of welding are introduced.

Assuming that the welder has mastered the art of arc striking, arc control and weld pool manipulation. The welder is moved to a more job like welding environment and there the welder is allowed to make a butt weld joining two plate sections. The most common thickness chosen for initial practices are the 1/8-inch-thick (3 mm) plate. As we said earlier, we will assume that the metal being welded is basic carbon steel plate.

We assume that the welder is familiar with welding safety, and is in possession of essential welding tools like chipping hammer, clamps for alignment, and of course those important PPEs.

2.11.3 Making of a SMAW Welder

The importance of training, and honing the skills to become a welder has been pointed out number of times in this book. The training activity must be planned and coordinated to achieve the full potential of the efforts made.

The following excesses are designed to make a new entrant to a certified welder who can weld a reasonable quality of weld, and progress to achieve of various levels of proficiency. The limit is only on person's dedication, and number of practice pieces welded, to master the level they are working at the time.

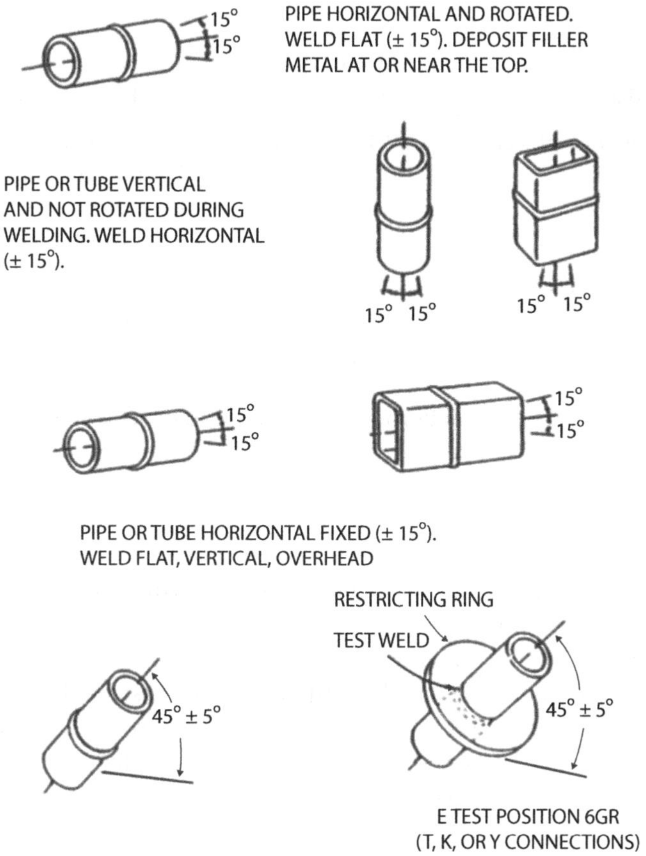

Figure 2.11.5 Above figure shows the permitted angular tolerance for specifically designated welding positions for pipe welding.

2.11.3.1 SMAW Welding Practice Step 1

For this very first exercise, the goal is to learn and master following steps.

- Learn to strike the arc,
- Learn to maintain the arc length,
- Learn to optimize the arc length for stable flow of arc, and good weld quality.
- Learn to sturdily progress the arc along the weld length, while maintaining the arc length.

The welder selects a 2.4 mm diameter E 6010 electrode and 1/8" thick plate. Following steps are performed.

The basic welding procedure specification (WPS) is as following.

Material	Edge preparation	Electrode type and diameter	Set-up type - position	Current and polarity
Carbon steel plate 1/8" thick.	Square butt weld	E 6010, Diameter 2.4 mm	1 G (flat position) For reference, see the figure above for the plates.	85 amperes, DC EP
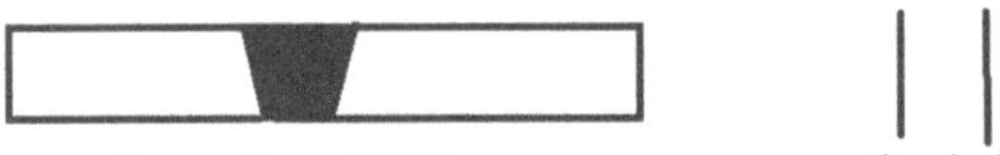 Square Grove Welding Symbol				
SMAW		Practice till 20 consecutive welds come out in acceptable range.		

- The welder sets the current on machine to 85 amps.
- The electrode is gripped in the electrode holder, which is connected to a DC welding machine's positive terminal.
- The other lead of the welding machine connected to the negative port, is clamped to the weld table on which the weld plate is securely positioned.
- The welder dons all his PPE, and lifts his helmet over his head to see the location where he needs strike the arc. Having inspected and ready to strike the arc he flips his head to lower the helmet on his face, while simultaneously striking the arc on the plate.
- As the arc is struck, and a stable arc length is established, the welder proceeds in the line of the joint direction. Initially the welder will practice leading arc technique, where the weld pool follows behind the arc.
- AS the weld is completed, the welder lifts the helmet hood, and chips the scale and slag off the weld and inspect its quality for uniformity of deposit, the speed of travel and for any other gross discontinuity that the welder needs to improve upon.
- Practice continues, for improvements leading to next step.

The welder must practice this exercise over and over again on at least 15 to 20 pieces of weld. To consolidate the gains towards the established goals of this exercise. Each time improving the ability to strike the arc, control the arc, and confidently manipulate the electrode. Once this task is mastered, they should move to the next step.

2.11.3.2 SMAW Welding Practice Step 2

The welder has learned the art of striking arc, maintain the arc length, and has achieved some control over the arc, and is able to manipulate in a straight line.

In this exercise the welder will take two pieces of 6-inch (152 mm) long and 2.5" (64 mm) wide 1/8" thick (3 mm) plates, and tack them for fillet weld practice, as shown in the following welding specification.

Material	Edge preparation	Electrode type and diameter	Set-up type - position	Current and polarity
Carbon steel plate 1/8" thick.	90° Fillet weld as shown I the figure below.	E 6010, Diameter 2.4 mm	1 F (flat Fillet weld)	85 amperes, DC EP
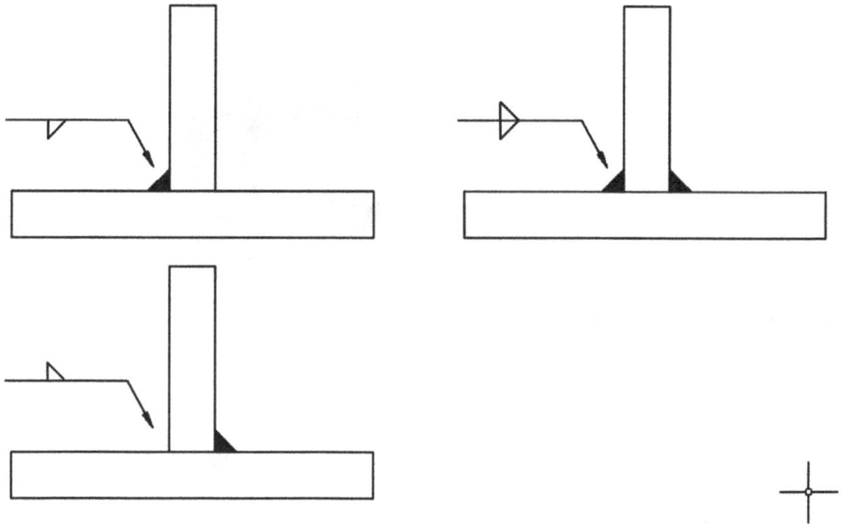				
SMAW		Practice till 10 to 15 consecutive welds come out in acceptable range.		

Initially, start with practice welding from one side, and inspect it for quality of weld, verticality of the welded members, and bend the weld by holding it in a vise to see penetration, as shown in the Figures 2.11.3.1 and 2.11.3.2 below.

Figure 2.11.3.1 Testing a fillet weld.

Figure 2.11.3.2 Testing a fillet weld using a hammer.

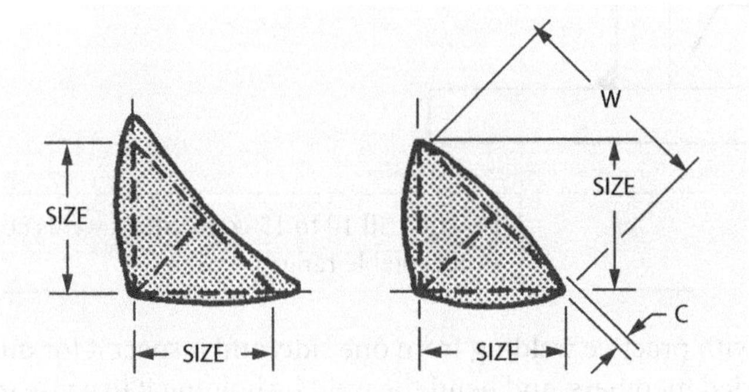

Figure 2.11.3.3 Size and nomenclature of fillet weld.

Figure 2.11.3.4 A single pass fillet weld.

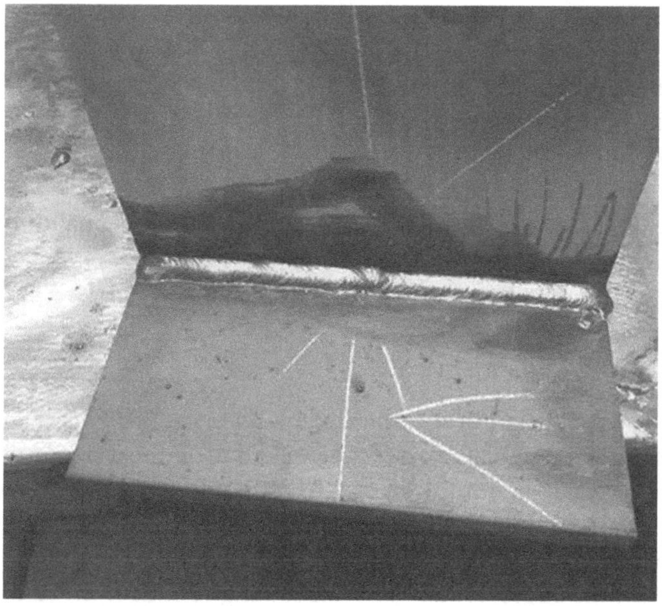

Figure 2.11.3.5 A single pass fillet weld with (arc termination) stop in the middle and restarted (arc re-initiation) from that point.

When initial task of welding clean and in a straight line is mastered, then move to welding from both sides as shown in the picture in the specification, and also start deposing two passes to achieve 3/32-inch (2.4 mm) leg size fillet weld.

See Figures 2.11.3.3 to 2.11.3.6 for understanding various aspects of fillet weld sizes, and other references.

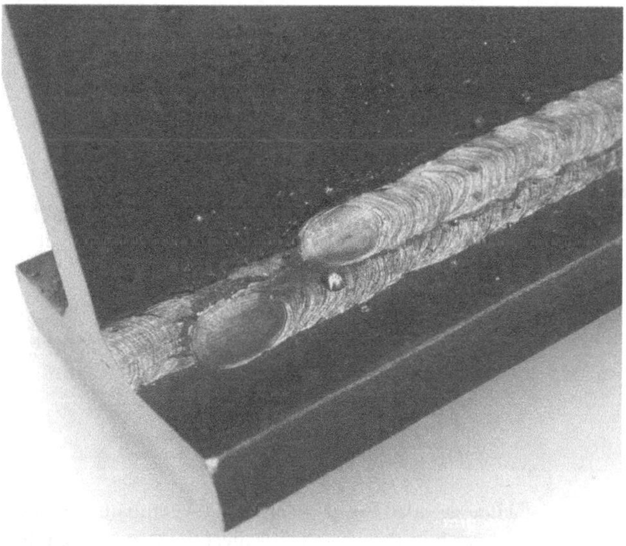

Figure 2.11.3.6 A multi-passes fillet weld-note the termination of arc start and stops are staggered.

Additional task to master fillet welds.

(i) Weld in Horizontal position,
(ii) Weld in vertical position,
(iii) Weld on thicker plate with minimum 0.23-inch (6 mm) fillet weld.

Inspect all welds for dimensional accuracy, uniformity, and quality of welds. Test all welds for strength and penetration.

2.11.3.3 SMAW Welding Practice Step 3

The welder has mastered the art of striking arc, maintain the arc length, and has good control over how the arc is manipulated across the weld line on the job.

For this exercise a 3/32" (2.4 mm) diameter E 6010 electrode is selected, and two plate sections 1/8 inch (3.2 mm) thick and of length 6" x width 3" are given for welding. The abutting edges of two plates are cut clean and squire, this is going to be squire edge butt weld.

The welder cleans the edges, for any possible fouling residue from cutting machine like, oil and dirt, or burr or sharp edges that may cause the weld to be defective. Welder uses file to remove the sharp edges and burr attached to the welding area. And aligns two pieces of plate side by side, keeping 1/16" (1.5mm) gap between the two pieces. Satisfied with the cleaning and alignment of two plates, the welder secures the two abutting plates firm in position, to the work table.

Since the plate thickness, and electrode diameter is same as practice plate in exercise 1 above, the welder uses the same electrical parameters as given in the procedure, the welder does not change the electrical stings on the machine. Welder uses 85 amperes, and DCEP

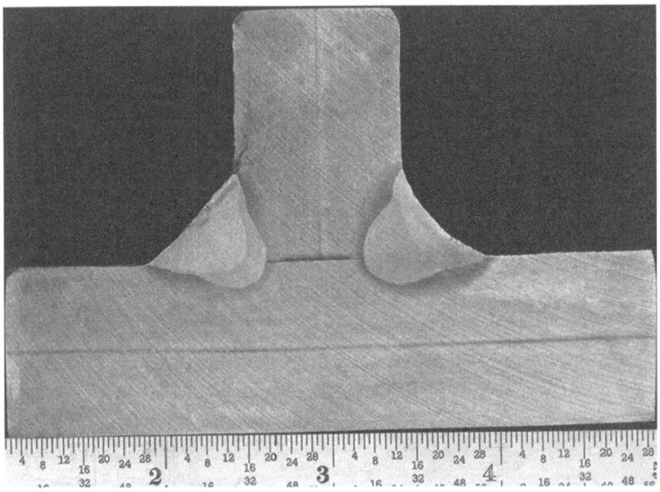

Figure 2.11.3.7 A micro-etch of a double sided two pass fillet weld – compare the weld with the nomenclatures figure given above, to see how these two welds meet the standard requirements.

to conduct this exercise. The welding procedure is reproduced below for easy reference. The WPS is to be used to weld several practice welds to master the technique.

Material	Edge preparation	Electrode type and diameter	Set-up type - position	Current and polarity
Carbon steel plate 1/8" thick.	Square butt weld	E 6010, Diameter 2.4 mm	1 G (flat position) For reference, see the figure above for the plates.	85 amperes, DC EP
 Square Grove Welding Symbol				
SMAW		Recommended Practice till 20 consecutive welds come out in acceptable range.		

Welder now dons, the PPE including the welding helmet. Inspects the weld line once again just before striking the arc, and then lowers the helmet and starts welding.

After completing the weld, the slag is chipped off, and it is inspected for the quality of the work. While the appearance of the weld bead, and the consistency of his weld speed was good.

2.11.4 Inspection of the Weld

The welder inspects for full penetration through the gap between two abutting plates, which is inconsistent, and finds a new problem with the weld. In this weld nearly halfway the length of the weld the two plates jammed the 1/16-inch gap that was kept ensure the penetration, this restricted the weld penetration. This jamming also had distorted the shape of the weld plates, this further restricted the weld penetration.

2.11.4.1 Appearance of the Weld

Also, at this point inspect the weld for current adjustment, if current is too high the weld would look like pear head pointing away from the direction of the arc travel.

If the current is too low, the arc strike and maintenance will be difficult however if some length of weld is made, the weld bead may be thin, very high in profile, and metal droplets appear over-lapping on each other. The following sketch shows the appearance of weld and the welding techniques that results in that kind of weld. The welder is expected to inspect their weld reach the quality of weld A in the sketch. Any other look of the weld would require either correction of current used, or the speed of arc travel or may be both, till a weld quality comparable weld A in the sketch is achieved, and consistently produced.

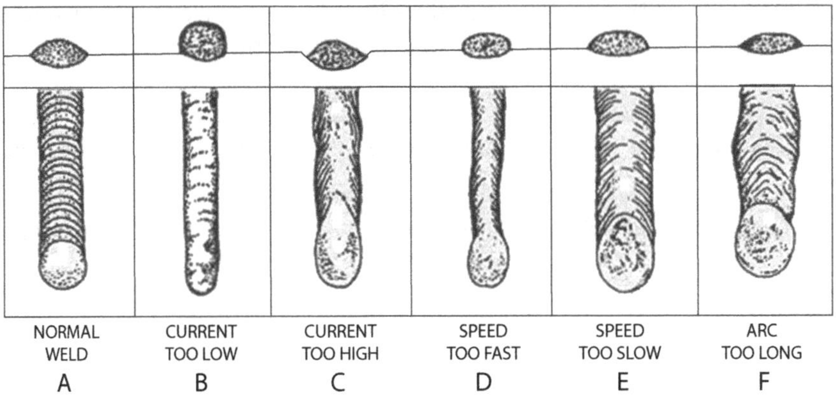

Figure 2.11.4.1 Weld appearances matched with arc current, and arc travel speed.

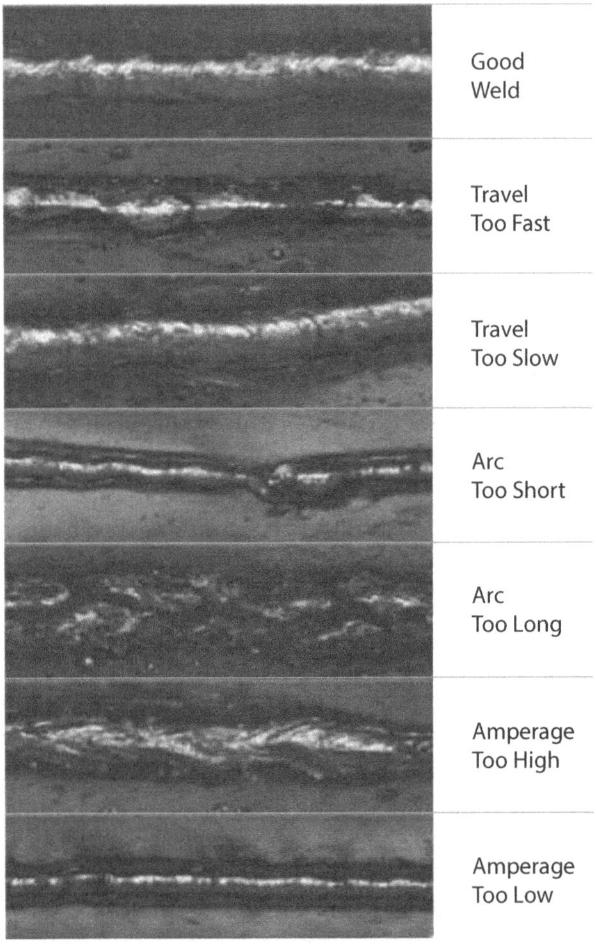

Figure 2.11.4.2 Pictures of the weld appearances and probable cause for the quality of weld produced.

2.11.5 Step 3 Practice 2

In this practice the welder uses same machine setting, and uses same electrode type and size, to weld same size and thickness of metal. But the current draw for welding is slightly lowered, to 80 amperes, for the same DCEP current.

The set-up also sees some important changes, the changes made are as follows. During the alignment, two edges are kept 1/16-inch (1.5mm) apart, however this time the gap is slightly tapered, opening wider towards the end of the weld that is closer to the welder.

And this time the plates are not lying flat, but slightly raised in the middle, uniformly along the length of the weld, the rise is about 3/32 - inch (about 2.5 mm) above the horizontal flat surface of the work-table. As shown in the figure below. The plates are secured and now ready to weld. See the scanned figure below, showing the pre welding offset, to control the distortion.

Weld is done, and the result is inspected.

Following is to be expected:

1. The weld had uniform penetration though the length of the weld.
2. Plates were not distorted, and lay nearly flat.
3. Weld profile is even, not too high, and each metal droplet is smoothly merged to the next one.

The offsetting of the weld off the horizontal plane compensated for the shrinkage, while the tapered gap opening also took care of both the shrinkage, and allowed gap for the penetration of weld-metal till the end of the joint.

The welder must repeat this practice on at least 10 to 15 times to gain full confidence in the process and acquire ability to manipulate arc and heat for more challenging tasks ahead.

Having successfully and repeatedly welded a square edge (square groove) weld, and number of practices, the welder is now ready to move to next stage ahead.

2.11.6 SMAW Welding Step 4

In this stage the welder uses ¼ -inch (6 mm) plate, with a single V-groove that has in groove-angle of 60°. Two pieces of 8" x 3" plates are taken, and each beveled as shown in the figure below. These two plates are fitted in the manner that they have gap opening of 1.2 mm that is slightly (not more than 1.5 mm wide) wider at the end of the weld joint, nearer to the welder.

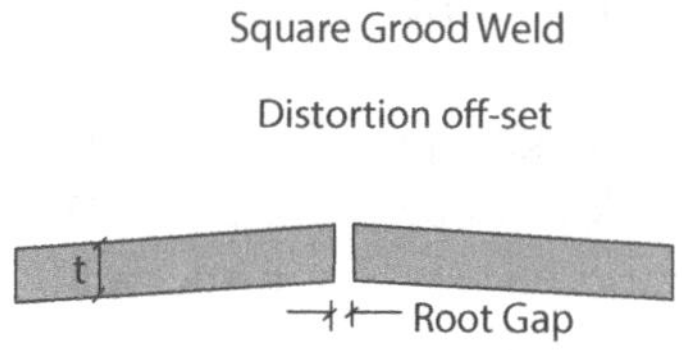

Figure 2.11.5.1 Offsetting the weld setup for distortion control.

Objective of this exercise is to make a practically acceptable welder. In this exercise the welder is required to use learned skills of previous steps. Welder will prepare weld pieces by preparing the weld edge preparation as shown in the WPS drawing. Set it up to eliminate distortions, and to achieve full and uniform penetration of weld.

The new WPS to follow for the welding practice is given below.

Material	Edge preparation	Electrode type and diameter	Set-up type	Current and polarity
Carbon steel plate 1/4" thick. (6 mm)	Single V-groove butt weld Bevel angle 30° Groove angle 60°	E 6010, Diameter 2.4 mm	1 G (flat position), Root opening: 1 to 1,2 mm Root face: 1 – 1.2 mm	70 - 85 amperes, DCEP (see note below)

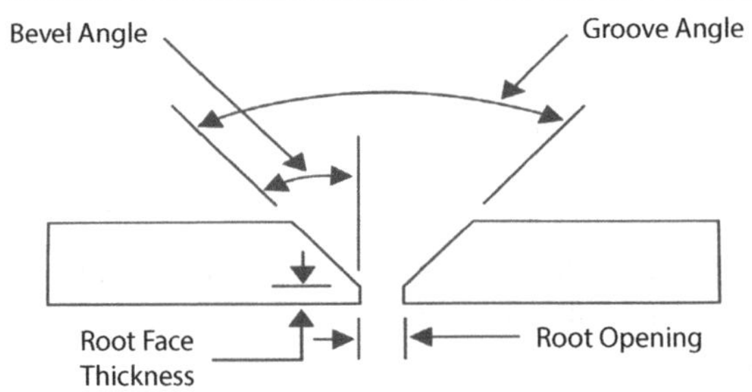

Welding Process: SMAW

Note: start with low amperes and adjust to suitable level before starting on the actual practice piece.	Practice till 20 consecutive welds come out in acceptable range.

2.11.7 SMAW Welding Step 5

In this stage the welder uses ½ -inch (12 mm) plate, with a single V-groove that has in groove-angle of 60°. Two pieces of 8" x 3" plates are taken, and each beveled as shown in the figure below. These two plates are fitted in the manner that they have gap opening of 1.2 mm that is slightly (not more than 1.5 mm wide) wider at the end of the weld joint, nearer to the welder.

Objective of this exercise is to make a practically acceptable welder. In this exercise the welder is required to use learned skills of previous steps. Welder will prepare weld pieces by preparing the weld edge preparation as shown in the WPS drawing. Set it up to eliminate distortions, and to achieve full and uniform penetration of weld.

This weld will require the welder to deposit more than one pass to complete the weld to an acceptable finish.

The new WPS to follow for the welding practice is given below.

<table>
<tr><th>Material</th><th>Edge preparation</th><th>Electrode type and diameter</th><th>Set-up type</th><th>Current and polarity</th></tr>
<tr><td>Carbon steel plate 1/2" thick. (12 mm)</td><td>Single V-groove butt weld Bevel angle 30° Groove angle 60°</td><td>E 6010, Diameter 3.2 mm</td><td>1 G (flat position) 1 G (flat position) Root opening: 1.2 to 1.5 mm Root face: 1 - 1.2 mm</td><td>90-110 amperes, DCEP (see note below)</td></tr>
<tr><td colspan="5">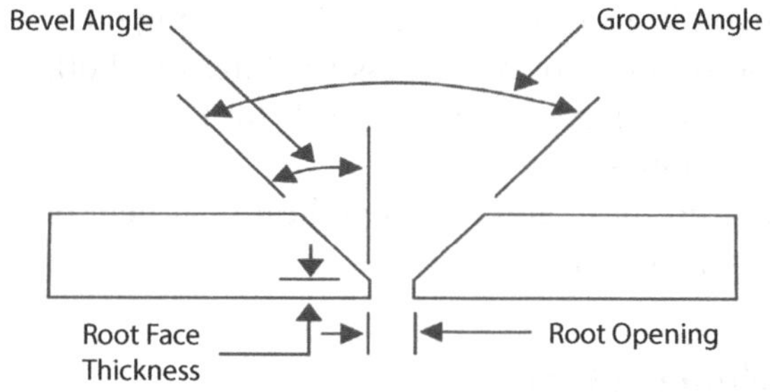

Welding Process: SMAW</td></tr>
<tr><td colspan="2">Note: start with low amperes and adjust to suitable level before starting on the actual practice piece.</td><td colspan="3">Practice till 20 consecutive welds come out in acceptable range.</td></tr>
</table>

The weld piece is setup as indicated in the sketch above, since this is a bit heavier plate to weld, the set up to counter distortion can be reduced to 1/32 inch (about 1 mm) above the flat surface of the work table.

This groove will be deeper, and will not fill in one weld pass. Welder will have to clean the first pass inspect and if needed use grinder to make suitable base for depositing next weld pass to fill and complete the weld with nice looking cap of the weld.

After successful welding of about 20 such weld specimen the welder is ready to hit employment market as basic structural fabrication welder. At this point the welder is able to do very basic welding work in a fabrication shop. And further work practice will improve their skills.

2.11.8 Set a Next Goal to Achieve

1. The next goal in this direction should be the ability to weld in different positions. While maintaining and improving the quality of weld. The skill to develop would be to strike, maintain, and manipulate the arc, and have good control over the molten weld-metal pool in all positions. Weld-Positions to practice are, Flat-horizontal, over-head and in incline positions. See weld positions and orientations further in the text.

Learning moment: Why it is difficult to maintain the position of the molten-weld-metal in various welding positions? **Hint** – Is earth's gravity playing with the weld?

2. The welder can now shift his electrode to low hydrogen E 7018 and practice getting used to the arc behavior of this type of electrode. More practice will be required.
   Leaning moment: Learn why the two different types of electrode have different arc behavior, and why the deposited welds do not look alike, and how weld slag is different. Note the difference in the sound these arcs make, and question why? All answers are there to find in the text. **Hint** – Has coating of these electrodes to do something?
3. Practice more on ½-inch thick, and similar welding on a ¾ inch thick plate for more understanding of multi pass welding. Use both E 6010 and 7018 electrodes to weld using the following WPS for E 7018 electrodes.
4. Move to welding pipes.
5. Learning about Pre-heat, and post weld heat treatment of welds. When it is necessary and why it is necessary?

2.11.9 SMAW Welding of Pipes

While the basic skills mastered during welding of the plate, is most essential skills to have in any welding activity, the pipe welding demands few more adaptations to those skills.

Welding pipe is a very generic term, it involves two very specific styles of welding (i) welding normally small diameter pipes for piping work (ii) welding pipeline, in that the pipe lies in the horizontal fixed position. These two basic job situations demand several adjustments to the welding technique and also in the selection of welding electrode(s). In this initial training we will concentrate on welding a smaller pipe and in 1G position whet the pipe lies in horizontal position and it can be rotated on its axis, to allow the welding around the pipes girth in 1G position.

Since the pipe is not a flat surface as a plate is, but has constituently changing surface, welding a pipe requires much better control of arc from the welder. Another challenge is that pipes in most cases do not allow good access to the internal of the pipe, to correct the weld if first attempt is not a good weld.

Cautions: Due to collected fumes, lack of oxygen, and also claustrophobic conditions, going inside a pipe to weld internally is very dangerous and should not be attempted.

2.11.9.1 Pipe Welding Step 1

For the first attempt on pipe welding, the welder may take any pipe of nominal size that is not less than 3-inch in diameter, and no less than 0.216 inch. A 3" nominal pipe's actual diameter is 3.5", for other sizes and corresponding wall thicknesses refer the pipe schedule table given table in the chapter 7 of this book.

Welder will weld in 1-G position, this is the basic and the easiest position to weld, because the welder is able to keep the arc in relatively fixed position and focuses on maintain the arc length and welding travel speed. See the sketch showing pipe welding positions, and note how the weld is positioned for 1-G. In the 1-G position the pipe is rotated, while the arc is held pointing downwards the weld groove. For rotating the pipe, often a motor driven rotator is used, which is often foot operated by the welder as the weld progresses, see the set of pictures of various types of pipe rotators used in the shop fabrication.

For smaller jobs like the welders' training, a hand moved simple rotating device can be used, it consists of simple rollers on which the pipe section is placed, and the welder operates it with the free hand as weld progresses. Note the weld piece is hot, and care must be taken to use protective gloves. In actual welding situation, the improvised tools, and wrenches are used to rotate the work. In more stablished shops, the rotating is done by the weld jigs, which give more regulated rotation and safe to use. Some of the most common types of pipe rotators are shown in Figures 2.11.9 to 2.11.11.

Now that the welder has selected the carbon steel pipe of 3-inch in diameter, that has the wall thickness of the Standard Schedule or schedule 40 which is 0.216 inch. Welder will weld in 1-G position, this is the easiest position to weld, because the welder is able to keep the arc in relatively fixed position and focuses on maintain the arc length and welding travel speed. See the sketch showing pipe welding positions, and note how the weld is positioned for 1-G. In the 1-G position the pipe is rotated, while the arc is held pointing downwards the weld groove.

The weld groove can be prepared either by machining if available or the welder should use grinder to prepare pipe edges as specified in the table below. Note to follow the sketch for root face thickness, and angle of the bevel.

The welder will set the machine to the initial current as shown in the following pipe welding WPS, along with other information given in the welding specification given below.

<table>
<tr><th>Material</th><th>Edge preparation</th><th>Electrode type and diameter</th><th>Set-up type-position</th><th>Current and polarity</th></tr>
<tr><td>Carbon steel pipe diameter 3-inch nominal</td><td>Single V-groove butt weld Bevel angle 30° Groove angle 60°</td><td>E 6010, Diameter 2.4 mm</td><td>1 G (Pipe flat position) Root opening: 1.2 to 1.5 mm Root face: 1 – 1.2 mm</td><td>90-110 amperes, DCEP (see note below)</td></tr>
<tr><td colspan="5">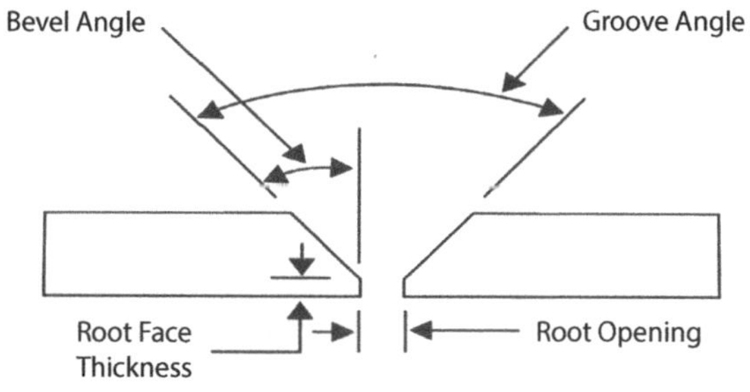

Welding Process: SMAW</td></tr>
<tr><td colspan="2">Note: start with low amperes and adjust to suitable level before starting on the actual practice piece.</td><td colspan="3">Practice till 20 consecutive welds come out in acceptable range.</td></tr>
</table>

Figure 2.11.9 This is a rotator with one end of the pipe held in a three-jaw, self-centering chuck the free end of the pipe rests on a free rotating roller, it can be raised or lowered to level the pipe to align the weld ends.

Once the pipes are set and aligned to maintain the root-opening (gap), welder may tack two pieces to hold them in place while weld is completed. For small diameter pipes three tacks placed equally spaced along the circumference of the pipe groove is sufficient, however for lager diameter four tacks may be necessary. If these tacks are made in the groove then, they should be removed by grinding as the weld progresses, alternatively if they are made outside the weld groove (as a bridge –linking two abutting pieces of pipe) they may be cut as the weld reached the take and weld progresses to the next tack location.

Figure 2.11.10 This rotator is similar to the one above except that the pipe end is placed on a motor driven set of rollers on one end, and the other end is on the set of idle rollers, which can be lowered or raised to align and level the weld joint.

Figure 2.11.11 A heavy-duty rotator.

Pictures of two bridge tacks are shown below, note that flange to pipe bridge tacks are made using pieces of material, whereas the pipe-to-pipe tack is made entirely of welding.

Figure 2.11.13 shows the flange to pipe tack using external piece of metal. These tacks will be removed as the welding reaches close to these tacks. This picture also shows very poor workmanship by the tack welder, there is arc strike on the pipe and flange, and the quality of tack is not very good.

After tacks are placed and surrounding surfaces are cleaned, free from weld slag and spatters, and all electrical connections are stablished and checked, the welder can start welding. Since this is a 1G welding position the welder will start on the top of the pipe and rotate the weld piece as the arc starts to go out of vertical down facing position to progress

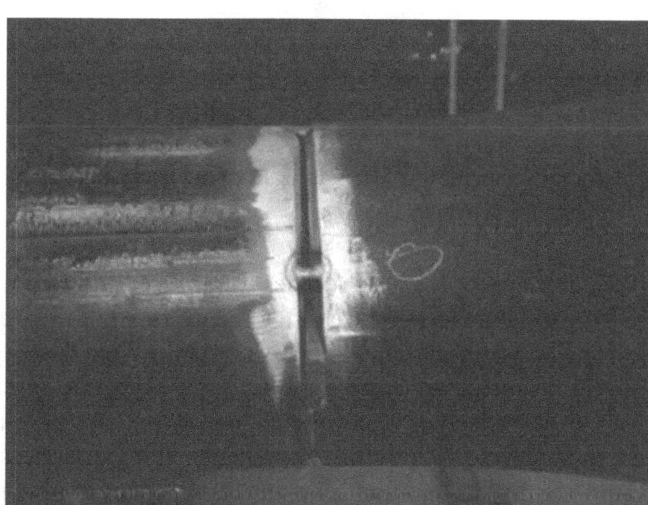

Figure 2.11.12 Weld tacks bridging two pieces of pipe.

Figure 2.11.13 Shows a removable tack.

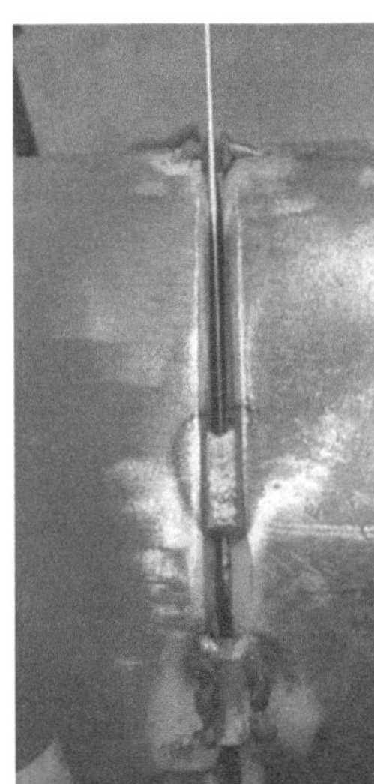

Figure 2.11.14 This picture shows both the bridge tack using external pieces of metal below, and just above that is the tack within the groove using welding.

the weld forward. Welder may stop as tacks reaches the arc area and remove those tacks by grinder, and progress the weld. It is a good welding practice to grind the weld where the arc is stopped to make a shallow groove to restart the arc. As the weld is completed, the scale is chipped off, and weld inspected for any anomalies – unacceptable anomalies are called defects, such as penetration or lack of it, undercuts, porosities, etc., and understand what caused those defects and try to correct in the subsequent welding practices.

The most common weld defects and their possible cause are given in the following table. This table is developed keeping in mind that the welding is done on basic mild steel material, and that the weld is not under any stress to develop cooling cracks. No metallurgical

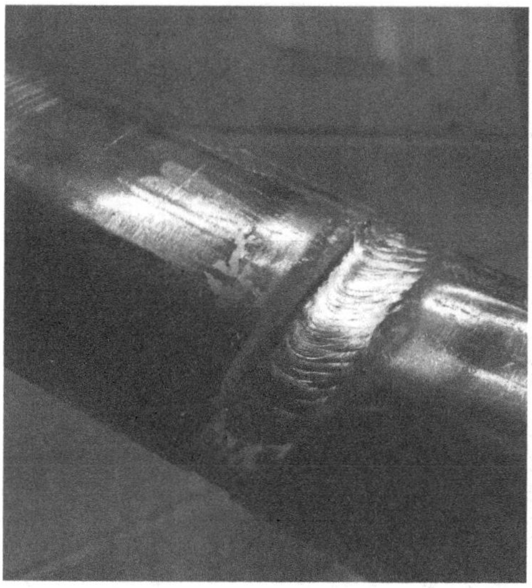

Figure 2.11.15 Typical CS pipe weld.

challenges are expected in welding of practice welds. However, conditions of welding electrodes, current draw, weld edge preparation and cleaning are all important factors, and they should all receive maximum attention.

The Table 2.11.10 below lists most common SMAW anomalies and their possible causes, and correction to remove those anomalies.

2.11.10 Pipe Welding Technique and Pipeline Welding

The pipe welding has a lot of common elements, that we have discussed in the basic learning to weld section of this chapter. Beyond these common features there are some points where the subject divides in two main parts the welding pipe (i) as in plant and welding, commonly referred as piping, and the other split is the (ii) welding a pipeline.

The approach in both cases is differ due to several factors related to the position of the pipe lay, and the demands of the material and the design specifications. Some of these are interdependent, that include the position of weld, then the choices available for the electrode, and the job's specific demand. In some cases, the welder may get a welding procedure where they may have to combine the approach to meet the engineering demand of the weld, most common example of the combination can be seen in thicker wall pipeline welding where often the root pass is deposited with vertical down hand technique but then switched over to vertical up technique for the rest of the weld. Here we will introduce some very common ways to weld both the plant piping, and pipeline welding and let the welder learn the complexities in their work environment and develop skills.

In welding arc manipulation, the control of the arc-heat, and weld pool is of the utmost importance, to produce a weld that is, (a) even on both sides of the bevel, and (b) has proper penetration through the metals being welded, (c) while managing the weld to be defect free, (d) and esthetically pleasing weld appearance is the ultimate goal. The techniques are developed around these principles.

Table 2.11.10 Common SMAW process anomalies and their suggested causes and corrections.

Type of weld discontinuity (defect)	Possible reasons	Correction
Porosity	Weld area not cleaned prior to welding, condition of the electrode.	Clean and remove dirt, oil, grease, and moisture from the weld area. Also see the electrode in use, if they are moist and have damaged coating, remove such electrodes from welding.
Slag inclusion in the weld (Very specific to SMAW process)	Improper weld-groove, improper weld end preparation prior to restart of weld during welding, leaving a grinding groove on the side of the weld groove that is not filled by weld metal, and slag is trapped in it.	Prepare good veld groove, of uniform shape and size. Grind restart area, and make a shallow groove to restart weld from behind the groove, prevent grinding damage to the side of the weld-groove.
Incomplete fusion	Low current, weld speed too fast.	Adjust current, and slow down welding speed. Allow arc to dwell just enough time to get desired level of fusion on the weld joint.
Excessive penetration/Burn through	High current, slow travel speed	Adjust current, adjust travel speed to get uniform melting under the arc.
Inadequate penetration (Lack of penetration)	Too tight root gap, low current, travel speed too fast.	Open the root gap, allow for the expansion of metal during the welding heat. Adjust current, slow down the travel speed.
Undercuts (a furrow or groove formed in the parent metal along the toe of the weld)	Excessive welding current, too long arc length, are two primary cause for undercut.	Lower the current, and maintain a stable arc length.
Profile of the weld	Improper welding technique, improper currents, poor alignment of weld pieces.	Adjust the current to suitable level, pay attention to the root opening and speed of travel of arc.
Cracks (Not likely to occur during welding of mild steel, in an unrestrained weld joints, as in the case of welding qualifications, or welding practices)	Cracks can occur due to metallurgical reasons of material being welded, or due to the restraint of the weld joints.	Ensure proper metallurgical steps like pre heat, Interpass and post weld heating steps are strictly followed as given the welding procedure.

2.11.10.1 Vertical Up Technique

SMAW process for welding in-plant piping often uses the Vertical Up technique. The technique uses (1) low current, (2) slow travel speed, (3) and often uses low hydrogen electrodes, this is possible, for among other reasons because of the slow speed and lower current, used for vertical up progression of weld.

This welding progression as the technique allows for heavier deposition rate, note that the slow speed and low current are factors that are responsible for this advantage. The technique may allow somewhat longer time to finish the weld, but that time delay is very short. Welds are often X-ray quality, and since low hydrogen electrodes can be used to produce a weld that has relatively low porosity, lower hardness and high ductility.

2.11.10.1.1 Procedure for Vertical-Up Welding Technique

The pipes to be welded often come coated with various types of protective coatings, or they may be a bit rusted. Before welding, all these foreign materials must be thoroughly removed and cleaned to bare steel surface. The beveled ends and the adjacent metal for about 1-inch further from the bevel, all around the circumference must be thoroughly cleaned. Cleaning is of utmost importance to achieve good quality weld that is free from defects. The bevel edge preparation for vertical up technique for a single V-grove weld is shown in the Figure 2.11.10.1 below, the figure also shows the welding electrode sizes that are commonly used for this type of weld joint set-up.

After cleaning the pipe bevel edges, the pipes are aligned and tacked at least four places, approximately equally spaced around the pipe circumference. These tacks are often removed as the welding progresses further. For the carbon steel plant piping often the root pass is done using a cellulosic electrode, this may or may not be in vertical-up progression, but often after the root-pass, the rest of the weld is switched to low hydrogen electrodes, which are easier to manage in the vertical-up progression.

For the root pass, the electrode is struck at the bottom (6 o'clock), of the joint, see Figure 2.11.10.2 below and arc is moved upwards, towards the top 12 o'clock position. The same processes are repeated on the other half of the pipe, on a larger pipes two welders may start together on each side of the pipe, but if a smaller diameter pipe is being welded then a single welder may weld one side after the other.

The penetration must be complete on both lips of the land of the weld bevel, and there should be some build-up on the inside of the joint, referred as bead. The inside bead should have even surface projecting approximately 1.6 mm or 0.0625 inch.

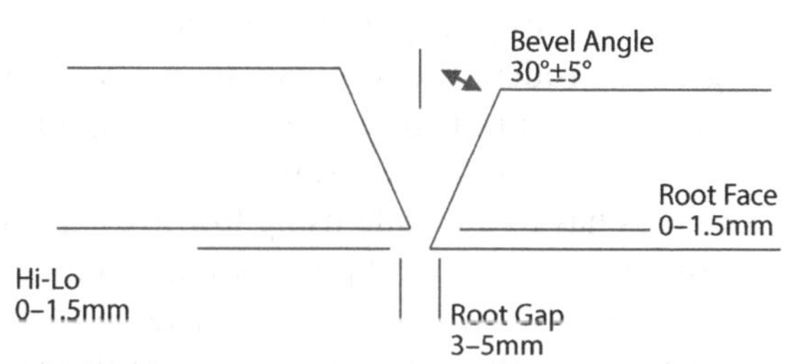

Figure 2.11.10.1 Bevel edge preparation for vertical-up pipe in 6G position.

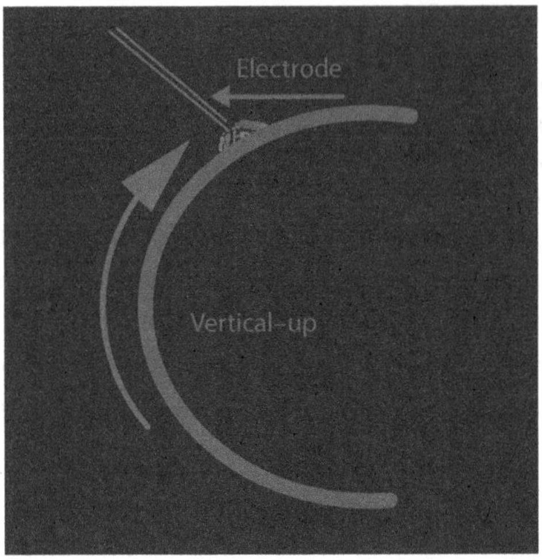

Figure 2.11.10.2 The vertical up progression - note the direction of electrode movement.

2.11.11 In-Plant Piping

As we have introduced above that the welding of piping in a plant conditions differ significantly from pipeline welding. This is due to several reasons, some of them are (1) the plant welding involves relatively smaller diameter pipes. (2) The piping in plants are not transportation lines, as such they are covered by a different set of design and construction codes. (3) The plant piping serves a variety of fluids that may require different properties from the weld than that is required of the pipeline transporting gas or liquid. (4) Plant piping provide some advantages to the welder that are not available to a pipeline welder, the position is one such advantage. This advantage changes several important variables for the pipeline welder.

For plant piping, the vertical up welding becomes necessary due to the use of low hydrogen electrodes. And the position of the weld or if the weld has backup ring the making of weld including the weld edge preparation becomes a little more complicated. We will discuss the simple bevel, welding from one side situation. Take note of the Figure 2.11.10.2 above, for the weld progression in vertical-up technique.

The root opening, and the land of the 30° to 37.5° bevel ranges from 0.094-inch to 0.125-inch (2.4 mm to 3.2 mm). In welding of carbon steel, and most of the low alloys it may be useful to use a cellulosic E XX-10 type of electrode for root pass to obtain desired penetration of the root pass, and then move to low hydrogen electrodes for rest of the fill and capping weld. In more stringent low hydrogen applications, the root pass also may be required to be done by low hydrogen electrodes. The root pass by E XX-10 electrode can be done in Vertical-Down progression, shown in Figure 2.11.11.1 below, note the difference in the direction of weld progression marked with the arrow.

In cases where the root pass has to be made using low hydrogen electrode as the pipe lays in fixed horizontal position (5G), the weld progression will be in upward progression (Vertical-Up) see Figure 2.11.10.2, this weld should be made in such a way that electrode arc is evenly maintained, chances are that if the current is low, the electrode would stick to the bevel-wall, or if the current is too high the electrode coating would over heat, making it

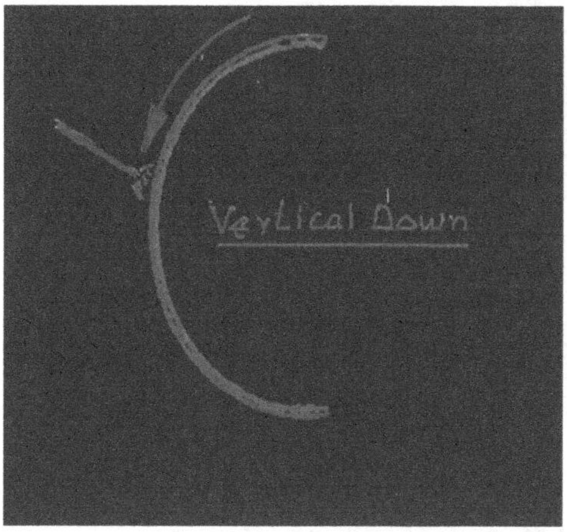

Figure 2.11.11.1 Vertical down progression.

unusable, or the arc control, and weld-pool maintenance will be very difficult to a point that welding will become near impossible.

In either case of cellulosic or low hydrogen electrode use, both edges of the joint bevel should be equally fused, while there is s even penetration of approximately 1.6 mm or 0.0626 inch. Care should be taken to stagger the start and stop of electrode. The weld cross section would look somewhat like as is shown in the Figure 2.11.11.2 below. Note the arrangement of various passes from root to cap pass.

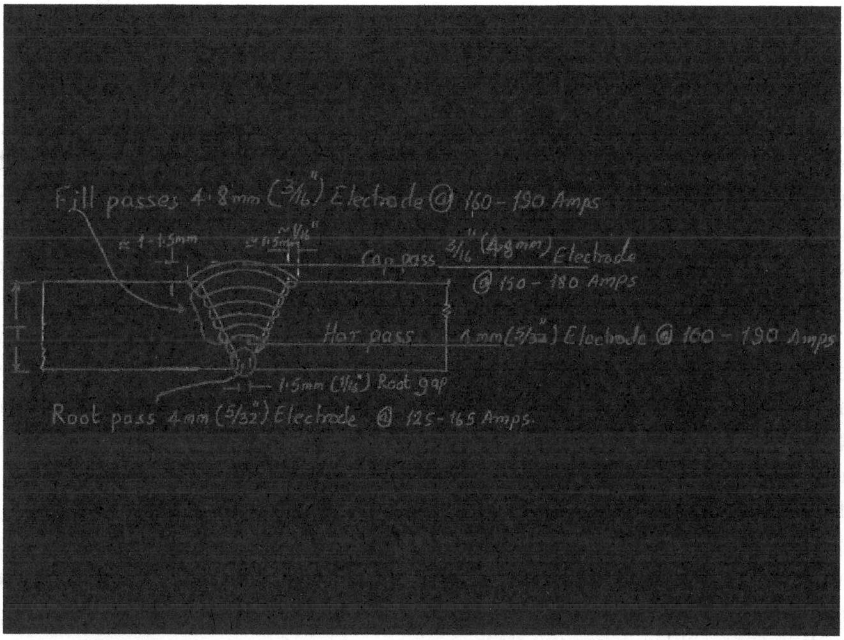

Figure 2.11.11.2 Weld profile of each pass.

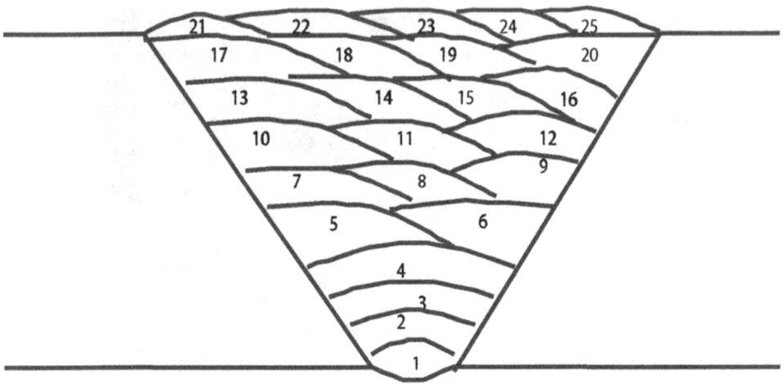

Figure 2.11.11.3 The sketch above shows a typical weld layers of several passes – note the sequencing numbers on each pass.

2.11.12 Pipeline Welding

For the pipeline welding the root pass with SMAW process is always done in the Vertical-down progression, and often the entire weld may be done in the vertical-up progression using E-XX10 type cellulosic electrodes. The weld bevel is different than the weld bevel for plant piping that is done in vertical-up progression. The vertical-down weld bevel is shown in the Figure 2.11.11.1 earlier in the section. The welding process goes by the following sequence, starting with root-pass. Figure 2.11.12 below gives a welding procedure details of a typical pipeline weld.

2.11.12.1 Making a Root Pass

The root pass is also referred as the stringer bead, it is made using a drag technique, in that the electrode coating rests on the side of the bevel, slightly above the landing of the weld edge preparation where the electrode makes contact. As the electrode strikes the arc and is drawn along the circumference of the weld in the bevel. The current is adjusted according to the size of the electrode chosen for welding, and also based on the quality of the fit up. As the electrode is dragged along the circumference at the bottom of the bevel, the arc creates a small key-hole just ahead of the arc, which is trailed by the melting weld-metal from the electrode and the melting steel from the sides of the bevel, filling the key-hole as the weld progresses. The continuation of size, and maintenance of the key-hole is the key to good root pass, if the hole closes it means that the current is low, if current is higher than necessary the key-hole will burn through, in both these situations the root pass weld will not be good.

Another difficulty can arise in the form of insufficient penetration, or lack of fusion, this could also be due to the low current, and poor fit-up. On the other hand, excessive root-gap, or higher current can cause burn through, and globular deposits would appear inside the pipe, often called grapes. Internal undercut, also called wagon tracks is also caused by excessive current, and or larger root-gap in the joint set-up.

Specifically, tailored electrodes are manufactured by leading manufacturers to address these conditions, however it is the welder's ability and dexterity of hand that matters the most even if the electrode is specifically tailored to eliminate some of these common defects in welding. The Table 2.11.12 below for details of some of the specific pipeline stringer-bead defects and their possible corrections.

Hot pass

After depositing the root pass the welder should quickly cleaned and inspect the weld, for any discontinuities in the weld, and remove any slag still adhered to the weld area with wire brush or light grinding. Grinding may also be used to improve and even-out the contour of the weld, for the even deposition of the next pass. This is as a preparation for next pass of weld called hot-pass. Often the hot-pass is expected to be started in shortest possible time after the root pass is completed. The time-delay between the root-pass and the hot-pass is often an essential variable in pipeline welding procedures.

Hot pass may require higher heat to remove defects like wagon tracks. For this, a whipping action of the arc is helpful, the quick passing of the arc over the wagon tracks helps in removing wagon tracks. Higher current is normally used but welder must ensure that they have full control over the arc, and not use very excessive current. The travel speed control is also important at this point, as excessive travel speed can result in pinholes.

Filler passes

After successful and satisfactory hot pass, the rest of the passes are required to fill the V-grove to the top before the final cap pass is deposited. The fill passes may be varied in numbers, and in some specific specifications may require low heat input depositions. But in the basic pipeline welding these passes are deposited with weave motion across the V-grove. Travel speed control like above in hot-pass is essential to avoid porosity and pin-hole type defects. If the current is excessive the welder will notice while welding the crowding of the arc. Even a slightly higher current than necessary will produce longer arc puddle, spatter, deep ripples, and on solidifying it will be noticeable. As a guidance, the correct current will give a crater of approximately 9 mm or 0.375 inch that is free of slag.

Stripper pass

Stripper passes are required to adjust the evenness of the deposited weld metal around the circumference. This is necessary for proper final capping pass. Stripper passes are small

Table 2.11.12 Weld defects and suggested changes that can correct them.

	Changes that can correct those defects possible					
Defects	**Current**	**Root gap**	**Land**	**Hi-Lo**	**Bevel angle**	**Polarity – change to**
Wagon Tracks	Decrease	Increase	Decrease	Decrease	Increase	Erratic (may be arc blow)
Internal Undercut	Decrease	Decrease	Increase	Decrease	Decrease	DC-
Lack of Fusion	Increase	Increase	Decrease	Decrease	Increase	DC+
Cracking	Decrease	Erratic	Decrease	Decrease	Increase	DC-
Hollow Bead	Decrease	Increase	Decrease	Decrease	Increase	DC-

Erratic means that there is no definitive result from the change.

length of weld that are sometimes required, mostly between 2 to 5 o'clock and 7 to 10 o'clock orientation of the weld.

Cap pass or Cover pass

These are the top passes, the final appearance of the weld, they are slightly above the adjacent pipe surface, normally about 1 mm to 1.5 mm (0.03 to 0.0625 inch) above the pipe surface. The overlap on the adjacent pipe metal beyond the grove edge on the pipe surface, by about 0.0625 inch or 1.5 mm. The weld should gradually blend into the both sides of the adjacent parent metal, without causing any abrupt change in the profile.

The Figure 2.11.12 below shows the weld profiles of each pass described above.

2.12 Welding Other Metals

Normally budding welders starts their career by training on carbon steel material, because training on a carbon steel plate or pipe is less costly affair than starting on any other material. After a welder has mastered the welding techniques, and understood the arc behavior, and gained confidence in handling the variance of welding variables, they are now ready to move if so desired, to weld other metals. However, some welders started their career on metals other than steel, and have done very good.

Welding other metals includes developing the skill to read and follow the instructions given in the welding procedures, as stated in the previous sections. Welding procedures can vary significantly with information and demands of the job. For example, they may require weld be done in a specific position, they may require tighter heat input control, the weld may be tested for its internal defects by X-ray or ultrasonic testing methods, or for its hardness, ductility and toughness, among other mechanical properties.

Alloy steel welding is very demanding, as they require very tight control on welding parameters, they may also require post weld heat treatment. In such cases the welder's job is not done until all inspections and all tests are completed and weld is finally accepted as sound to the specified requirements.

So, it is important for new welders to develop and master skills and habits that will make them most desired welder for welding expensive materials, and with minimum possibility of defects and repairs, and rejections.

2.12.1 SMAW Welding Aluminum

Steel is often considered the 'default' metal to for structural construction, and therefore for welding. During welding the steel gives stage-wise indication of what is happening to it by the application of heat, primarily by change of color, forming weld-pool etc. The human brain is trained to expect somewhat similar temperature indications from other materials including aluminum when heated during welding aluminum. Aluminum does not give these visual cues. Therein lies the fundamental challenge of welding aluminum.

Welding aluminum is different than welding steel because of one very basic property, unlike steel, aluminum when heated does not change color, this lack of color change does not give heat perspective to the welder, especially to the new welder, welder does not know what is happening with the material as it is heated, and it simply collapses under the heat. Low melting point of aluminum brings the catastrophe even earlier than the welder realizes.

Aluminum is much greater conductor of heat that changes the behavior of heat dwell on the parts to be welded.

Before we start on the subject of welding aluminum, it is important to know how aluminum is available in the market for welding and fabrication. Understanding of various grades of aluminum makes the welding it a much easier task.

2.12.2 Aluminum Alloys and Their Characteristics

There are seven series of wrought aluminum alloys, and as a welder, or a welding engineer, or manager of aluminum fabrication, it is imperative to know these alloys and their differences and understand their applications and characteristics.

2.12.2.1 1xxx Series Alloys

Aluminum in 1xxx grade classification is non-heat treatable, their ultimate tensile strength varies from 10 to 27 ksi, this series is often referred to as the pure aluminum series because it is required to have 99.0% minimum aluminum. They are weldable. However, because of their narrow melting range, they require certain considerations in order to produce acceptable welding procedures. When considered for fabrication, these alloys are selected primarily for their superior corrosion resistance such as in specialized chemical tanks and piping, or for their excellent electrical conductivity as in bus bar applications. These alloys have relatively poor mechanical properties and would seldom be considered for general structural applications. These base alloys are often welded with matching filler material or with 4xxx filler alloys dependent on application and performance requirements.

2.12.2.2 2xxx Series Alloys

Aluminum under 2xxx classification is heat treatable, and their ultimate tensile strength ranges from 27 to 62 ksi.

These grades are alloys of aluminum and copper where copper is added ranging from 0.7 to 6.8%, these are high strength aluminum alloys, higher strength makes these grades most useful for aerospace and aircraft applications. They have excellent strength over a wide range of temperature. Some of these 2xxx grade aluminum alloys are considered non-weldable by the arc welding processes, because of their susceptibility to hot cracking and stress corrosion cracking. However, others are successfully arc welded, with the correct welding procedures. These base materials are often welded with high strength 2xxx series filler alloys designed to match their performance, but can sometimes be welded with the 4xxx series fillers containing silicon or silicon and copper, dependent on the application and service requirements.

2.12.2.3 3xxx Series Alloys

Alloys in the 3xxx series are non-heat treatable, their ultimate tensile strength is relatively low ranging from 16 to 41 ksi. These are aluminum, manganese alloys, where the added manganese is between 0.05 to 1.8%. The alloy has moderate strength, but it has very good corrosion resistance, good formability, and are suited for use at elevated temperatures. They are commonly used for manufacture of cooking utensils, but in the industrial sector they are used, for the equipment where heat is a factor, such as the heat exchangers in vehicles and power plants. Their moderate strength, however, often precludes their consideration for structural applications. These base alloys are welded with 1xxx, 4xxx and 5xxx series filler alloys, dependent on their specific chemistry and particular application and service requirements.

2.12.2.4 4xxx Series Alloys

These aluminum alloys comprise of some heat treatable and some non-heat treatable grades. Their ultimate tensile strength varies from 25 to 55 ksi. These are alloys of aluminum and silicon, where the silicon additions range 0.6 to 21.5%.

As noted above the series consists of both heat treatable and non-heat treatable alloys. The aluminum alloyed with the Silicon, reduces the melting point and improves its fluidity when molten. These characteristics are desirable for filler materials used for both fusion welding and brazing. Consequently, this series of alloys is predominantly found as filler material. Silicon, independently in aluminum, is non-heat treatable; however, a number of these silicon alloys have been designed to have additions of magnesium or copper, which provides them with the ability to respond favorably to solution heat treatment. Typically, these heat treatable filler alloys are used only when a welded component is to be subjected to post weld heat treatments.

2.12.2.5 5xxx Series Alloys

Aluminum alloys in 5xxx series are non-heat treatable, their ultimate tensile strength varies from 18 to 51 ksi. Alloys of aluminum and magnesium where the alloying magnesium is between 0.2 to 6.2%. The alloys in series have the highest strength of the non-heat treatable alloys. In addition, this alloy series is readily weldable, and for these reasons they are used for a wide variety of applications such as shipbuilding, transportation, pressure vessels, bridges and buildings. These magnesium-based aluminum alloys are often welded with filler alloys, which are selected after consideration of the magnesium content of the base material, and the application and service conditions of the welded component. Alloys in this series with more than 3.0% magnesium are not recommended for elevated temperature service above 66°C (150°F) because of their potential for sensitization and subsequent susceptibility to stress corrosion cracking. Base alloys with less than approximately 2.5% magnesium are often welded successfully with the 5xxx or 4xxx series filler alloys. The base alloy 5052 is generally recognized as the maximum magnesium content base alloy that can be welded with a 4xxx series filler alloy. Because of problems associated with eutectic melting and associated poor as-welded mechanical properties, it is not recommended to weld material in this alloy series, which contain higher amounts of magnesium with the 4xxx series fillers. The higher magnesium base materials are only welded with 5xxx filler alloys, which generally match the base alloy composition.

2.12.2.6 6XXX Series Alloys

These alloys are heat treatable, and have the ultimate tensile strength between 18 to 58 ksi. These aluminum magnesium and silicon alloys contain alloying element up to 1%. Often the Silicon content is about 0.40 to 0.7 x actual magnesium content. Rest being the aluminum.

These alloys are widely used throughout the welding fabrication industry, predominantly in the form of extrusions, and incorporated in many structural components. The addition of magnesium and silicon to aluminum produces a compound of magnesium-silicide, which provides this material its ability to become solution heat treated for improved strength. These alloys are naturally solidification crack sensitive, and for this reason, they should not be arc welded autogenously, that is, without filler material. The addition of adequate amounts of filler material during the arc welding process is essential in order to provide dilution of the base material, thereby preventing the hot cracking problem. They are welded with both 4xxx and 5xxx filler materials, dependent on the application and service requirements.

2.12.2.7 7XXX Series Alloys

These are heat treatable alloys. Their ultimate tensile strength ranges from 32 to 88 ksi. The series is alloys of aluminum with 0.8 to 12.0% zinc. This makes these alloys one of the highest strength aluminum alloys. These alloys are often used in high performance applications such as aircraft, aerospace, and competitive sporting equipment. Like the 2xxx series of alloys, this series incorporates alloys which are considered unsuitable candidates for arc welding, and others, which are often arc welded successfully. The commonly welded alloys in this series, such as 7005, are predominantly welded with the 5xxx series filler alloys.

2.12.3 The Aluminum Alloy Temper and Designation System

Further from the 7 series of aluminum alloys discussed above, they are identified and designated by number system, that indicates their series, their specific heat treatment, and other methods used to impart strength, and their hardening level to the specific alloy.

Since we are discussing about welding of aluminum and its various alloys, it is necessary to learn how the material is designated and identified. The Aluminum Association Inc. https://www.aluminum.org is registers, maintains, and allocate new designations to aluminum alloys. It is estimated that there are over 400 wrought aluminum, and wrought aluminum alloys and over 200 aluminum alloys in the form of castings and ingots registered with the Aluminum Association.

The alloy chemical composition limits for all of the registered alloys are contained in the Aluminum Association's two books.

1. **Teal Book** entitled "International Alloy Designations and Chemical Composition Limits for Wrought Aluminum and Wrought Aluminum Alloys" and
2. The **Pink Book** entitled "Designations and Chemical Composition Limits for Aluminum Alloys in the Form of Castings and Ingot".

These two publications can be extremely useful to the welding engineer when developing aluminum welding procedures, and when the consideration of chemistry and its association with crack sensitivity is of importance.

Aluminum alloys can be categorized into a number of groups based on the particular material's specific characteristics, theses could be its ability to respond to thermal and mechanical treatment, or the primary alloying element added to the aluminum alloy. When we consider the numbering and identification system used for aluminum alloys, the theses characteristics are identified through these numbers. The wrought and cast aluminums have different systems of identification; the wrought having a 4-digit system, and the castings have a combination of a 3-digit and 1-decimal place system.

2.12.4 Wrought Alloy Designation System

So far, the discussion has been about various grades of aluminum, we need to further extend this body of knowledge about those four digits in the alloy designation system. These 4-digits together tell a detailed story about that specific alloy. This is tabulated in Table 2.12.1 and described here in. The numbering XXXX, for the wrought aluminum alloy identification system consist of following;

The first digit (Xxxx) indicates the principal alloying element, which has been added to the aluminum alloy and is often used to describe the aluminum alloy series, i.e., 1000 series, 2000 series, 3000 series, up to 8000 series as shown in Table 2.12.1.

The second single digit (xXxx), if different from 0, indicates a modification of the specific alloy, and the third and fourth digits (xxXX) are arbitrary numbers given to identify a specific alloy in the series. Example: In alloy 5183, the number 5 indicates that it is of the magnesium alloy series, the 1 indicates that it is the 1st modification to the original alloy 5083, and the 83 identifies it in the 5xxx series.

The only exception to this alloy numbering system is with the 1xxx series aluminum alloys (pure aluminums) in which case, the last 2 digits provide the minimum aluminum percentage above 99%, i.e., Alloy 1350 would mean that the alloy has 0.50% more purity above the 99% aluminum, the meaning of this alloy 1350 in pure aluminum group, contains 99.50% of aluminum purity and this alloy has had three modification.

2.12.5 Cast Alloy Designation

The cast alloy designation system is based on a 3 digit-plus decimal designation xxx.x, for example 356.0. The first digit (Xxx.x) indicates the principal alloying element, which has been added to the aluminum alloy as shown in Table 2.12.5 below.

The two successive digits (xXX.x) together are arbitrary numbers given to identify a specific alloy in the series. The number following the decimal point is a binary system of identification where .0 is casting and 0.1 or 0.2 signify that the metal is an ingot. If a capital letter prefix is used it indicates a modification to a specific alloy.

Example: Alloy - A356.0 the capital A (Axxx.x) indicates a modification of alloy 356.0. The number 3 (A3xx.x) indicates that it is of the silicon plus copper and/or magnesium series. The 56 (Ax56.0) identifies the alloy within the 3xx.x series, and the .0 (Axxx.0) indicates that it is a final shape casting and not an ingot.

Table 2.12.1 Aluminum alloy designation system.

Alloy designation	Grade, primary alloying element	Secondary alloying element	Weldability and heat treatment
1xxx	99.000% Pure Aluminum		Weldable and requires no PWHT
2xxx	Copper	Magnesium may be added.	Weldable and PWHT required
3xxx	Silicon, Copper and Manganese		Weldable and PWHT not required
4xxx	Silicon		Weldable and PWHT not required
5xxx	Magnesium		Weldable and PWHT not required
6xxx	Magnesium	Silicon	Weldable and PWHT is required
7xxx	Zinc	Copper and magnesium may be added.	Weldable and PWHT is required
8xxx	Tin		
9xxx	All other alloys		Weldability and heat treatment to be determined.

Table 2.12.5 Cast aluminum designation and numbering system.

Alloy series	Principal alloying element
1xx.x	99.000% minimum Aluminum
2xx.x	Copper
3xx.x	Silicon Plus Copper and/or Magnesium
4xx.x	Silicon
5xx.x	Magnesium
6xx.x	Unused Series
7xx.x	Zinc
8xx.x	Tin
9xx.x	Other Elements

2.12.6 The Aluminum Temper Designation System

From the description and Table 2.12.5, presented above, one can see that there are different series of aluminum alloys, what is important to note is the considerable differences in their characteristics. The first point to recognize, after understanding the identification system, is that there are two distinctly different types of aluminum within the series described above. These are the Heat Treatable Aluminum alloys that can be imparted strength through the treatment of heat and cooling cycle. The other group consist of the non-heat Treatable Aluminum alloys. This distinction is particularly important when considering the effects of arc welding on these two types of materials.

The 1xxx, 3xxx, and 5xxx series wrought aluminum alloys are non-heat treatable and are strain hardenable only. The 2xxx, 6xxx, and 7xxx series wrought aluminum alloys are heat treatable and the 4xxx series consist of both heat treatable and non-heat treatable alloys. The 2xx.x, 3xx.x, 4xx.x and 7xx.x series cast alloys are heat treatable. Strain hardening is not generally applied to castings.

The heat treatable alloys acquire their optimum mechanical properties through a process of thermal treatment, the most common thermal treatments applied are the Solution Heat Treatment and Artificial Aging.

Solution Heat Treatment is the process of heating the alloy to an elevated temperature to around 482°C (about 990°F) in order to put the alloying elements or compounds into solution. This is followed by quenching, usually in water, to produce a supersaturated solution at room temperature. Solution heat treatment is usually followed by aging. Aging is the precipitation of a portion of the elements or compounds from a supersaturated solution in order to yield desirable properties. The aging process is divided into two types: aging at room temperature, which is termed natural aging, and aging at elevated temperatures termed artificial aging. Artificial aging temperatures are typically about 160°C (about 320°F). Many heat treatable aluminum alloys are used for welding fabrication in their solution heat treated and artificially aged condition.

The non-heat treatable alloys acquire their optimum mechanical properties through Strain Hardening. Strain hardening is the method of increasing strength through the application of cold working. The Temper Designation System addresses the material conditions called tempers. The Temper Designation System is an extension of the alloy numbering system and consists of a series of alpha-numeric identification, which follow the alloy designation number that we previously discussed, and are connected by a hyphen. Examples: 6061-T6, 6063-T4, 5052-H32, 5083-H112. Table 2.12.6 below details the letters used for temper designations, and meaning of those letters.

Further to the basic temper designation, shown in the Table 2.12.6 above, there are two subdivision categories.

(i) addressing the "H" Temper – Strain Hardening, and
(ii) The other addressing the "T" Temper – Thermally Treated designation.
(iii) H Temper – Strain Hardened

- The first digit after the H indicates a basic operation:

H1 – Strain Hardened Only.
H2 – Strain Hardened and Partially Annealed.

Table 2.12.6 Temper designation letters and meaning.

Designating letter	What the letter means
F	As fabricated – Applies to products of a forming process in which no special control over thermal or strain hardening conditions is employed
O	Annealed – Applies to product which has been heated to produce the lowest strength condition to improve ductility and dimensional stability
H	Strain Hardened – Applies to products which are strengthened through cold-working. The strain hardening may be followed by supplementary thermal treatment, which produces some reduction in strength. The "H" is always followed by two or more digits (see table 4)
W	Solution Heat-Treated – An unstable temper applicable only to alloys which age spontaneously at room temperature after solution heat-treatment
T	Thermally Treated - To produce stable tempers other than F, O, or H. Applies to product which has been heat-treated, sometimes with supplementary strain-hardening, to produce a stable temper. The "T" is always followed by one or more digits (see table 5)

H3 – Strain Hardened and Stabilized.
H4 – Strain Hardened and Lacquered or Painted.

- The second digit after the H indicates the degree of strain hardening:

HX2 – Quarter Hard,
HX4 – Half Hard
HX6 – Three-Quarters Hard
HX8 – Full Hard
HX9 – Extra Hard

- Subdivisions of Temper – Thermally Treated

T1 - Naturally aged after cooling from an elevated temperature shaping process, such as extruding.
T2 - Cold worked after cooling from an elevated temperature shaping process and then naturally aged.
T3 - Solution heat treated, cold worked and naturally aged.
T4 - Solution heat treated and naturally aged.
T5 - Artificially aged after cooling from an elevated temperature shaping process.
T6 - Solution heat treated and artificially aged.
T7 - Solution heat treated and stabilized (overaged).
T8 - Solution heat treated, cold worked and artificially aged.
T9 - Solution heat treated, artificially aged and cold worked.
T10 - Cold worked after cooling from an elevated temperature shaping process and then artificially aged.

Additional digits indicate stress relief.

Examples:

TX51 or TXX51 – Stress relieved by stretching.

TX52 or TXX52 – Stress relieved by compressing.

Aluminum is an active metal and it forms oxides while being heated for welding, this makes it harder to develop a weld-metal pool suitable for welding aluminum. When combined with the metal's high heat conductivity and low melting point, it is very easy for a new welder to completely melt the aluminum pieces involved in the process. As a result, and to prevent such situation, two very important steps need to be taken.

1. First step to arc welding aluminum is to clean the base metal of any oxides or solvent oils, and prevent oxide formation during welding. And
2. The second step is to be mindful of aluminum's behavior under heat,

    i. Aluminum does not change color,
    ii. Aluminum does not show molten pool,
    iii. Aluminum has high conductivity of heat,
    iv. Aluminum has very low melting point as compared to the steel.

On the positive side, if the welding technique is mastered, the welding of aluminum is less energy intensive, and therefore easier to weld than steel.

Another important point to note is that most of the welding machines are tailored to weld steel, they are mostly calibrated and set with those parameters. So, it is important to know the features of the welding machine, and if required reprogram the machine to suite aluminum welding.

SMAW process is the least expensive method for aluminum welding, whereby the shielding is provided by the coating around the electrode itself. And since it does not use shielding gases as other processes like GTAW or GMAW use, it has some advantages over other welding processes used for welding aluminum, especially when it comes to weld in relatively open area where protection from wind may be of particular concern.

SMAW electrode do create some slate while welding, and this requires considerable cleanup at the end of the job.

2.12.6.1 Aluminum Welding Electrodes

Number of electrodes producers supply electrodes in various diameter and lengths. Common diameters are 3/32" (2.4 mm) and 1/8" inch (3.2 mm). Electrodes are available in 12-inch, 16-inch and 18-inch lengths. For welding with 1/8" electrode with 120 amps to maximum 160 amps works very well for flat downhill position on plate.

Alloy 4043 is one of the oldest and most widely used welding and brazing alloys. E 4043 electrode can be classed as the general-purpose welding electrode that can be used for welding a variety of aluminum grades, like, 3003, 3004, 5052, 6061-T4, 6061-T6, 6063-T6 and 2014-T6.

The alloy contains about 4% to 6 % silicon in it, silicon is a wetting agent, and it allows the weld metal fluidity and this property makes the electrode use very welder-friendly. The silicon additions result in improved fluidity (wetting action) to make the alloy a preferred

Figure 2.12.6 Aluminum fillet weld-bend testing.

choice by welders. The weld metal is less sensitive to weld cracking, and produces brighter, almost smut free welds.

2.12.6.2 Electrical Parameters

Welding is done using DC with electrode positive current (DCEP).

Electrode positive polarity (DCEP) helps constant removal of oxide layers that form during welding, on the base metal, while the electrode flux keeps oxides from forming on the electrode metal and the molten pool. The flux protects the weld as it cools and forms a protective barrier, which works great even in breezy conditions. The key to welding aluminum is (a) to move faster than the welding speed used for welding steel. (b) No weave movement, since the aluminum flows better than steel, and helped by presence of silicon the fluidity is even better.

AC welding electrodes are also marketed, these electrodes have AC suffix at end of the electrode class marking. Some welders find alternating current very helpful, the alternating current functions in somewhat similar manner to remove oxides from the surface as the welding progresses.

2.12.7 SMAW Welding of Stainless Steel

Welding stainless using SMAW process somewhat similar to welding basic carbon steel, however there is significant metallurgical issues that are different from, the carbon steel and those need to be understood to make a sound stainless steel weld.

The term stainless steel is very generic, and somewhat misleading. It is essential to understand some fundamental metallurgical aspects of "Stainless-steel".

There are number of groups, and within those groups may be different grades of steels that may be generally classified stainless steels. The following basic discussions are included for those who wants to be more educated about the material they intend to weld.

2.12.8 Introduction to Stainless-Steels

Stainless steels are iron base alloys that contain a minimum of approximately 11% chromium (Cr), this is an important number that is needed to create a passivating layer of chromium-rich oxide to prevent rusting on the surface. Several stainless-steel grades are produced to address specific demands of the environment that they are expected to protect the material, for this purpose other elements are also added to the steel. Nickel, Copper, Titanium, Aluminum, Silicon, Molybdenum, Niobium, Nitrogen Sulfur and Selenium are some of the commonly used elements that are alloyed to impart the required properties to the specific steel grade.

Stainless steels are marketed in various shapes and sizes and in various finishes. However, for industrial application and ease of understanding we limit this to the following.

Bars are available in all grades and come in rounds, squares, octagons, or hexagons of 0.25 inch (6 millimeter) in size.

Wire is usually available up to 0.5 inch (13 millimeter) in diameter or size.

Plate is defined as rectangular shapes of more than 0.1875 inch (5 millimeter) thick and over 10 inches (250 millimeter) wide.

Strip are defined as rectangular shapes of less than 0.185 inch (5 millimeter) thick and less than 24 inches (610 millimeter) wide.

Sheet are defined as rectangular shapes of less than 0.1875 (5 millimeter) thick and more than 24 (610 millimeter) wide.

Further processing is done to produce specific shapes like pipes tubes and structural shapes.

2.12.8.1 Cutting Stainless Steel for Fabrication

Cutting operation is usually necessary to obtain the desired blank shape or size. This is done to trim the part to final size. Mechanical cutting is accomplished by a variety of methods, including straight shearing by guillotine knives, circle shearing by circular knives horizontally and vertically positioned. Blanking by metal punches and dies to punch out the shape by shearing. Nibbling is a process used for cutting by blanking out a series of overlapping holes and is ideally suited for irregular shapes, only some stainless steels can be saw cut by high-speed steel blades.

Normally stainless steel cannot be cut using flame cutting. Another cutting method that is used is the Plasma jet cutting, to make a cut this process uses an ionized gas column in conjunction with an electric arc passing through a small orifice the force of the gas and high heat generated by the gas plasma, melts the metal and makes the cut.

2.12.8.2 Finishing

Surface finish is an important requirement for stainless steel products, depending on the end application. The surface finish is a very important property and it is specified on the fabrication drawing. The main reasons to consider for specifying the surface finish could include one or all of the following.

1. The appearance.
2. Process convenience.
3. Corrosion protection.

4. To facilitate lubrication - often rougher surface is specified.
5. Surface condition specific to facilitate further manufacturing steps.

Stainless steels are produced in variety of different metallurgical class, and in different grades. Each of these classes, and grades have very specific properties to offer, and may behave differently during fabrication and welding. Broadly speaking Stainless steels are of following types,

- Austenitic Stainless steels
- Martensitic Stainless steels
- Ferritic Stainless steels

There are number of grades within these groups.

There are also stainless steels, that are classified as super austenitic and an entirely different class that is termed as Duplex steel that has number of different grades within.

Knowing each class and grade is an important part of being a good "Stainless Steel" welder.

2.12.9 Fabrication of Stainless Steel

After the stainless steel in its various forms are packed and shipped to the fabricator or end user, a variety of secondary processes are needed to make it useful for specific service. Further shaping is accomplished using a variety of secondary processing that may include rolling forming, press forming, forging, press drawing, and extrusion, welding cutting, additional heat treating, machining, and cleaning processes.

2.12.9.1 Why Use Stainless Steel

Stainless steel is chosen for a project based on any single or a combination of following specific properties.

1. Resistance to corrosion
2. Resistance to oxidation at higher temperatures.
3. Good mechanical properties at room temperature.
4. Good mechanical properties at low temperature.
5. Good mechanical properties at high temperature.
6. Aesthetic values - Good Appearance.

Stainless steels are corrosion resistant material which relay on surface passivity for resistance to corrosion attack. Use of these materials is governed by the oxidizing characteristics of the environment. For more oxidizing conditions, stainless steel is superior in performance to several more-noble metals and alloys, available for fabrication by welding.

2.12.10 General Welding Characteristics

All the Chromium-nickel (300 Series) austenitic stainless steels with the exception of high Sulphur or selenium added free-machining grade (AISI 303) are easily welded. The welded joints are tough and ductile in "as welded" condition. These welds if used in non-corrosive or mildly corrosive services do not require any post weld heat treatment.

In welding a temperature gradient is achieved ranging from room temperature to molten steel. The area that is heated in this process rage from 425°C to 900°C (800°F to 1 650°F) becomes sensitized as carbides are precipitated. This carbide precipitation may affect the life of equipment under severe corrosive conditions, therefore annealing the welded parts is recommended to restore optimum corrosion resistance. This annealing process is called solution annealing. The process of solution annealing consists of heating the material up to a temperature above sensitizing temperature, generally 1 100°C (or about 1 960°F) and holding it long enough for the carbon to go into solution. After this, the material is quickly cooled to prevent the carbon falling out of the solution. Solution annealed material is in its most corrosion-resistant and ductile condition.

It is not always possible to solution anneal the weldments. This could be for various reasons like its size or other post fabrication process etc. If for any reason the welded part cannot be annealed then extreme care should be taken in welding stainless steel and either a low carbon grade of stainless steels that have less than 0.03% carbon or AISI 321 or 347 grade of steel should be selected.

The grades 321 and 347 are stabilized alloys. They contain Titanium and Columbium respectively. The ratio of these elements is dictated by the percentage of carbon in these steels for example the minimum amount of Titanium in Grade 321 is about 5 times that of carbon in the steel. Similarly, Columbium in Grade 347 is about 10 times that of carbon content of the steel. But accurate ratio of carbon to Titanium or Columbium has to be designed in the steel, based on the specific requirement of the service environment including welding requirements of the project. When these steels are heated during welding and material reaches in the sensitizing range carbide precipitation occurs, as in any other grades of stainless steel, except that due to high affinity of carbon to these elements, the carbide of titanium and Columbium is precipitated, thus leaving Chromium free from Intergranular corrosion. In some very specialized conditions Grade 321 may be further heat treated by heating to 815°C – 900°C range for 2 to 4 hours and air cooled to secure complete carbide precipitation as stable titanium carbides. This heat treatment is some time called stress relief treatment.

In low carbon (less than 0.03%) grades like 304L and 316L the carbon is so low that during welding the heat does not precipitate carbides. The use of these grades of steel is limited to service temperature below 425°C to 870°C (800°F to 1 600°F).

Welds in other corrosion resistant steels like, ferritic and martensitic stainless steels are not as ductile and tough as in austenitic steels discussed above. Ferritic alloy type 405, 430, 442, and 446 are more readily weldable. The martensitic grades like 403 and 410 are more weldable than types 420 and 440.

2.12.10.1 Protection Against Oxidation

A welding process must protect the molten weld metal from the atmosphere during arc transfer and solidification. Fluxing may be required to remove the chromium and other oxides from the surface and the molten weld metal. Gas shielded processes do not require fluxing since the shielding gas prevents oxidation.

2.12.11 Welding and Joining Stainless Steel

The practical welding techniques make it easy to weld in the flat and horizontal position. However, welding in vertical uphill position is tougher to master. This is because stainless steels have lower conductivity, as a result the electrode heats up quickly and then maintaining the arc is difficult. Even if the weld is completed the appearance of the weld bead is not uniform, it looks peaked to the crown. Experienced welders recognize the situation beforehand, and control the arc by lowering the amperage. This specific situation is caused due the lower thermal conductivity of the stainless-steel property.

During the welding, the weld deposition rate changes very soon from the rate at the start, this is because of the hot electrode starts to melt faster than when it was colder, at the start of the weld.

Some welders use weave technique to spread out the weld metal, but not all engineering specifications accept this method due to the metallurgical changes that happens due to the longer arc-dwell time.

Good welding technique is to start at the minimum possible level of amperage, and keep control of the arc during the welding by adjusting the amperage.

A new welder training to weld stainless steel, should practice on the same machine that is actually to be used for the test welding. This will allow familiarization with the accuracy of the output amperage, and also the adjustments response of the machine. In other words, the welder should familiarize with the machine before actual welding.

2.12.12 Importance of Cleaning Before and After Welding

The high chromium content of stainless steels promotes the formation of tenacious oxides that must be removed for good welding results. Surface contaminants affect stainless steel welds to a greater extent than they affect carbon steel welds.

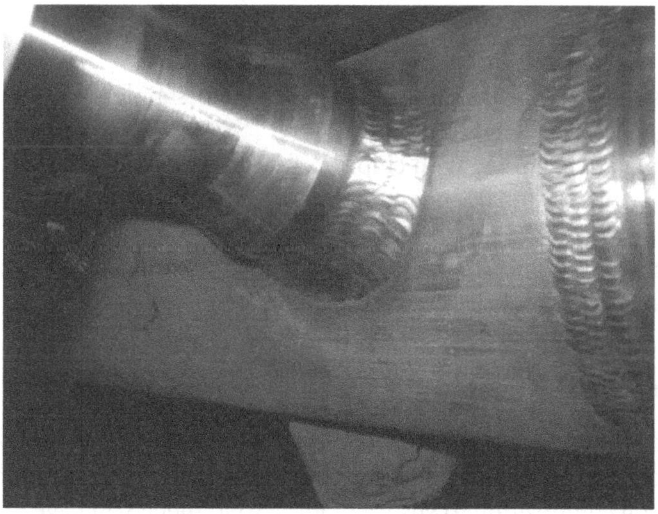

Figure 2.12.12 Typical stainless-steel pipe weld, and weld-o-let on the header.

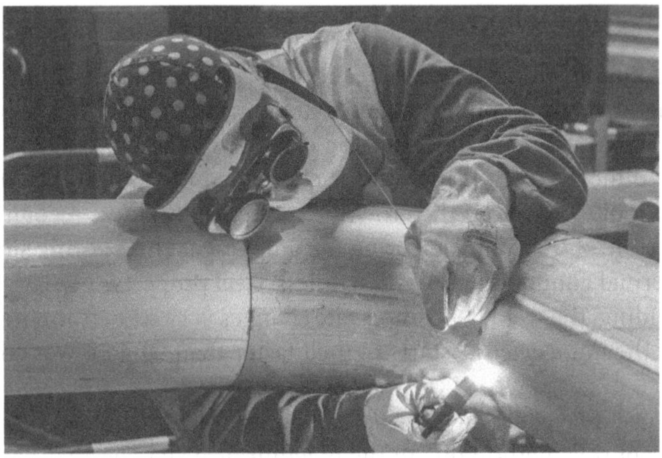

Figure 2.12.13 Pipe is assembled and prior to welding, the welder is tacking them with the GTAW process.

The surfaces to be joined must be cleaned prior to welding. An area surrounding the weld joint for at least 12 mm (0.5 inch) is cleaned, far wider area if thicker plates are being welded. As a general rule of thumb cleaning a band of metal about 1.5 times the plate thickness will be considered good practice, it would avoid contaminations.

Special care in surface cleaning is required for gas shielded welding because of the absence of fluxing.

Carbon contamination can adversely affect the metallurgical characteristics and corrosion resistance of stainless steel. Pickup of carbon contaminants or embedded particles must be prevented.

Suitable solvents are used to remove hydrocarbon and other contaminants such as cutting fluids, grease, oil, waxes, and primers. Light oxide films can be removed by pickling or by carefully selected mechanical means of cleaning. Acceptable pre-weld cleaning techniques include:

1. Stainless steel wire brushes that are used only for stainless.
2. Blasting with clean sand or grit.
3. Machining and grinding with chloride-free cutting fluid.
4. Pickling with 10% to 20% nitric acid solution.

Thorough post weld cleaning is required to remove welding slag. The surface discoloration is best removed by wire brushing or mechanical polishing.

2.12.13 Filler Metals

Covered electrodes and bare solid and cored wire are available to weld most of the grades. The chemical composition of allweld metal deposits vary slightly from the corresponding stainless-steel metal composition to ensure that the weld metal will have the desired microstructure and be free from cracks.

Covered electrodes are available with either lime or titania coverings. Lime type covering (EXXX-15) are suitable for DCEP (electrode positive, reverse polarity) only and EXXX-16 electrodes are suitable for AC or DCEP. Type EXXX–15 electrode coverings produce deeper penetration and Type EXXX –16 electrode coverings produce a smoother surface finish when used with DCEP current polarity system.

Covered electrodes must be stored in sealed containers or holding ovens at temperature range of 100°C to 125°C (200°F to 250°F).

2.12.14 Austenitic Stainless Steels

As we have noted, there are various types of stainless steels, some metallurgically very distanced from other while there are others that are in somewhat similar group but are made different by specially added elements. These precise additions are made for specific properties beyond the basic properties of its group.

Such additions are of essential concern for welding as the resultant weld metal must meet the parent metal properties as closely as possible.

2.12.14.1 Metallurgical Concerns Associated with Welding Austenitic Stainless Steels

The properties of austenitic stainless steel as value to industry are, high ductility, excellent toughness, strength, corrosion resistance, weldability, and excellent formability and castability. Because of these properties, austenitic stainless steels are the most commonly used material from the family of stainless steels. There is a virtual continuum of austenitic alloys containing Fe, Cr, Ni, and Mo.

The distinction between highly alloyed stainless steels and lower-alloyed nickel-base alloys is somewhat arbitrary. Nickel alloys must satisfy either (a) Cr >19; Ni>29.5; Mo >2.5, or (b) C r>14.5; Ni >52; Mo >12 over their entire composition range, those that do not meet this criterion e.g., alloy 20, UNS N08020 are classified as stainless steels.

Unified Numbering System (UNS) alloy numbers starting with a prefix "S" are grouped with the austenitic stainless steels discussed in this section whilst super austenitic stainless steels, defined in this report as alloys with *FPREN* greater than 30.0 are discussed in the section entitled "Super austenitic Stainless Steels. The alloys that begin with prefix "N" are grouped with the nickel-based alloy.

2.12.14.2 Mechanical Properties of Stainless Steels

The lower-alloyed austenitic stainless steels, such as type 304 and 316 (UNS S30400 and S31600), possess yield strengths around 30 to 40 ksi (210 to 280 MPa) in the annealed condition. Some higher-alloyed austenitic stainless steels with nitrogen have higher yield strengths.

Cold working often increases strength, especially in higher-alloyed austenitic stainless steels.

Cold deformation during fabrication, although less severe than that applied during temper rolling, can produce martensite in some austenitic stainless steels, thereby increasing their susceptibility to hydrogen embrittlement. Fabrication processes can also induce residual stresses that may help increased prospects of stress corrosion cracking (SCC).

Many of the common austenitic stainless steels can be readily welded using matching filler metals. Higher alloy grades are normally weldable, but non-matching, over-alloyed nickel-base filler metals are used.

Generally, these alloys are readily weldable whether for longitudinal seam welded pipe or girth welds, etc., via a range of processes (SAW, GTAW, GMAW, SMAW, and PAW etc.). They are usually welded with matching composition filler metal. For some of the molybdenum-containing grades, over-alloyed filler with an extra 1-3% molybdenum and higher nickel content are specified. Normally, argon is used for both shielding and backing gases. Austenitic stainless steels typically require care in welding and adherence to good stainless-steel welding practice.

Since these alloys are in austenitic phase and they do not have phase transformation on cooling they do not require Preheat or Post weld heat treatment (PWHT). Except in some specific cases where solution annealing may be specified after welding and hot working.

Welding technology for the typical austenitic stainless steels is common practice using standard consumables like, ER308L, etc. Welding of higher-strength (650-690 MPa UTS), 200-series austenitic stainless steels can be done with standard 308 L-type fillers if matching the strength of the base metal is not critical. Use of duplex ER2209 filler metal (this is a bare wire filler metal, not an electrode) is one way of matching or exceeding the strength of the base metal, but toughness and embrittlement concerns restrict use of this approach to service temperatures of about - 40 to 315 °C (-40 to 600 °F). For cryogenic applications, such as liquefied natural gas (LNG) equipment, use of less-standard fillers such as E16-8-2 or 316 L Mn, or use of nickel-base fillers such as UNS N06022, are often of selected to take advantage of necessary strength and cryogenic toughness.

2.12.15 Welding of Austenitic Stainless Steels

At the beginning of this chapter, we discussed general requirements of welding and addressed fundamental essentials like sensitization control, difference in welding stainless steel and carbon steel and importance of weld hygiene. We take those discussions further in the subsequent paragraphs.

The austenitic steels have high coefficients of thermal expansion and low thermal conductivity and are particularly susceptible to distortion during welding. They have better ductility and toughness than carbon steels and excellent notch toughness even at cryogenic temperatures. They are stronger than carbon steels above 500°C (1 000°F) and have good oxidation resistance.

When austenitic stainless steels are joined to carbon steel, construction codes often mandate PWHT in the temperature range of about 550-675°C (1 025-1 250°F) for relief of residual stresses. These heat treatments can adversely affect intergranular corrosion and stress corrosion cracking (SCC) resistance of the stainless steel. In these situations, use of a low-carbon grade type 304-L or stabilized grade like type 347-L is recommended. It may however be noted that the service temperature range of 304-L and 347-L is limited to –40°C to 315°C. If PWHT is one of the limiting factors imposed on the design, then other alternatives must be thought, one of them is to butter the carbon steel as described below.

A buttering layer of austenitic stainless-steel electrode/filler wire is deposited to the carbon steel. Often the selection is based on the available chromium in the as deposited weld metal, after compensating for the dilution, if the resulting weld metal is close to the austenitic level then the buttering is completed, the most common interface electrode for welding and buttering austenitic steel and carbon steel is E 309 grade of consumable. Once the buttering is completed, the buttered carbon steel is heat treated as required. After the post weld heat treatment (PWHT) is carried out to relive stresses in carbon steel, then the stainless-steel member is welded on to the PWHT buttered section of carbon steel.

Heat input ranges and inter-pass temperatures are not especially important for the austenitic stainless steels. Interpass temperatures up to 150 °C (300 °F) is usually permissible.

After welding, there is usually a heat tint in the weld/HAZ area, and it is usual to remove this in some applications where feasible. The heat tint is often removed by manual (but not mechanical) brushing, by mechanical abrasives, such as a flapper wheel, or by a suitable pickling paste or gel. The inside of small-bore pipe welds, flowlines, and clad line pipe are difficult to clean, this requires that welding procedure uses inter gas as backing to keep the inside surface oxide free.

Weld deposit microstructures are very different from wrought metals with the same composition. 100% austenitic structure in welds is prone to cracking. Some amount of ferrite is essential to control cracking of these welds. In austenitic welds small pools of delta ferrite often form and carbides may also be present. The weld metal ferrite control is essential; Schaeffler and DeLong diagrams are used to predict as-welded microstructures. These diagrams are also useful in selecting the electrode for keeping control on the ferrite in austenitic steel weld metal.

2.12.16 Super-Austenitic Stainless Steels

Material Properties and Applications

Like the austenitic stainless steels, the super-austenitic stainless steels are highly ductile; they have excellent toughness, high strength, outstanding corrosion resistance, good weldability, and excellent formability. The super-austenitic stainless steels are normally used where greater resistance to corrosion, especially protection from chloride pitting and crevice corrosion, is needed. Super-austenitic stainless steels are defined as austenitic, iron-based alloys that have PREN greater than 40.

The higher PREN values are achieved primarily by adding nitrogen (N) to these alloys, and upper working temperature limits of 400°C (750°F) are generally imposed by industry codes to prevent Σ (sigma) or X (chi) phase embrittlement.

Many of the super-austenitic stainless steels, especially those containing nitrogen, possess higher yield strengths in the annealed condition than the standard austenitic stainless steels.

These alloys are generally available in most product forms (bar, wrought plate, castings, pipe, forgings, etc.), and are usually supplied in the solution-annealed condition. Specialized parts (fittings, fasteners, etc.) of these grades are not generally inventoried by stockiest, but often custom manufactured. As an alternative suitable nickel alloys are often selected. Castings are solution annealed to homogenize the as-cast, cored, dendritic structure.

The super-austenitic stainless steels are generally used in the solution-annealed and rapid-cooled condition. Prolonged heating in the temperature range of about 510 to 1 070 °C (950 to 1 960 °F) can cause precipitation of carbides, nitrides, or intermetallic phases. This precipitation increases susceptibility to intergranular corrosion, IGSCC, and chloride pitting and crevice corrosion. These alloys cannot be strengthened by heat treatment.

The higher alloy contents of the super-austenitic stainless steels give them greater resistance to the formation of martensite during cold working. Thus, they show reduced susceptibility to hydrogen embrittlement (HE) as compared to the austenitic stainless steels. The higher alloy contents of the super-austenitic stainless steels also give them greater resistance to stress corrosion cracking (SCC) when compared to the austenitic stainless steels. Thus, fabrication induced residual stresses are less likely to cause SCC in these alloys.

2.12.17 Welding and Joining of Supper-Austenitic Stainless Steels

These alloys are easily weldable, and by a range of processes like, SAW, GTAW, GMAW, SMAW, etc. All these processes have been discussed in this book. Because in part these alloys rely on molybdenum to provide corrosion resistance properties, the segregation of molybdenum that occurs during solidification of welds can impair the corrosion resistance of welds. To counter this effect the normal practice is to use over-matching composition filler metal. The over-alloyed fillers typically contain about 1.5 times the molybdenum to that of the base metal. To keep these high levels of molybdenum in solid solution, nickel-based fillers are used. This ensures that even the solute-depleted dendrite cores will have local PREN values meeting or exceeding the PREN of the base metal. Examples of such filler metals include UNS N06625, N06022, and N06686 wires.

From the above description of weldability of these molybdenum alloyed steels the autogenous welding is not normally recommended for super-austenitic stainless steels, although it has been performed successfully in thin sections of < 2 mm and with special gases. Some specialized post weld solution annealing can also restore autogenous welds to corrosion resistance levels approaching that of the base metal. Normally, argon is used for shielding gas, but the addition of small amounts of nitrogen is considered more beneficial. Backing gases can be argon or nitrogen. As with high alloy stainless steels, care is typically taken in welding super-austenitic, and adherence to good stainless-steel welding practice is generally considered to be good practice. Suitable joint design, inter-pass temperatures and low heat inputs is the path to successful welding of these alloys. Pre and post weld heat treatment are not required for super-austenitic stainless steels.

During welding, heat input ranges and inter-pass temperatures is very carefully monitored and controlled. The maximum permissible heat input and inter-pass temperature increase with section thickness. The values for these parameters generally decrease as the alloy content increases. Specialist publications for suitable values for a specific joint should be consulted. If heat input or inter-pass temperatures is kept too high, the risk of precipitating sigma or chi phases in the HAZ or weld metal increases. These intermetallic phases are rich in chromium and molybdenum, thus leaving chromium depleted area around them, which is responsible for reducing the localized corrosion resistance in these metals.

The austenite stainless steels contain a combined total chromium, nickel, and manganese content of 24% or more, with the chrome generally above 16%. Nickel and manganese stabilize austenite to below room temperature.

Ferrite content is designated by ferrite number (FN). Ferrite is difficult to measure accurately although automated equipment is now available. The importance of ferrite in weld microstructure cannot be understated, as it increases resistance to hot cracking. Ferrite provides sites with good ductility for interstitial or tramp elements to distribute. However, excessive ferrite can also lower corrosion resistance and high temperature properties of material.

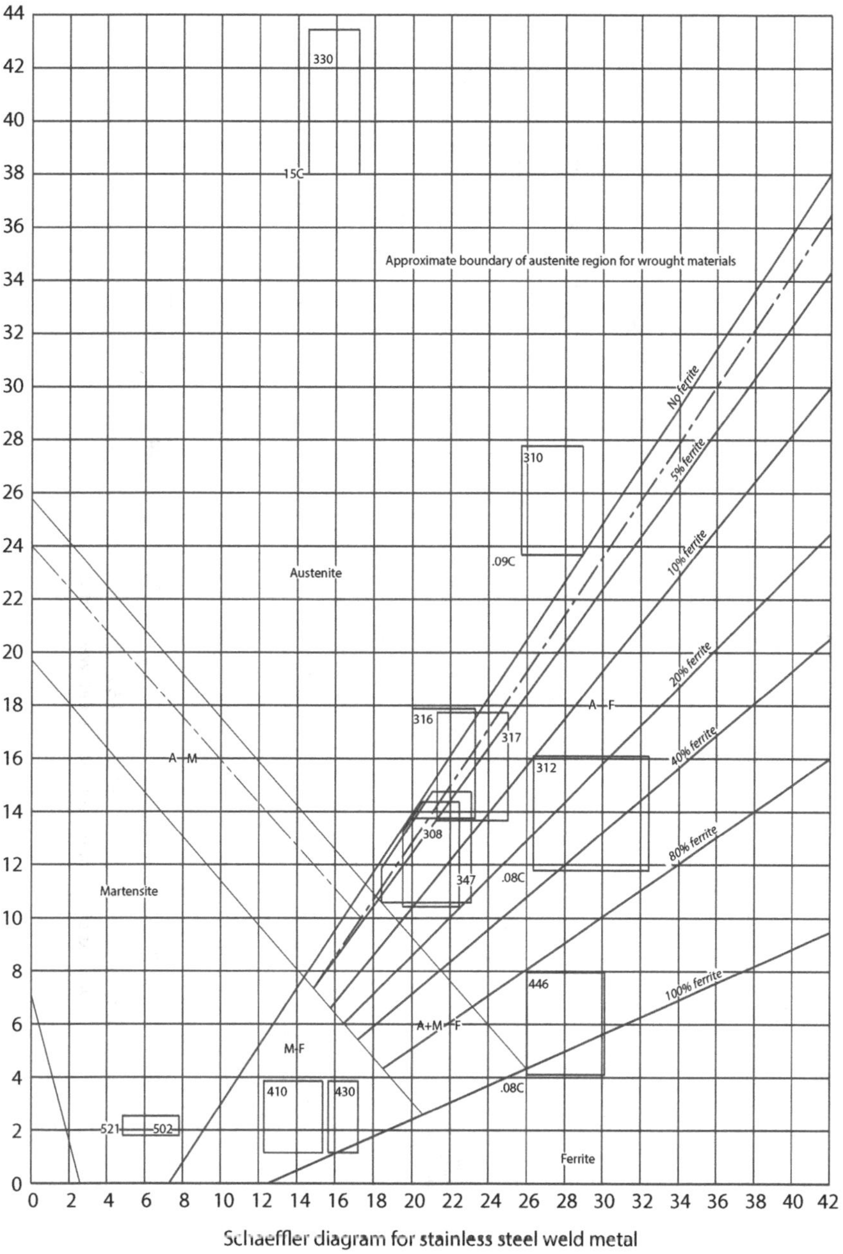

Figure 2.12.18 Schaeffler diagram.

Welding parameters and technique have a significant effect on the amount of ferrite formed and retained in a weld and they must be controlled to reduce desired properties in the weld.

2.12.17.1 Difficulties Associated with Welding Stainless Steel

Austenitic stainless-steel welding appears to be similar to the normal carbon steel welding. But little more in-depth observation will reveal that it is anything but similar to the carbon

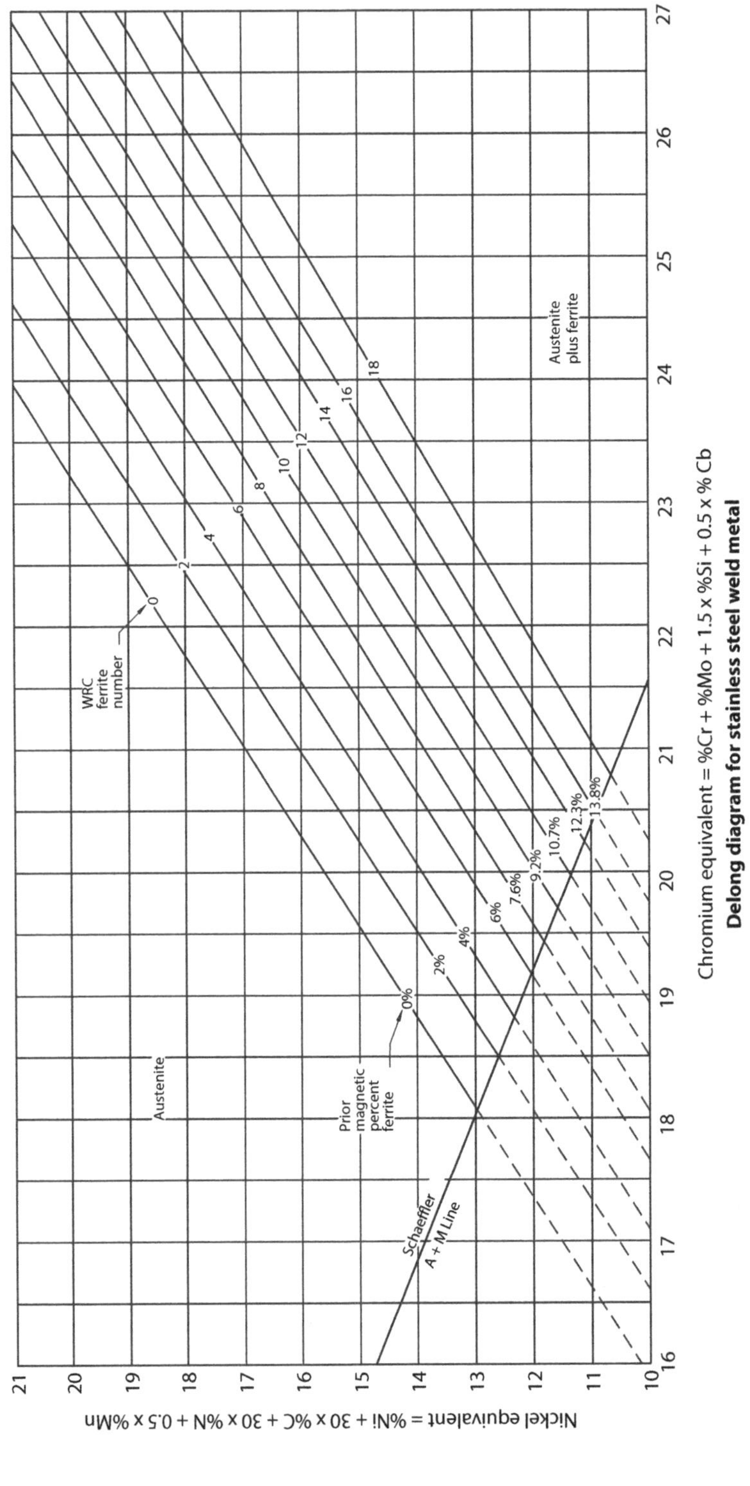

Figure 2.12.19 DeLong diagram.

steel welding. This is basically due to the metallurgical difference between the two types of steel, Stainless steel do not undergo the normal phase changes associated with carbon steels. While ferritic steels are austenitic and nonmagnetic at elevated temperatures, they transform to ferrite, pearlite, martensite and other phases as they are cooled through the transformation range.

In contrast when stainless steel is cooled, all or nearly all of the material retains the austenite structure at room temperature. No phase changes occur, and no hardness increase is associated with cooling. This property of austenitic steels reduces the need to pre heat or post heat.

When these steels are welded two very specific points need to be considered.

1. Carbide precipitation or sensitization.
2. Micro fissuring, ferrite content and Sigma phase formation.

The corrosion resistance of austenitic steels depends on the addition of various alloys of which Chromium is of primary importance.

In our earlier discussion on stainless steel welding, we introduced terms like sensitization, sigma phase etc., we discuss them in little more details. When austenitic steel is heated to a temperature range called "sensitization range" that is 425°C to 870°C (800°F to 1 600°F), some of the chromium in the solution can combine with any carbon that is available and form a chromium rich precipitate called chromium carbide, thus reducing the chromium in the steel, now less chromium is available in the alloy to carry its primary duty that is to resist corrosion. This reduction of chromium by forming Chromium carbide is called sensitization.

The steel in such venerable conditions is easily attacked by acidic environment. Under certain conditions, austenitic welds are subject to intergranular corrosion. A narrow band of metal in the heat-affected zone is always heated to the sensitizing range. The amount of precipitation that occurs is a function of the carbon content. Carbon levels higher than 0.03% are particularly susceptible.

Many metallurgical solutions are available to overcome the formation of sensitized area in austenitic welds; we discuss them in subsequent paragraphs.

Since we have discussed the importance of austenite in steel for corrosion resistance, we may make it clear that while pure austenite has excellent mechanical and corrosion resistance properties, its ability to absorb impurities without cracking during solidification is severely limited. During the cooling process the low meting impurities are forced out to the grain boundaries. Excessive amounts of such grain boundary accumulations weaken the material at the grain boundary. They create grain boundary flaws called "micro-fissures". This condition is of significance in welding because in relation to the parent metal a very small area is associated with welding. One method to reduce such micro-fissuring is to disperse these impurities among the disconnected gain boundaries that surround the island of second phase. This can be accomplished by modifying the chemical composition of steel that would allow creation of islands of ferrite in the welds. Ferrite has enormous capacity to absorb impurities, and ferrite islands are dispersed throughout the microstructure. The presence of ferrite has potential to very slightly reduce the corrosion resistance of steel, however it can certainly prevent micro fissuring, which is more serious and can lead to catastrophic failures. On the other hand, too much ferrite is also detrimental to the material. It can cause other problems called "Sigma phase" developed within the welding temperature range. This is a very brittle constituent and is caused by very evenly dispersed ferrite, even a small amount of sigma phase will embrittle large areas of stainless steel.

It is clear that both minimum and maximum limitations on ferrite phase are desirable in stainless steel welding to prevent micro-fissuring and sigma phase embrittlement.

Since carbon rapidly decreases the corrosion resistance and changes the properties of austenitic welds, it must be carefully controlled. Filler metals are usually chosen to match the base metal composition.

It is established that the Ferrite content must be appropriate for the weldment's service requirements. Many electrodes are developed to produce deposits containing ferrite limits within the range of 4 to 10 "ferritic number". There are methods that can measure ferrite in the weld, one of them is magnetic ferrite gage as described in AWS A 4.2.

2.12.18 Martensitic Stainless Steels

Martensitic stainless steels are metallurgically different from austenitic stainless steels discussed above. The weldability differs slightly and different grades of martensitic steels may present different challenges depending on the alloying elements.

2.12.18.1 Properties and Application

Martensitic stainless steels are Fe-Cr-C alloys that are capable of the austenite-martensite transformation under all cooling conditions. Compositions for most of martensitic steel alloys are covered by number of specifications, such as ASTM A 420 or API 13 Cr L80 and 420 M with additional small amounts of Ni and/or Mo. Although 9Cr1Mo is not strictly a martensitic stainless steel; it is often included in this alloy group, especially because of challenges associated with welding of 9-Cr-Mo steel is in many ways similar to this group of materials.

The martensitic stainless steels are generally used in the quenched and tempered, or normalized and tempered condition. For services where hydrogen evolution or presence of Sulphur is expected as in sour gas services in oil and gas industry a maximum hardness of 22 HRC is specified by most of the specifications, and for most of the alloys. Some of the alloys like type 410 and 420 develop quench crack if quenched in water, so they are quenched only in oil, or polymer, or air-cooled before tempering.

Some alloys like type 410, type 415, and J91540 (CA6NM) receive a second temper treatment called "Double tempering" at a temperature lower than the first tempering temperature, to reduce the untampered martensite in type 410, type 415, and J91540 (CA6NM). Double tempering has not been shown to improve resistance to stress corrosion cracking in type 420 tubular products and for 9Cr1Mo tubular or forgings.

The mechanical properties of typical base metal strength (SMYS) are grouped as 414 MPa (60 ksi), 517 MPa (75 ksi), 552 MPa (80 ksi), and 586 MPa (85 ksi), with hardness controlled to the maximum 22 or 23 HRC, and often have specified to maximum yield strength of 95 to 100 ksi (660 to 690 MPa). For sour service applications the tubular products are generally used according to the API Specification 5CT, or L80, strength level; forgings and castings are generally specified with hardness not exceeding 22 on Rockwell C scale. Higher strengths are used in sweet service; however, corrosion resistance and ductility are adversely affected as the strength of steel is increased.

2.12.18.2 Welding Martensitic Stainless Steels

Weld design strength levels range from 414 MPa (60 ksi) upward, but they can be different than the parent metal; for example, a 552 MPa (80 ksi) mandrel could be welded with a duplex or austenitic stainless filler metal that results in a lower weld joint strength, provided this has been considered to meet the design and operation demands.

The martensitic stainless steels are easy to work with, including welding, the welding processes used include SMAW, GMAW (MIG/MAG), FCAW, GTAW (TIG), SAW, EBW, and laser beam welding (LBW) Typical welding consumables include 410 Ni Mo (matching weld metal), or 2209, 309LSi (overmatching consumables), while some limited application of autogenous welding is also practiced. These alloys are not used in the as-welded condition in more demanding environments like, sour service.

Extreme care is typically required when these alloys are welded, because they are susceptible to high hardness. Tubing and casing are generally not welded.

When welding type 410, high pre-heat temperatures are used. The alloys classified as type 410, type 415 (F6NM), and J91540 (CA6NM) are tempered again as a post weld heat treatment after welding to ensure that they have maximum specified strength and hardness. These alloys have been welded using nominally matching filler metals. The use of non-matching austenitic types consumables can increase the risk of fusion boundary cracking in sour service, this increase in fusion boundary cracking is irrespective of the hardness limits in the weld area.

These alloys are known for moderate corrosion resistance, heat resistance up to 535°C (1 000°F), relatively low cost, and the ability to develop a wide range of properties by heat treatment.

If left in as-welded condition, the intergranular (sensitization) cracking is common occurrence in both sweet (CO_2 containing) and sour conditions. These problems also arise as a result of poor PWHT cycles, where the treatment has been ineffective in refining structure and reducing HAZ hardness.

They are capable of air hardening from temperatures above 815°C (1 500°F) for nearly all section thicknesses. Maximum hardness is achieved by quenching from above 950°C (1 750°F). They lack toughness in the as-hardened condition and are usually tempered.

Martensitic alloys can be welded in any heat treat condition. Hardened materials will lose strength in the portion of the heat affected zone. With a carbon content of 0.08% and 12% Cr (Type 410), the heat affected zone will have a fully martensitic structure after welding. The steep thermal gradients and low thermal conductivity combined with volumetric changes during phase transformation can cause cold cracking.

The hardness of the heat-affected zone depends primarily on the carbon content and can be controlled to some degree by developing an effective welding procedure. As the hardness of the heat affected zone increases, its susceptibility to cold cracking become greater and its toughness decreases.

Weldability is improved when austenitic stainless steel filler is used because it will have low yield strength and good ductility. This also minimizes the strain imposed on the heat-affected zone.

Martensitic steels are subject to hydrogen-induced cracking like low alloy steels. Covered electrodes used for welding must be low-hydrogen and maintained in dry condition.

Preheating and good inter-pass temperature control is the best means to avoid cracking. Preheating is normally done in the 200°C to 315°C (400°F to 600°F) range. Carbon content, joint thickness, filler metal, welding process, and degree of restraint are all factors in determining the pre-heat, heat input, and post weld heat treatment requirements.

Post weld heat treatment is performed to temper or anneal the weld metal and heat affected zone with aim to decrease hardness and improve toughness, and to decrease the residual stresses associated with welding. Subcritical annealing and annealing are performed. When matching filler metal is used, the weldments can be quench hardened and tempered to produce uniform mechanical properties.

Types 416 and 416Se are free machining grades that must be welded with caution to minimize the hydrogen pickup. ER312 austenitic filler metal is recommended for welding type 416 and 416Se alloys, since it can tolerate the sulfur and selenium additions.

Type 431 stainless can have high enough carbon to cause heat affected zone cracking if proper preheat, preheat maintenance, and slow cooling procedures are not followed.

2.12.19 Welding Ferritic Stainless Steels

Ferritic stainless steels are another metallurgical variation the stainless-steel grades. As the name suggest these steels are corrosion resistant and retain the ferrite as primary grain structure.

2.12.19.1 Properties and Application

Ferritic stainless steels are Fe-Cr-C alloys with ferrite stabilizers such as aluminum (Al), columbium (Cb), molybdenum (Mo), and titanium (Ti) to inhibit the formation of austenite on heating. Therefore, they are non-hardenable. In annealed conditions, Lower-alloy ferritic stainless steels have mechanical properties somewhat similar to the low-alloy austenitic stainless steels like type 304. The typical yield strength is in the rage of 30 to 50 ksi (205 to 345 MPa). Alloys with increased chromium, molybdenum, and nickel content have higher strengths.

In the high-chromium-containing alloys such as UNS S44626, the welding procedure typically developed to minimize interstitial pickup during welding and to retain material toughness. These alloys are predominantly utilized as thinwalled tubing products.

These alloys generally exhibit a loss of toughness with increasing section thickness, and a maximum thickness has been established for each alloy depending on the toughness requirements. In high-chromium-containing alloys, the interstitial contents have been carefully controlled for this purpose.

First generation ferritic alloys (Types 430, 422, 446) are subject to intergranular corrosion after welding and exhibit low toughness.

Second generation ferritic alloys (Types 405 and 409) are lower in chromium and carbon and have powerful ferrite formers and carbide formers to reduce the amount of carbon in solid solution. Although they are largely ferritic, some martensite can form as a result of welding or heat treating. Ferritic alloys are low cost, have useful corrosion resistance with low toughness properties.

Recent improvements in melting practice have resulted in third generation ferritic alloys with very low carbon and addition of nitrogen e.g., Types 444 and 26-1 steel. Stabilizing with powerful carbide formers reduces their susceptibility to intergranular cracking after welding, improves toughness, and reduces susceptibility to pitting corrosion in chloride environments and to stress corrosion cracking.

The most important metallurgical characteristic of the ferritic alloy is the presence of enough chromium and other stabilizers to effectively prevent the formation of austenite at elevated temperature. Most grades do form some small amount of austenite since interstitials are present.

Since austenite does not form and the ferrite is stable at all temperatures up to melting, these steels cannot be hardened by quenching. The small amounts of austenite which may be present and transform to martensite are easily accommodated by the soft ferrite.

Annealing treatment at 760°C to 815°C (1 400°F to 1 500°F) is required to restore optimum corrosion resistance after welding. Ferritic stainless steels cannot be strengthened appreciably by heat treatment. These steels are generally used in the annealed condition. The cooling rate from the annealing temperature chosen depends on the particular alloy. The importance of proper heat treatment is emphasized by the fact that the higher-chromium-containing alloys are subject to embrittlement by sigma or alpha prime phase if not properly heat-treated.

All ferrites if heated above 927°C (1 700°F) are susceptible to severe grain growth, due to this the material toughness is reduced and it can only be restored by cold working and annealing.

2.12.20 Welding Ferritic Steel

Types 430, 434, 442, and 446 are susceptible to cold cracking when welds are made under heavy restraint. A 150°C (300°F) preheat can minimize residual stresses that contribute to cracking. These steel grades are also susceptible to intergranular corrosion.

Filler material selection would include any of the three available options.

1. Matching compositions,
2. Use of austenitic stainless steels consumables, and
3. Use of nickel alloy consumables.

Matching fillers are normally used only for Types 409 and 430. Austenitic stainless steels electrode or filler wire matching E 309 or E312 grade or nickel alloys are often selected for dissimilar welds.

The need for preheating is determined by the chemical composition, desired mechanical properties, thickness, and conditions of restraint. High temperatures can cause excessive grain growth and heat affected zone cracking can occur in some grades. Low 150°C (300°F) and inter-pass temperatures are usually recommended.

If post weld heat treatment is deemed necessary, it is done in the 700°C (1 300°F) to 843°C (1 550°F) range to prevent excessive grain growth. Rapid cooling through the 538°C (1 000°F) to 371°C (700°F) range is necessary to prevent embrittlement.

2.12.21 Precipitation Hardening (PH) Stainless Steels

2.12.21.1 Properties and Application of Precipitation Hardening Steels

Precipitation hardening (PH) stainless steels can develop high strength with simple heat treatments. They have good corrosion and oxidation resistance without the loss of toughness and ductility that is normally associated with high strength materials. Precipitation hardening is promoted by alloying elements such as copper (Cu), titanium (Ti), columbium (Cb), and aluminum (Al). Submicroscopic precipitates formed during the ageing treatment increase hardness and strength.

Martensitic PH steels provide a martensitic structure which is then aged for additional strength. Semi austenitic precipitation hardened steels are re-heated to form martensite and also aged. Austenitic precipitation hardened steels remain austenitic after cooling and strength is obtained by the ageing treatment.

As a group, the precipitation hardened steels have corrosion resistance comparable to the more common austenitic stainless steels. Corrosion resistance is dependent on the heat treatment and the resulting microstructure. Welding can reduce corrosion resistance by over-aging and sensitization.

Precipitation hardened steels tend to become embrittled after exposure to temperatures above 300°C (580°F), particularly if heated for long periods of time in the range of 370°C to 427°C (700°F to 800°F) temperature. After welding, the maximum mechanical and corrosion resistance properties can be obtained by solution heat treatment followed by ageing. For some applications, only ageing treatment is sufficient.

Martensite precipitation hardened steels are often fabricated in the annealed or overaged condition to minimize restraint cracking. Solution heat treatment and ageing is performed after fabrication.

2.12.22 Welding Precipitation Hardened (PH) Steels

The Semi-austenitic precipitation hardened steels are welded in all conditions. Austenitic conditioning and ageing are performed after welding for maximum mechanical properties. Austenitic precipitation hardened steels are difficult to weld because of cracking problems. Matching, nickel alloy, or austenitic filler materials are used. The selection of suitable filler metal is dependent on the post-weld heat treatment and final property requirements.

The following are the key points that must be kept in mind for selection of material as well as welding of all stainless steels discussed thus far.

- Thermal expansion, thermal conductivity, and electrical resistivity have significant effects on the weldability of stainless steels.
- The relatively high coefficient of thermal expansion and low thermal conductivity of austenitic stainless steel require better control of distortion during welding.
- Low thermal conductivity for all stainless steels indicate that less heat input is required.
- The weldability of the martensitic stainless steels is affected mainly by hardenability that can lead to cold cracking.

Table 2.12.23 Stainless steel welding electrodes and heat treatments.

AISI type	Recommended heat treatment		Common recommended electrode for welding
	Pre-weld	**Post weld**	
301, 302	Not required if steel temp is above 15°C	Rapid cooling from temperature between 1065°C to 1150°C (1950°F to 2100°F), if service condition is moderate to severe corrosive.	308
304	As above	Rapid cooling from temperature between 1010°C to 1095°C (1850°F to 2000°F), if service condition is severe corrosive.	308
304L	As above	Not required for corrosion resistance	308L, or 347
309, 310	As above	Not required for corrosion resistance, because steel is usually at higher temperature in service.	309, 310
316	As above	Rapid cooling from temperature between 1065°C to 1150°C (1950°F to 2100°F), if service condition is severe corrosive.	316
316L	As above	Not required for corrosion resistance	316L
317	As above	Rapid cooling from temperature between 1065°C to 1150°C (1950°F to 2100°F), if service condition is severe corrosive.	317
317L	As above	Not required for corrosion resistance	317L
321,347	As above	Not required for corrosion resistance	347

(*Continued*)

Table 2.12.23 Stainless steel welding electrodes and heat treatments. (*Continued*)

AISI type	Recommended heat treatment: Pre-weld	Recommended heat treatment: Post weld	Common recommended electrode for welding
Ferritic and Martensitic Steels			
403,405	150 to 300°F Light gauge sheet need no pre heat	Air Cool from 1200°F/1400°F (650°C/760°C)	410
410	As above	Air Cool from 1200°F/1400°F (650°C/760°C)	410
430	As above	Air Cool from 1400°F/1450°F (760°C/785°C)	430
			Can be welded with 308, 309 or 310 without pre heat.
442	As above	Air Cool from 1450°F/1550°F (785°C/840°C)	446
446	300°F to 500°F	Rapid cooling from temperature between 840°C to 900°C (1550°F to 1650°F)	446
501	300°F to 500°F	Air Cool from 1325°F/1375°F (715°C/745°C)	502
502	300°F to 500°F Light gauge sheet need no pre heat	Air Cool from 1325°F/1375°F (715°C/745°C)	502 Can be welded with 308, 309 or 310 without pre heat.

- Welded joints in ferritic stainless steels have low ductility as a result of grain coarsening related to the absence of an allotropic transformation.
- The weldability of the austenitic stainless steels is governed by their susceptibility to hot cracking.
- The precipitation hardening stainless steels have welding difficulties associated with transformation (hardening) reactions.
- Stainless steels which contain aluminum or titanium can only be welded with gas-shielded processes.
- Joint properties of stainless-steel weldments will vary considerably as a result of their dependence on welding process and technique variables.
- Suitability for service conditions such as elevated temperature, pressure, creep, impact, and corrosion resistance must be carefully evaluated. The complex metallurgy of stainless steels must be accounted for.

2.13 Welding and Fabrication of Duplex Stainless Steels

This alloy group was developed over past 30 years; the development progress has resulted in a range of compositions including lean 22% chromium (Cr) and 25% chromium, listed in the Table 2.13 below. These alloys have high strength, good toughness, good corrosion resistance, good weldability, and formability, all of which ease manufacturing. These alloys combine the strength characteristics of ferritic stainless steels and the corrosion resistance of austenitic stainless steels. They have higher resistance to environmental corrosion than austenitic stainless steels. Dual phase alloying requires relatively lower Ni and Mo contents than single-phase austenitic alloys.

The alloys with higher F_{PREN} values, this is possible as a result of adding nitrogen in the alloy.

The duplex stainless steel contains up to 22% chromium. The key property that is of value to industry is the material's pitting resistance F_{PREN}, which is typically in the rage of 35 to 40.

The chromium content of super duplex steel is up to 25% and its pitting resistance F_{PREN} is typically in the rage of 40 and 45.

2.13.1 Mechanical Properties

The mechanical properties of the different types of duplex stainless steel are shown in Table 2.13.1 below. The mechanical properties of the cast versions of these alloys (e.g., UNS J93380, J92205, etc.) are lower than their wrought counterparts. ASTM A 995 "Standard Specification for Castings, Austenitic-Ferritic (Duplex) Stainless Steel, for Pressure-Containing Parts" details the compositions and mechanical properties of cast duplex alloys that are used for pressure-containing parts.

The duplex stainless steels used by the oil and gas industry have a roughly 50/50 austenite/ferrite, in general the duplex steel of various types would present a phase balance within the range 35% to 65% ferrite. They have adequate toughness at low temperatures, the alloy is commonly used to temperatures as low as minus 60°C (–76 °F).

Table 2.13 Nominal compositions of some of duplex steels.

		Nominal composition (wt%)							
Type	UNS no.	Fe	Cr	Ni	Mo	N	Cu	W	F_{PREN}
Lean	S32101	Bal	21	1.5	0.5	0.16	0.5	-	25
	S32304	Bal	23	4	0.3	0.16	0.3	-	26
	S32003	Bal	20	3	1.7	0.16	-		>30
Standard	S31803	Bal	22	5	3	0.16	-	-	35
	S32205	Bal	22	5	3.2	0.16	-	-	35
25 Cr	S32550	Bal	25	6	3	0.2	2	-	37
Superduplex	S32750	Bal	25	7	3.5	0.27	0.2	-	>40
	S32760	Bal	25	7	3.5	0.25	0.7	0.7	>40
	S32520	Bal	25	7	3.5	0.25	1.5	-	>40
	S39274	Bal	25	7	3	0.26	0.5	2	>40

Table 2.13.1 Nominal mechanical properties of duplex stainless steels.

Type	0.2% Proof stress (MPa)	Tensile strength (MPa)	Elongation (%)
Lean Duplex	450	620	25
Standard Duplex	450	620	25
25 Cr Duplex	550	760	15
Superduplex	550	750	25

Super duplex stainless steel (UNS S32760) has been successfully used up to minus 120 °C (–184 °F), but this requires a well-developed welding procedure and closely monitored welding parameters during the production process.

On long exposure to temperatures above 320°C (608°F) and up to about 550°C (1 022°F), the ferrite decomposes to precipitate alpha prime. This phase causes a significant loss of ductility; hence, duplex stainless steels are not normally used above 300°C (572°F).

In oil and gas service applications these alloys have fared very well in both sour and sweet environmental conditions.

2.13.2 Heat Treatment

Generally, these alloys are used in the annealed or annealed and cold worked condition. Prolonged heating at temperatures between 260 and 925 °C (500 and 1 700 °F) can cause the precipitation of number of phases, including sigma, which reduces toughness and can reduce SCC resistance. Any prolonged heating below the minimum solution-heating temperature is to be normally avoided. Low-temperature toughness generally decreases with decreasing cooling rates in annealing.

Cold-worked alloys are usually not welded because the mechanical strength of the weld would be lower than the base metal. Annealed alloys are easily welded. The weld filler metal is chosen to produce a desired volume fraction of ferrite and austenite. Hence, fabrication using autogenous (without filler) metal can result in welds that are poorer in mechanical and corrosion-resistant properties. The welding procedure is typically developed to control and balance the ferrite/austenite phase, this is essential to prevent deleterious phases or intermetallics.

These alloys are readily weldable, by SMAW, SAW, GTAW, GMAW processes. Other processes are also successfully used. Where the weld is to be heat treated after completion it is usual practice to weld with matching composition filler metal.

In as welded application, it is normal to use an over alloyed filler metal with an extra 2 to 2.5% nickel (Ni). This helps in getting the austenite/ferrite phase balance of about 50/50, if the weld is cooled rapidly. The lean duplex grades are welded with the filler metal used for 22% Cr duplex stainless steels. Except for thin sheets of up to 2 mm thickness autogenous welding is normally not recommended for duplex stainless steels.

Normally, argon gas is used for both shielding and backing gases, and welding does not begin until the oxygen level is dropped below 0.1%. As with high alloy stainless steels, care is to be taken in welding duplex alloys, and adherence to good stainless-steel welding practice discussed earlier in this chapter a good practice. Good joint design, control of inter-pass temperatures and keeping low heat inputs are other essential variable for good welding. Preheat and post weld heat treatment are not required for duplex stainless steels.

Maximum permissible heat input and inter-pass temperature increase with section thickness. The values for these parameters generally decrease as the alloy content increases. If heat inputs or inter-pass temperatures are too high, there is a risk of precipitating sigma (Σ) or chi (X) phases in the heat affected zone (HAZ) or weld metal. These are intermetallic phases, rich in chromium and molybdenum that leave a denuded area around them, which reduces the localized corrosion resistance. Sigma and chi phases also reduce impact toughness properties. In many applications especially in some oil and gas applications, the low temperature toughness is compromised for the corrosion resistance properties.

After welding, there is usually a heat tint in the weld and heat affected zone (HAZ), and it is normal to remove this, by manual brushing, by mechanical abrasives, or by a suitable pickling solutions or gels.

While developing welding procedures it is common and advised to include a corrosion test for example testing according to ASTM G 48, (http://www.astm.org/) as part of the weld qualification procedure.

The corrosion test sets important weld parameters; hence it is essential that the qualified parameters of welding are followed very closely during production welding. The experience tells us that sometimes "less experienced" welders have difficulty passing the corrosion test. Although the weld made by these less experienced welders would meet the mechanical requirements, it may not meet the corrosion tests as specified above. This increases the importance of welders/operators' qualification test and production weld parameters monitored by inspectors. In a very limited way this problem is resolved with use of 2% nitrogen gas along with argon as shielding gas. The reasons for the corrosion test failure can be due to the development of third phases, which are the result of poor supervision and control over the heat input and the inter-pass temperature.

Duplex Stainless-steel welds usually have lower impact toughness than their parent metals. The welding process used often affects the level of toughness, the GTAW welds being the toughest and SAW being the poorest. The weld-metal toughness is a function of both the heat input and the type of flux used. The experience suggests that a minimum of 70-Joule Charpy impact toughness in the parent metal ensures adequate toughness in a duplex weld, and it is easily achieved when correctly welded. To improve the low temperatures toughness requirements, especially for very low temperature services, it is worth considering the use of a nickel alloy filler metals, taking into account that other properties are not compromised for example, the nickel alloy weld must have the same strength as the parent duplex stainless steel. A practical example of above would be the selection of C-276 (UNS N10276) filler metal to improve the impact toughness of cast super-duplex (UNS J93380) at minus120 °C (-184 °F) service.

Some specifications for duplex material that are used in subsea environments with cathodic protection (CP) require maximum austenite spacing. In a weld, this cannot be controlled and the result cannot be changed by any heat treatment. However, duplex welds usually have a fine microstructure and meeting a maximum austenite spacing of 30 μm, is usually not difficult. Although the welding in itself does not necessarily degrade the resistance of duplex stainless steel against HISC, the presence of higher stress and stress raisers like weld toe, poses a significant problem when uncoated duplex stainless steels or steels with defective coating are exposed to CP under mechanical stress. Failures have occurred as a result of this effect, and guidance to avoid them can be sought from industrial specifications.

2.14 SMAW Welding Nickel Alloys

Nickel is a very versatile material with excellent weldability. Nickel and nickel alloys are major corrosion resistant materials in use in chemical and petrochemical industries. They are also used in other industries like, marine engineering, aeronautical and automobile making.

Nickel is an element, with the symbol Ni and atomic number 28, as shown in the periodic table below.

In appearance, it is a silvery-white lustrous metal with a slight golden tinge. Nickel belongs to the transition metals and it is hard, and ductile. Nickel is slow to react with air because of passivation, an oxide layer forms on the surface and prevents further corrosion.

Nickel is slow to oxidize in air at room temperature and is considered corrosion resistant. It is used for plating iron and copper alloys like brass, and coating chemistry equipment, and manufacturing alloys that retain a high silvery polish, one such alloy is called German silver, which contains about 60% copper, 20% nickel and 20% zinc.

Nickel like iron, cobalt, and gadolinium is ferromagnetic at room temperature. That property allows its use to make permanent magnets. The metal is valuable in modern times primarily to make various alloys, primary among them are various grades of stainless steels, and cupro-nickel alloys.

Nickel 200 is the purest of nickel commercially available in wrought condition. The metal is weldable. Among the various alloys of nickel most of them are commercially named with number identifiers, some of them are discussed below for introduction purpose.

Nickel and Nickel alloys are welded by number of welding processes including SMAW process. The wrought nickel alloys can be welded under conditions similar to those used to weld austenitic stainless steels. Cast nickel alloys, particularly those with a high silicon content, present difficulties in welding.

Figure 2.14 Nickel alloy plate being welded.

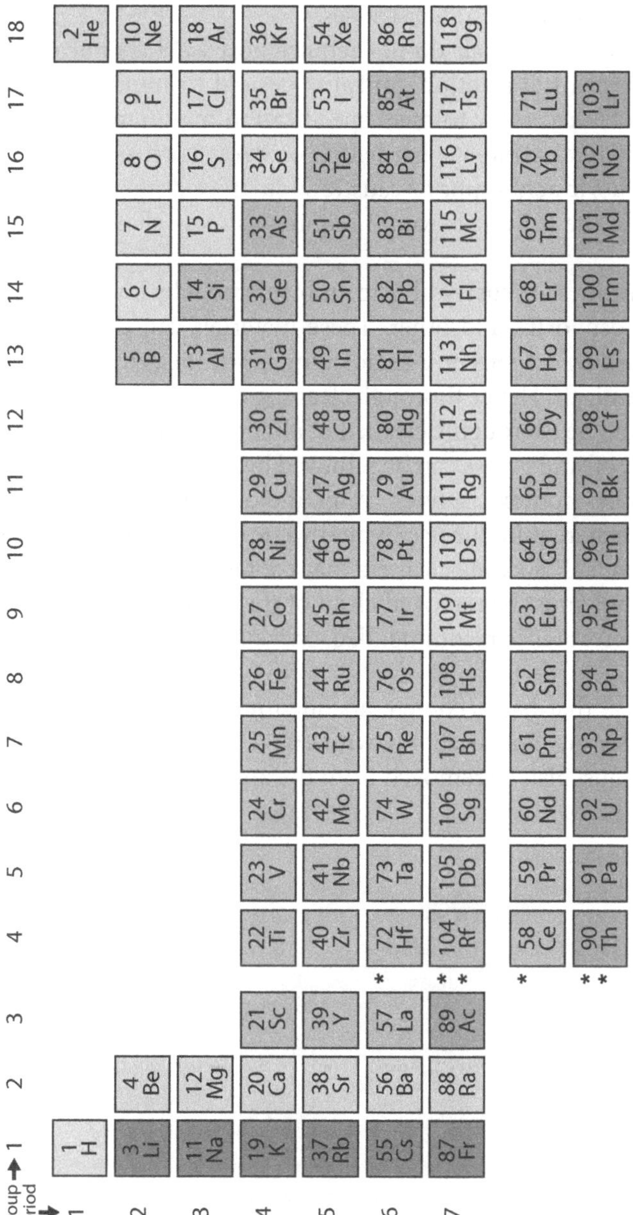

Figure 2.14.1 Nickel is in 10th group and 4th period in the periodic table, its atomic number is 28.

The most widely employed processes for welding non-age-hardenable (solid-solution-strengthened) wrought nickel alloys are shielded metal arc welding (SMAW), gas-tungsten arc welding (GTAW), and gas-metal arc welding (GMAW).

Nickel alloys are usually welded in the solution-treated condition. Precipitation-hardenable (PH) alloys should be annealed before welding if they have undergone any operations that introduce high residual stresses.

Most of the time these materials do not require post weld chemical or heat treatment, however in some cases a full solution anneal is desired to improve corrosion resistance. Heat treatment may be necessary to meet specification requirements, such as stress relief of a fabricated structure to avoid age hardening or stress-corrosion cracking (SCC) of the weldment in hydrofluoric acid vapor or caustic soda. If welding induces moderate-to-high residual stresses, then the PH alloys would require a stress-relief after welding and before aging.

Nickel and nickel alloys are susceptible to embrittlement by low-melting-point elements like, lead, sulfur, phosphorus, and other. These materials can exist in contaminants like grease, oil, paint, marking crayons or inks, forming lubricants, cutting fluids, shop dirt, and other processing chemicals. Hence thorough cleaning of parts to be welded is very essential.

Work-pieces should be thoroughly cleaned of all foreign material before they are heated or welded. Shop dirt, oil and grease can be removed by either vapor degreasing or swabbing with acetone or another nontoxic solvent. Paint and other materials that are not soluble in degreasing solvents may require the use of methylene chloride, alkaline cleaners, or special proprietary compounds. If alkaline cleaners that contain sodium carbonate are used, then the cleaners themselves must be removed clean of the material, prior to welding. Spraying or scrubbing with hot water is recommended. Marking ink can usually be removed with alcohol.

Processing material that has become embedded in the work metal can be removed by grinding, abrasive blasting, and swabbing with 10% HCl solution, followed by a thorough water wash. Oxides must also be removed from the area involved in the welding operation,

Figure 2.14.2 Typical nickel welding electrodes – note the electrode identification making on the electrode.

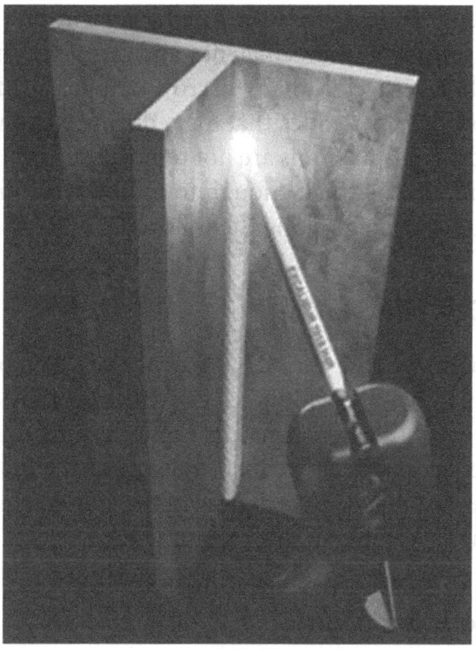

Figure 2.14.3 Nickel alloy welding (note the fillet weld in upward progression).

primarily because oxides get imbedded in the weld as inclusions, because the oxide have much higher meting point than the base metal melting points. Oxides are normally removed by grinding, machining, abrasive blasting or pickling.

Nickel alloys, both cast and wrought and either solid-solution-strengthened or precipitation-hardenable, can be welded by the SMAW process.

2.14.1 Welding of Precipitation Hardenable Nickel Alloy

The PH alloys, alloy 718 is the primary of them are very good weldability, but they also have higher susceptibility of cracking, both in the base-metal, and heat affected zone, hence they require special welding procedures for successful welding.

Cracks also occur post weld and during the operation if the service temperature is greater than the aging temperature, and residual stresses developed during welding or stresses induced during the precipitation. Before welding these alloys, a full-solution anneal is usually performed. After welding, the appropriate aging heat treatment is performed. To further improve alloy properties, a full anneal after welding, followed by a post-weld heat treatment, can be incorporated in the welding procedure.

Any part that has been subjected to severe bending, drawing or other forming operations should be annealed before welding. If possible, heating should be done in a controlled atmosphere furnace to limit oxidation and minimize subsequent surface cleaning.

A generic welding procedure for most commonly used, and weldable nickel and nickel alloys with SMAW electrode is given below. However, the information is not sufficient and more detailed procedure must be developed by the Welding engineer responsible for the project welding.

Heat input during the welding operations should be held to a moderately low level in order to obtain the highest possible joint efficiency and minimize the extent of the HAZ. For multiple-bead or multiple-layer welds, many narrow stringer beads should be used, rather than a few large, heavy beads. Any oxides that form during welding should be removed by abrasive blasting or grinding. If such films are not removed as they accumulate on multiple-pass welds, then they can become thick enough to inhibit weld fusion and produce unacceptable laminar type oxide stringers along the weld axis.

2.14.2 Welding of Cast Nickel Alloy

Cast nickel alloys can be joined by the SMAW processes. For optimum results, casting should be solution annealed before welding to relieve some of the casting stresses and provide some homogenization of the cast structure. Light peening of solidified metal after the first pass will relieve stresses and, thus, reduce cracking at the fusion line, or the interface of the weld metal and the cast metal. The peening of the subsequent passes is of little, if any, benefit. Stress relieving after welding is also recommended.

2.14.3 Nickel - Chromium Alloys

- Alloy 600,
  This alloy has nominally about 72% nickel and 16% Chromium, among other alloying elements. It is resistant to oxidation at high temperatures.
  Weldability: Has good weldability.
  Recommended SMAW Electrode: E NiCrFe-3
- Alloy 601,
  This alloy has nominally about 60% nickel and 25% Chromium, 1% Aluminum, Iron among others. It has higher strength and is excellent resistant to oxidation at high temperatures.
  Weldability: Has good weldability.
  Recommended SMAW Electrode: E NiCrFe-3
- Alloy 617,
  This nickel, chromium, molybdenum, and cobalt alloy is metallurgically stable alloy, that serves well as corrosion resistant in wide range of corrosive environment, and in high temperature environment, while maintain it strength.
  Weldability: Has good weldability.
  Recommended SMAW Electrode: E NiCrCoMo-1
- Alloy 625,
  This Nickel, chromium, and molybdenum alloy is designed to be an excellent corrosion resistant material by adding niobium, which stabilizes the structure matrix and the strength of the material at high temperature applications. One of the most significant properties is the high resistance to pitting in vary corrosive environment.
  Weldability: Excellent weldability.
  Recommended SMAW Electrode: E NiCrMo-3

- Alloy 718,
  This precipitation hardenable, nickel and chromium alloy with iron and niobium, and molybdenum presents excellent creep resistance properties, and resists cracking after welding.
  Weldability: Excellent weldability.
  Recommended SMAW Electrode: E NiCrFe-3

2.14.4 Nickel - Copper (Cupro-Nickle Alloys)

- Monel alloy 400,
  This alloy with nominal, about 60% nickel, and 30% copper, is excellent material for service in sea water, and other acids like sulfuric acid, and hydrofluoric acid environment.
  Weldability: Good weldability.
  Recommended SMAW Electrode: E NiCu-7
- Monel alloy 401,
  This copper and nickel alloy with nominal, about 45% nickel, and reminder copper, is excellent material for electrical application service.
  Weldability: Not considered.
- Monel alloy 450,
  This cupro-nickel alloy with nominally 70% copper, and 30 % nickel is resistant to biofouling and sea water corrosion.
  Weldability: Superior weldability.
  Recommended SMAW Electrode: AWS Class, E CuNi
- Monel alloy K-500,
  This precipitation hardenable alloy of nominally 60% nickel, and 30% copper with other alloying elements is a version of Monel 400 discussed above, but it has greater hardness, and strength.
  Weldability: Not considered.
  Recommended SMAW Electrode: Not recommended

2.14.5 Nickel - Iron - Chromium Alloys

- Alloy 800,
  This Nickel- Iron- Chromium, with good creep properties is excellent alloy for service in oxidizing and carburizing environment in high temperature atmosphere service.
  Weldability: Good
  Recommended SMAW Electrode: E NiCrFe-2
- Alloy 825,
  The nickel-iron chromium with added molybdenum and copper has excellent resistance to both oxidizing and reducing acids. Resists SCC and has good pitting and crevice resistance.
  Weldability: Very Good,
  Recommended SMAW Electrode: E NiCrMo-3

- Alloy 902
  The nickel-iron chromium is designed for precipitation hardening by adding aluminum and titanium.
  Weldability: Not considered.
- Alloy 330
  The nickel-iron chromium with added silicon for increased resistance to oxidation, is good alloy in high temperature service in both oxidizing and carburizing environment.
  Weldability: Good.
  Recommended SMAW Electrode: E NiCrFe-1
- Alloy 020
  The nickel-iron chromium with added copper and molybdenum and stabilizer niobium. The alloy has good resistance to general corrosion, and localized corrosion forms like pitting and crevice corrosion occurring in chlorides and sulfuric, nitric, and phosphoric acids.
  Weldability: Good.
  Recommended SMAW Electrode: E NiCrFe-1

2.15 Minimizing Discontinuities in Nickel and Alloys Welds

The discontinuities including the metallurgical issues encountered in the arc welding of nickel include can be listed as the following.

1. Porosity
2. Susceptibility to high-temperature embrittlement by sulfur and other contaminants
3. Cracking in the weld bead, caused by high heat input and excessive welding speeds
4. Stress-corrosion cracking in service.

2.15.1 Porosity

From welders' point of view, cleaning of the parent metal, and surrounding area can reduce the possibility of porosity. However, from the metallurgical angle more reactions are possible that can cause porosity, gases that are either intentionally present in the weld area, or present due to the service environment of the material being welded can react in welding heat and cause discontinuities. Oxygen causes oxidation, carbon dioxide is a reducing agent, nitrogen allows formation of nitrides, or hydrogen reacts with atmospheric oxygen to form water vapor that cause porosity, all of these gases can cause porosity in welds. In the SMAW processes porosity can be minimized by using electrodes that contain deoxidizing or nitride forming elements, such as aluminum and titanium. These elements have a strong affinity for oxygen and nitrogen and form stable compounds with them. Presence of deoxidizers in either type of electrode serves to reduce porosity. In addition, porosity is much less likely to occur in chromium-bearing nickel alloys than in non-chromium-bearing alloys.

2.15.2 Weld Cracking

Hot shortness of welds can result from contamination by sulfur, lead, phosphorus, cadmium, zinc, tin, silver, boron, bismuth, or any other low-melting-point elements, which form intergranular films and cause severe liquid-metal embrittlement at elevated temperatures. Many of these elements are found in soldering and brazing filler metals.

Hot cracking of the weld metal usually results from such contamination. Cracking in heat-affected zone is often caused by intergranular penetration of contaminants from the base-metal surface. Sulfur, which is present in most cutting oils used for machining, is a common cause of cracking in nickel alloys.

Weld metal cracking also can be caused by heat input that is too high, as a result of higher current and slower travel speed. Welding speeds have a large effect on the solidification pattern of the weld. High welding speeds create a tear-drop molten weld pool, which leads to uncompetitive grain solidification at the center of the weld. At the weld centerline, residual elements will collect and cause centerline hot cracking or lower transverse tensile properties.

In addition, cracking may result from undue restraint. When conditions of the high restraint are present, as in circumferential welds that are self-restraining, all bead surfaces should be slightly convex. Although convex beads are virtually immune to centerline splitting, concave beads are particularly susceptible to centerline cracking. In addition, excessive width-to-depth or depth-to-width ratios can result in cracking may be internal (that is subsurface cracking).

2.15.3 Stress Corrosion Cracking

Nickel and nickel alloys do not experience any metallurgical changes, either in the weld metal or in the HAZ that affects normal corrosion resistance. When the alloys are intended to contact substances such as concentrated caustic soda, fluorosilicates, and some mercury salts, however, the welds may need to be stress relieved to avoid stress corrosion cracking. Nickel alloys have good resistance to dilute alkali and chloride solutions. Because resistance to stress-corrosion cracking increases with nickel content, the stress relieving of welds in the high-nickel-content alloys is not usually needed.

2.15.4 Effect of Slag on Weld Metal

Because fabricated nickel alloys are ordinarily used in high-temperature service and in aqueous corrosive environments, all slag should be removed from finished weldments. If slag is not removed in these types of application, then crevices and accelerated corrosion can result. Slag inclusions between weld beads reduce the strength of the weld. Fluorides in the slag can react with moisture or elements in the environment to create highly corrosive compounds.

2.16 Review Your Knowledge

To review the knowledge, take the test on the question set below, these questions are based on the content of this chapter of the book.

After wring the answers, check the chapter for correctness and where necessary read related topics again, and correct your answers.

After a couple of hours re-take the test, till you answered all questions correctly.

1. What types of AC we1ding machines are discussed in the book?
2. What are different types of DC welding machines?
3. What is the normal no-load voltage range in DC welding arc?
4. What is the method to stabilize the AC arc?
5. Explain the DCEN and DCEP, in each circuit what direction the electrons flow?
6. State the importance of having tight connections to the work, and electrodes?
7. What are the dangers of arc flash?
8. What are different electrode coatings?
9. State the variables that are considered in the selection of specific DC arc polarity?
10. What is the resistance of a welding circuit when the arc voltage is 20 volts and the current is 130 amperes? Use Ohm's law and step by step calculation.
11. How a weld is properly ended?
12. Describe what you understand by the electrode designation E 309 and E 7024-H8R? What is the meaning of letter, number combination H8R?
13. What is the weld position that allows easiest manipulation of welding arc?
14. Justify the choice of smaller diameter electrodes for welding.
15. Give factors responsible for running a good weld bead.
16. Describe the weld profile of a bead that is welded with long arc length.
17. How is the weld restarted in the middle of the plate?
18. AS a welder, what points you need to keep in mind when welding aluminum?
19. What is porosity, and how to control it? List all that may cause the porosity.
20. What is the difference between pre-heat and Interpass temperatures?
21. What is the difference in welding carbon steel and aluminum?
22. How is welding carbon steel different from welding stainless steel?
23. What do you understand by Post weld heat treatment, and why is it necessary for PH alloys?
24. What is additional knowledge that a welder should have to successfully weld Nickel alloys?
25. What are special precautions taken for welding alloy steels, as compared to welding carbon steel?

3

Gas Tungsten Arc Welding

3.1 Synopsis

The chapter describes the GTAW welding process, the various equipment types, their accuracy in delivering required power for welding, and control methods. The chapter has a section on the important aspects of tungsten electrode including the importance of precise grinding of the non-consumable electrode for effective use of the arc initiation and control.

The chapter focuses on how to make new welders to improve their efficiency in GTAW welding.

3.2 Keywords

Gas tungsten arc welding, power sources, AC power, sine wave, square wave, triangular wave, DC power, transformers.

3.3 Introduction to Gas Tungsten Arc Welding Process

The Gas tungsten arc welding process, is also termed as GTAW process. It is one of the several electric arc welding processes that is very commonly used to join several types of metals, normally seeking high quality and esthetically pleasing welds. The Figure 3.3.1 below depicts the very common image of the GTAW process.

The GTAW process uses a non-consumable tungsten electrode to initiate arc that provides the heat to melt the metal being welded, and this electrode heat is also used to melt the welding wire, often introduced in to the weld pool by other hand of the welder, of course in the mechanized version of the process this work is done by the machine wire-feeder. Welds can also be done without adding this filler material, in which case the weld is called an autogenous weld. The atmospheric contamination of the molten metal in the weld pool is protected by creating an envelope of inert gas environment. Argon (Ar) and Helium (He) are the two inert gases that are used.

Generally, the process gives a sound weld, which is aesthetically pleasing, because of this attribute, the process finds much use in food industry, and for fabricating medical equipment and tools and equipment.

The GTAW process, is relatively expansive, and requires training to master it, but it is also a very well used process across the industrial spectrum.

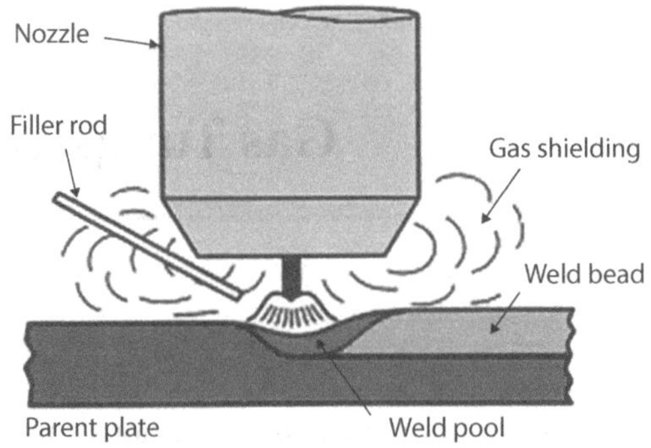

Figure 3.3.1 Typical GTAW welding.

The main disadvantage of GTAW is that it produces the slowest metal deposition rate of all the electric arc welding processes. The emphasis of the process is to make welds that are relatively defect free, and near perfect in appearance, to achieve these objectives the process is associated with lower welding current, which translates in longer welding time. The operator needs to learn hand coordination to manipulate precise movements of the torch which is held in one hand with adding filler metal with the other hand as shown in the Figure 3.3.2 below, along with the controlling of current with a foot pedal.

The operator also needs to learn how to properly setup the GTAW machine, select and maintain the tungsten electrode for more details see paragraph 3.11 in this chapter, preparation, spark intensity, upslope, downslope, pulsing rate, peak intensity, background current, high frequency and proper grounding of the job, all of these factors are very important issues for successful GTAW welding. Combined with lower deposit rates, it's easy to see how the GTAW process has a great following in industries such as aerospace, and any other industry where quality of the weld is much more important than the time and cost.

The biggest advantage of GTAW process is that it delivers high quality welds, on almost any weldable metal or alloy. Another major advantage of a GTAW process is that filler metal

Figure 3.3.2 A GTAW welder, note the welding torch, and the filler wire in each hand.

can be added by the operator, to the weld pool independently of the arc current, this gives the welder full control over the weld pool, and the weld speed management. This is in total contrast with the other arc welding processes, where the rate of filler metal addition controls the arc current. GTAW process has very low spatter, and produces no slag to clean up after the weld is completed.

3.4 Process Description

Gas Tungsten Arc Welding (GTAW) process uses a non-consumable tungsten electrode which must be shielded with an inert gas, the Figure 3.4.1 below shows a schematic of GTAW process where a cut section of the welding torch is shown, the figure shows the positioning of tungsten electrode, and how the shielding gas envelops the weld pool and the hot electrode from the atmospheric contamination. The details of GTAW torch are further described and discussed in this chapter. In the GTAW, an electric arc is established between a non-consumable tungsten electrode and the base metal. The arc zone is covered under the canopy of the inert gas, typically argon, which protects the tungsten and molten metal from oxidation and provides an easily ionized path for the arc current to travel across the electrode tip to the job. GTAW produces high quality welds on almost all metals and alloys. Because it can be controlled at very low amperages, it is ideally suited for welding on thin metal sheets and foils.

The process is also called Argon-arc or Heli-arc these unofficial terminologies are created in the field primarily based on the type of inert gas used for weld shielding. Heli-arc welding uses Helium gas, whereas Argon gas is used in argon-arc welding. The term GTAW

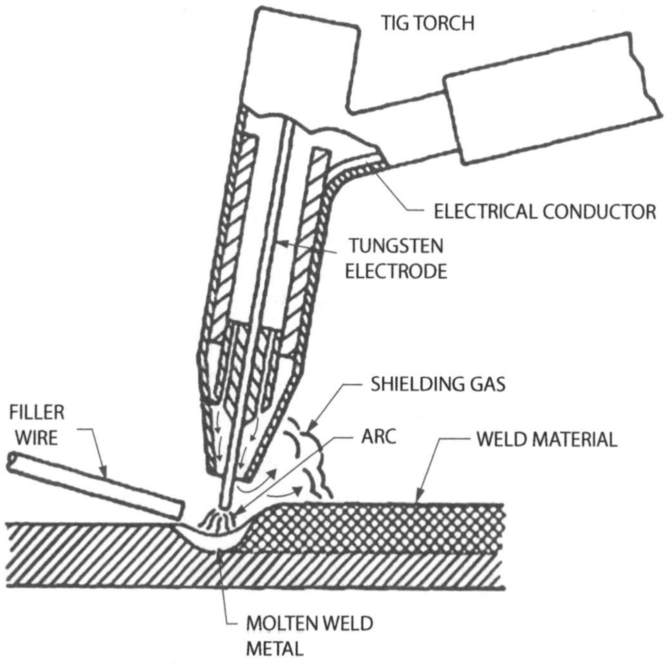

Figure 3.4.1 Typical GTAW welding process with details of the welding torch.

is a generic name, it encompasses both gas versions of the process since it only refers to the inert gas as opposed to any specific gas. Another term very commonly used is tungsten inter gas (TIG) welding. For this chapter we will try to keep the discussion going through the term GTAW, which is an American welding society (AWS) used terminology, and an acceptable term throughout most of the world.

3.5 How the Process Works

The welding arc is established between the tip of the tungsten electrode and the work, the heat generated melts the metal being welded and the consumable filler metal is added either manually or if the process is mechanized, by some mechanical process. The inert gas shielding protects the molten metal, which is cooling just behind the arc and also the tungsten electrode, and the ionized gas also helps to establish the required arc characteristics.

The process may use direct current with positive or negative polarity attached, to the tungsten electrode, though in most application electrode is attached to the negative polarity. Alternative current (AC) is also used to produce different effects on the welding. These details are discussed further in this chapter.

In the process description earlier in the chapter we have said that the Argon and Helium are the two inert gases used for this process. Choice of gas and type of current and polarity depends on the type of material being welded, and quality of weld desired. For example, use of helium gas will result in deeper penetration, and if helium gas is used with DC current then the process would provide deepest penetration of the weld, for more details refer tables and figure in section 3.11 of this chapter. A typical GTAW welding setup is shown Figure 3.5.1 below.

Use of AC current with Argon shielding helps remove oxides from such materials as aluminum, and stainless steels, and other metals that have passivation films on the surface giving

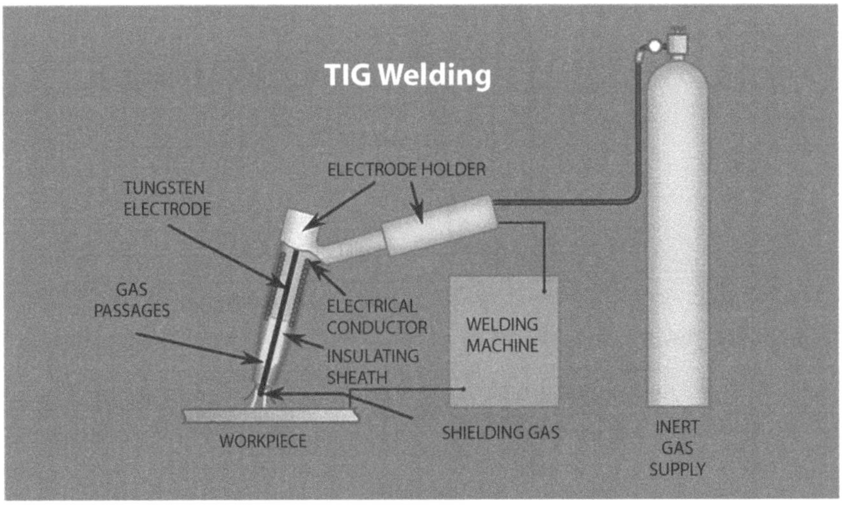

Figure 3.5.1 A typical GTAW set-up with positions of gas cylinder, welding machine, electrode holder and work-piece.

them corrosion resistance properties, but these films are not good for welding as they have very high melting temperatures. The process of cleaning during the welding is shown in Figure 3.5.2 below. Note that the current cycle alternates as cleaning and heating phase within each cycle.

The process uses constant current welding power. High frequency oscillation is generally provided for AC power sources. High frequency attachments with DC process allow for "no-touch starting" of the arc, a distinct advantage for producing high quality welds.

There are variations of GTAW process that includes different types of automation and they may also include improved deposition rate. These systems are in market with different trade names from verity of the manufacturers. In the automation of the GTAW the variants could be a system that can be used to weld pipe girth weld, in which the tungsten electrode, wire feed unit and gas supply connection are often mounted in one unit. That unit rotates on a rack and pinion, or rail and the wheel, mechanism to go around the weld. The early use of these systems included welding in tight spaces like boiler tubes and tube sheet welding etc. however more advance systems are in use now for other orbital wilding like pipeline and in more adopted forms for cladding of valve and pump internals etc.

The primary advantage of the GTAW process is the good quality weld through the slow deposition rate. The slow rate is also deemed as the disadvantage of the process. To mitigate this disadvantage to some extent few methods have been devised, and added to the process. To increase the deposition rates various options are available, there can be multiple head wire feed units that feed more than one wire to increase the deposition rate but more often used option is the hot wire process. The Hot-Wire option uses an independent source to heat the wire to a level just below the wire's melting temperature. The hot wire is introduced to the weld just at the tip of the molten pool created by the tungsten electrode. The process allows for the limit on the arc energy utilization.

Other derivations have also emerged as proprietary development, K-TIG® is one such process that increase both the deep-penetration and increases the productivity of the process. More information about the process is included further in the chapter.

The pulsed version of GTAW power sources often used direct current (DC) pulsed power source. These are often specialized applications power sources of proprietary nature designed for automatic girth welders or for cladding of special parts where the efficiency of the process is of higher demand.

The pulsed current system alternatively outputs high and low currents; at a nominal value of about 6Hz. normally the natural frequency of steel is considered to be between 6Hz to 7Hz depending on the type and grade of steel. In a pulsed current system, the current is

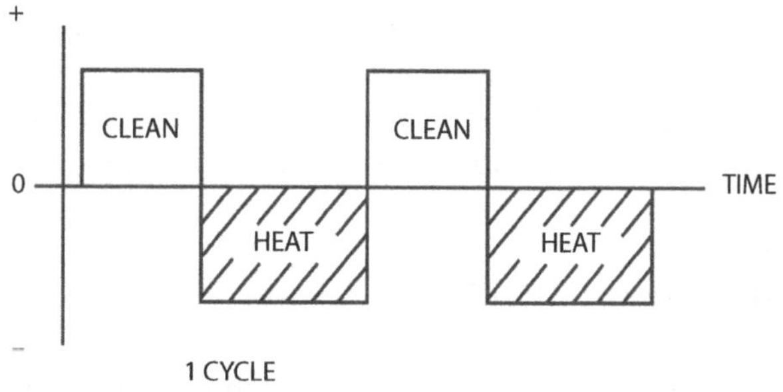

Figure 3.5.2 The cleaning process by the current cycle.

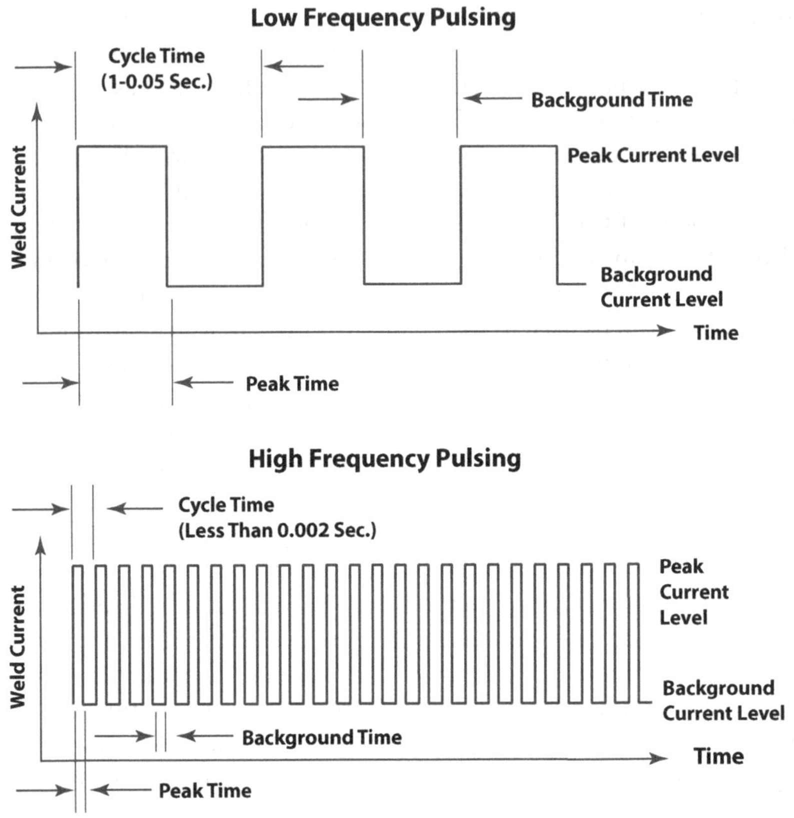

Figure 3.5.3 High and low frequency currents in pulsing.

synchronized with the rate of wire feed and voltage control of the machine. During the high current state, the pulsed current melts the metal and creates a molten pool to which the filler-wire is added to accomplish the weld. In the lower current phase of pulse, the back low ground current gives the time to cool and allows control over the weld pool and its management. When more variations are added like effect of oscillation, heating current then the synchronization of pulse is required and this can bring down the frequency to very low numbers like 1 or 2 pulse per second. The other variable like thickness of material where higher current is required to heat and melt, the demand on the time scale of the background current is critical, it is required that the time of the high current phase is matched with the lower cooling current phase. The Figure 3.5.3 depicts the high and low frequency current pulsing, note the difference in the cycle-time and the width (time) of the peak within the cycle-time.

Since most of these newer variants are patent protected, more details on any of the specific equipment can be obtained from specific equipment or system manufacturers.

3.6 Process Advantages and Limitations

GTAW process produces superior quality welds that are free from most defects, by the superior quality it is implied that the welds are free from slag inclusions, free from any inclusions, and if properly maintained including tungsten inclusion, it is also free of spatter process. The process can be used with or without filler metal (autogenous), as required. It allows excellent control of the root pass weld penetration. It allows precise control of the welding variables. It can be used to weld almost all metals. That makes the GTAW process an excellent process for root pass welding, with good control over the heat source, and filler metal additions, that is required for making a successful root pass weld. It can use relatively inexpensive power supplies, but more modern high-end machines with better control systems can be expensive.

Deposition rates for this process are low, compared to other processes the welder requires more skill to be able to produce a quality weld. The weld area must be protected from wind and drafts to maintain the inert gas envelop over the weld zone. Tungsten inclusions can occur if the electrode is allowed to contact the weld pool. Weld contamination can occur if proper shielding is not maintained or if the filler metal or base metal is contaminated.

As stated in the introduction, the GTAW uses a non-consumable tungsten electrode. It's held in a torch assembly, as shown in Figure 3.4.1 and also in Figure 3.5.1 above. The electrode initiates and maintains the arc. In manual GTAW, the heat of the arc commonly is controlled by one of the following two ways,

(i) through the use of a foot pedal or
(ii) By a static setting on the power source.

The foot pedal provides variable control of the welding amperage, or heat, however most modern machines have electronic controlled static setting on the power source. Theses electronic control systems are discussed and explained in some detail for the in the chapter.

The torch assembly is either water- or air-cooled. A solenoid on the power source activates the flow of shielding gas to the torch. Some power sources provide an additional solenoid, which switches the water flow to the torch, while others don't have a solenoid for the torch coolant so the cooler can circulate coolant constantly through the torch, providing cooling all the time.

High frequency (HF) may be required to start the arc for non-touch-start applications. Most upscale GTAW power sources provide HF capability, but not all do. DC high frequency output circuit shown in Figure 3.6.1 below.

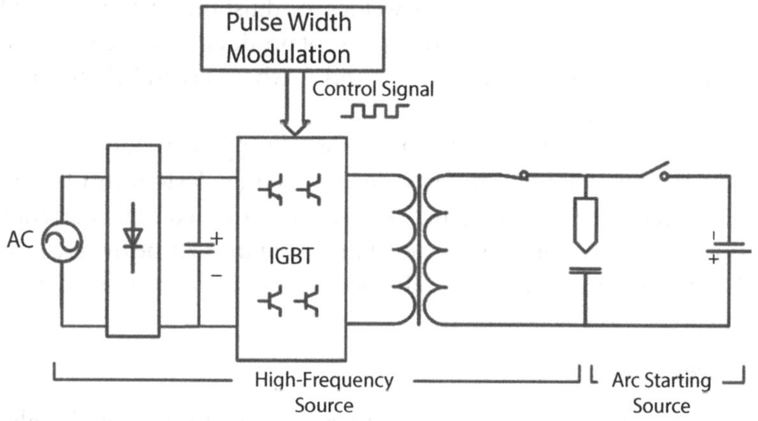

Figure 3.6.1 DC HF output circuit.

Note that the HF circuit is activated at the initiation of the arc, once the arc is stablished, the HF mode is disconnected, and normal arc mode takes over for the maintenance of the arc and continuance of welding.

3.7 Power Sources

As stated in the introductory paragraphs we have learned that the GTAW process can use either AC or DC current. Also, a single welding machine may be able to deliver both AC and DC current.

While some newer machines offer technology and flexibility, achieving high-quality welds with a less advanced, less expensive power source is always possible. Technology has its place for efficiency of time, cost, and general consistency of quality, but it can't always beat what a skilled welder can bring to the table. Many simple machines are affordable and can provide years of service, and skilled welder can use them effectively.

A plug and go option are not possible for buying a power source. However, a comprehensive understanding of the process and the materials to be welded is imperative and can lead to a satisfying purchase of optimum machine suitable for the task at hand.

Transformer transformers, rectifier, and generator driven machines are also used for GTAW process. While the transformer machines are capable of producing AC current only, the generator type machines can produce both AC and DC currents, and often the DC power source is fitted with rectifiers to produce rectified AC current.

A GTAW power source as shown in Figure 3.7.1 below, is unlike constant-voltage power sources, it is a constant-current machine, and will maintain a constant current. The current in this respect translates in to the heat required to weld, despite changes in the voltage, which translates in to the arc length.

A constant-voltage power source does just the opposite, which is required for gas metal arc welding (GMAW), see Chapter 4 of this book for more about GMAW process.

The primary requirement for the GTAW process machine is that it should be able to supply constant current. In a Constant current machine, the volt-ampere curve has a steep slope, as shown in the Figure 3.7.1. Wherein we can see that a change in voltage shown by red dotted lines, will only result slight change in output voltage.

The constant current machines allow the welder to adjust the current output by changing the arc length. More modern machines are capable of producing near constant current. This is depicted in the curve with the dotted line 3 and 4, where no real change in amperage at OCV of the machine is seen.

This is stability of current output, and maintenance during the welding is possible with the use of electronic circuits that use silicon-controlled rectifiers (SCR) to rectify the current output. The transformer rectifier type of welding machines delivers such constant current with the help of several SCRs in the circuit. The curves 3 and 4 in the Figure 3.7.1 below, are possible because of the rectified current output derived from the electronic circuitry, attributed to the SCRs.

3.7.1 AC Power Sources

Alternating current welding machines are either the transformer or the alternator type machines.

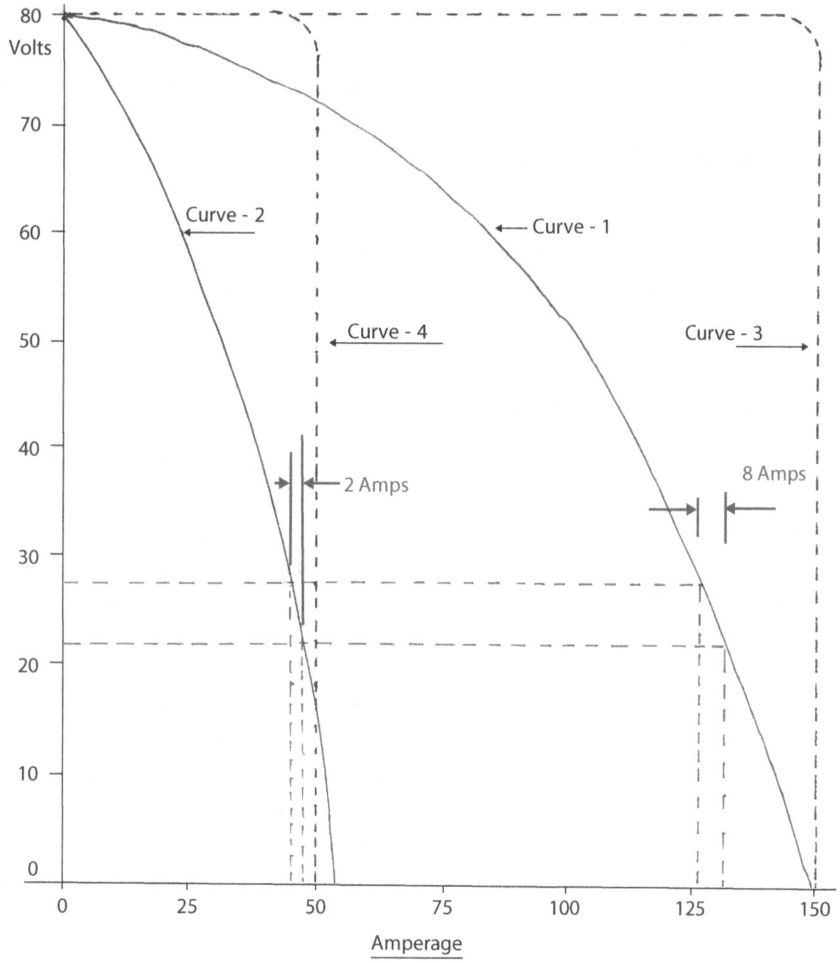

Figure 3.7.1 The graph.

The transformers reduce the high voltage and low amperage supply electricity to high current and low voltage welding power. Transformers are constructed on a soft iron core, which is laminated. The iron core is used to build electro-magnet, to develop magnetic field. The iron core is wound with a copper wire that carries the current. There are two such coils, they are called primary and secondary coils. Primary coils receive the primary (supply) current and the secondary coil outputs the changed (transformed) current for the welding. Primary coils have more copper wire winding turns, and the wire is thinner compared to the secondary coil, this is because the primary coil carries lower amperage - less current. The significance of more turns than secondary coil, is that the transformer is reducing the voltage and conversely increasing the amperage – the current. These transformers are aptly called the stepdown transformers because they are reducing the voltage. Number of coil turns indicate the force of the magnetic field created. In an alternating current this magnetic field collapse as the alternating cycles reverses through its cycles, at this point the magnetic field collapses. Magnetic field is directly proportioned to the number of the turns the coil has. Figure 2.2.2 in previous chapter details of primary and secondary coils and other major components of and AC transformer.

The secondary coil has fewer turns of relatively thicker diameter copper wire. These two coils are not connected by any means, they are set apparat from each other in one container, in such a way that the magnetic field of the primary coil transfers to the secondary coil. The precise spacing of two coils plays very important role, if they are distanced too far apart, not enough field from the primary coil will cross to the secondary core, reducing the current output.

The action of reversal and rebuilding of magnetic field induces the current in the secondary coil. This successive action at 120 times per second, causes the current to flow from primary to the secondary coil.

3.7.1.1 *The Alternator Type AC Welding Machines*

These machines are different from transformer type welding machines, in that these machines generate their won electricity through a rotor, which may be driven by any hydrocarbon fuel diesel or gasoline for example, or even by an electric motor driven by supply power source. The rotor is wrapped with multiple coils of copper wire and housed in a casing that has magnetic field. Thus, the rotating coils on the rotor create alternating current for welding.

The electrical parameters are very important variables for welding, hence the necessity to have better control on the current output cannot be over emphasized. The control in this context means the accuracy of current available for welding in various positions, and continuity of the available current for welding. The control for transformer type AC power sources is done either through the mechanically moving some parts of the system or through an electronic device fitted in the circuit. We can discuss them here for good understanding of the function of AC power system. The first three methods of control discussed below use the manual or mechanical means to vary and control output. It should be noted that the output AC is not limited by the frequency, it can be at any frequency.

Alternating current can also be obtained by a method called Dual Source with inverter switching. This machine uses inverter to deliver AC as the main welding current, a second power supply unit is included within the machine's circuit that by switching, supplies the direct current for welding. The inverters are discussed in more details further in this chapter.

3.7.1.2 *Movable Coil Movable Core (Reactor)*

In this type of control system, the transformer has three coils. The usual two coils, the primary coil and the secondary coil are fixed, these two coils produce the current for welding, a movable third coil is introduced. This movable third coil is also referred as the reactor; it controls the current output of the machine.

The copper-coil wound is to create a counter voltage, also called bucking voltage. If the movable coil is turned on then its inductive voltage increases the resistance of the secondary coil of the transformer, reducing the current output. When the inductive reactor (coil) is turned out, the output from the secondary coil of the transformer is in its full capacity. Figure 2.2.4 in the previous chapter shows a schematic of a movable coil reactor, the position of the reactor coil causes the inductive reactance of the secondary output coil resulting in the variance in current output.

Apart from the above some other transformers use the tapping devices to the secondary coil to mark stepped stages for the control of current output. Though these devices are useful for welding, the current output of tapping devices are not as precise as that of the movable coil and consequently not as accurate. The accuracy of the current derived from the electronic control methods is discussed further in the chapter.

3.7.1.3 Magnetic Amplifier Method of Current Control

Yet another method of current control is the magnetic amplifier approach. This method uses the welding current coils and diodes in series with control coils. The load coil is used to assist the control coil to increase the magnetic field of the cores. The high magnetic field in the cores cause an inductive reactance in the secondary welding current. The increased reactance decreases the welding current from the transformer. Figure 2.2.5 in Chapter 2 of the book depicts the magnetic amplifier transformer output control, the diode allows the current to flow in one direction, and this allows a remote-control operation possible.

The more modern transformers use electronic devices to control the current output. These devices are capable of providing finer control, and they are more prevalent in modern welding equipment. These are also referred as the solid-state devices. The best example of a solid-state devices is the diode, and solid-state rectifier or SCR. Both these electronic devices are discussed further in this chapter.

3.7.1.4 AC Inverters for GTAW Process

The inverter welding units, are the units that convert the utility power to the output that is suitable for welding. These inverters first rectify the utility AC power to DC; then they switch (invert) the DC power into a stepdown transformer to produce the desired welding voltage or current. The switching frequency is typically 10 kHz or higher. Although the high switching frequency requires sophisticated components and circuits, it drastically reduces the bulk of the step-down transformer, as the mass of magnetic components (transformers and inductors) that is required for achieving a given power level goes down rapidly as the operating (switching) frequency is increased. The inverter circuitry can also provide features such as power control and overload protection. The high frequency inverter-based welding machines are typically more efficient and provide better control of variable functional parameters than non-inverter welding machines.

The use of insulated gate bipolar transistors (IGBTs) see Figure 3.6.1 HF circuit that allows an inverter-based machines are to control by a microcontroller, so the electrical characteristics of the welding power can be changed by a software in real time, even on a cycle-by-cycle basis, rather than making changes slowly over hundreds if not thousands of cycles. Typically, the controller software will implement features such as pulsing the welding current, providing variable ratios and current densities through a welding cycle, enabling swept or stepped variable frequencies, and providing timing as needed for implementing automatic spot-welding; all of these features would be prohibitively expensive to design into a transformer-based machine, but require only program memory space in a software controlled inverter machine. Similarly, it is possible to add new features to a software-controlled inverter machine if needed, through a software update, rather than through having to buy a new more modern welder.

3.7.2 Other Control Methods

Among the other control methods, none is more versatile, and useful as the variations of AC waveform.

Through the new arc controls and waveform options available to choose from, it is much easier to tailor the current wave forms and arc characteristics for specific challenges and different type of welding demands. Variables like material type, material thickness, weld penetration, weld bead width, etc., can be easily manipulated with the selection of wave forms.

This option allows the welder to have easier control over the arc characteristics, weld puddle behavior, and weld bead profile, penetration, and appearance. Such controls include those for:

- AC waveform shaping.
- Independent control of amperage values during the electrode negative (EN) and electrode positive (EP) portions of the AC cycle.
- Adjustable AC output frequency.
- Extended balance control.

The Figure 3.7.2 below shows different types of wave form that are available to choose from for welding.

i. Figure A indicates the Squire wave,
ii. Figure B is soft squire wave,
iii. Figure C is typical sine wave, and
iv. Figure D is triangular wave

3.7.2.1 Wave Forms

Current wave forms are modified to derive desired properties from the welding current. The following paragraphs discuss them in detail.

3.7.2.1.1 Advanced Square Wave

The alternate current wave is shown in the Figure 3.7.2-(A), this advanced square wave waveform offers fast transitions between electrode negative (EN) and electrode positive (EP) modes for a responsive, dynamic, and focused arc with better directional control. It forms a fast-freezing puddle with deep penetration and fast travel speeds.

3.7.2.1.2 Soft Square Wave

Soft square wave is shown in the Figure 3.7.2-(B), it provides a smooth, soft, delivers a "buttery" arc, with a fluid puddle and good wetting action. The puddle is more fluid than with advanced square wave and more controllable than with sine wave.

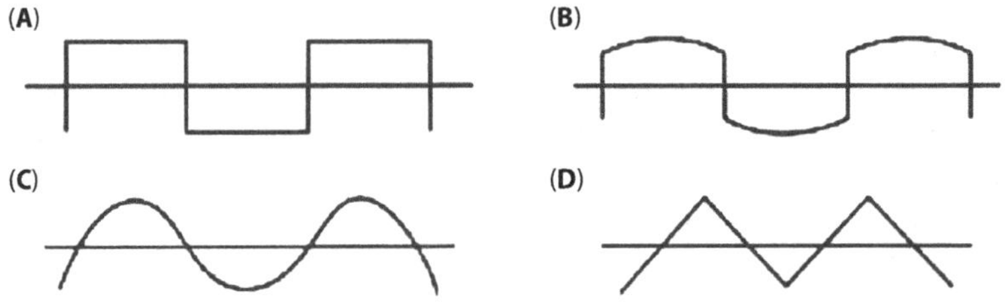

Figure 3.7.2 Four AC wave forms.

3.7.2.1.3 Sine Wave

The alternate current wave is shown in the Figure 3.7.2-(C) is the sine wave. The sine wave offers a soft arc with the feel of a conventional power source. It provides good wetting action and it is much quieter than other waves. Its fast transition through the zero amperage point also eliminates the need for continuous high frequency.

3.7.2.1.4 Triangular Wave

This wave form of the alternate current is shown in the Figure 3.7.2-(D). The triangular wave offers peak amperage while reducing overall heat input into the weld. This leads to quick puddle formation, low weld distortion, and fast travel speeds. It is especially good for welding thin aluminum.

3.7.2.2 Independent Amperage Control

Independent amperage, also referred as the amplitude, control allows the independent setting of electrode positive (EP), and electrode negative (EN) amperages. This precise control system allows better control of the heat input into the work and even takes heat off the electrode. The EN portion of the cycle controls the level of penetration, and the EP portion affects the arc cleaning action.

Figure 3.7.2.2 below shows, compares, and describes the effect of independent AC Amperage Control. At the top of the figure, we can see that the current with greater EN than EP creates a narrow bead with deeper penetration and no visible cleaning action. These attributes are ideal for fillet welds and automated applications.

At the bottom of the figure, we see the effect of current with greater EP than EN gives the operator a wider bead with less penetration and clearly visible cleaning action, ideal for buildup work, like cladding.

The example of the above control method: when welding a thick piece of aluminum, the operator can use 350 amperes of EN into the weld, and only 175 amperes of EP into the tungsten. This allows faster travel speeds, faster filler metal deposition, deeper penetration, and the potential to eliminate preheating. Case studies on the effect of the use of GTAW inverters with independent amperage control suggest that fabricators can cut production time, by as much as two-thirds.

Increasing EN while maintaining or reducing EP also permits the use of a smaller-diameter tungsten. This takes heat off of the tungsten, and more precisely directs it into the weld. Fabricators who have used this option have reported that this has allowed them to purchase thinner-diameter electrodes, which are less expensive than the thicker variety.

3.7.2.3 Adjustable AC Output Frequency

Another way of welding-current control is the adjusting the AC frequency, the number of times per second that the direction of the electrical current completes a full cycle. This method gives welders excellent control over bead appearance and penetration profile as is shown in the Figure 3.7.2.3 below. While conventional GTAW technology limits AC frequency to 50 or 60 hertz (Hz), new GTAW inverters gives a wide band of frequency to be adjusted anywhere from 20 to 400 Hz.

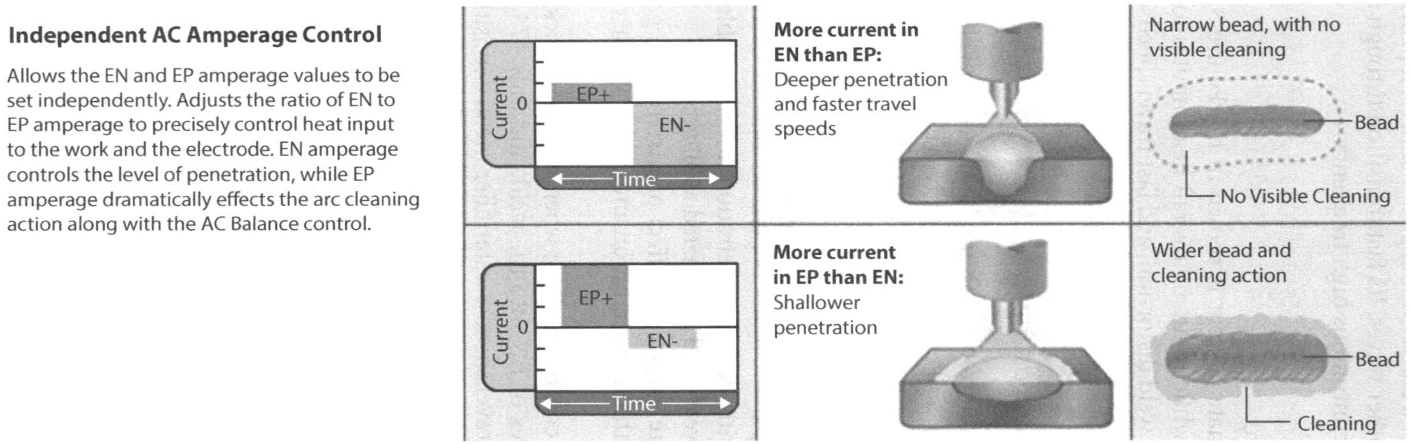

Figure 3.7.2.2 Effect of Independent AC amperage control on weld penetration and weld bead profile.

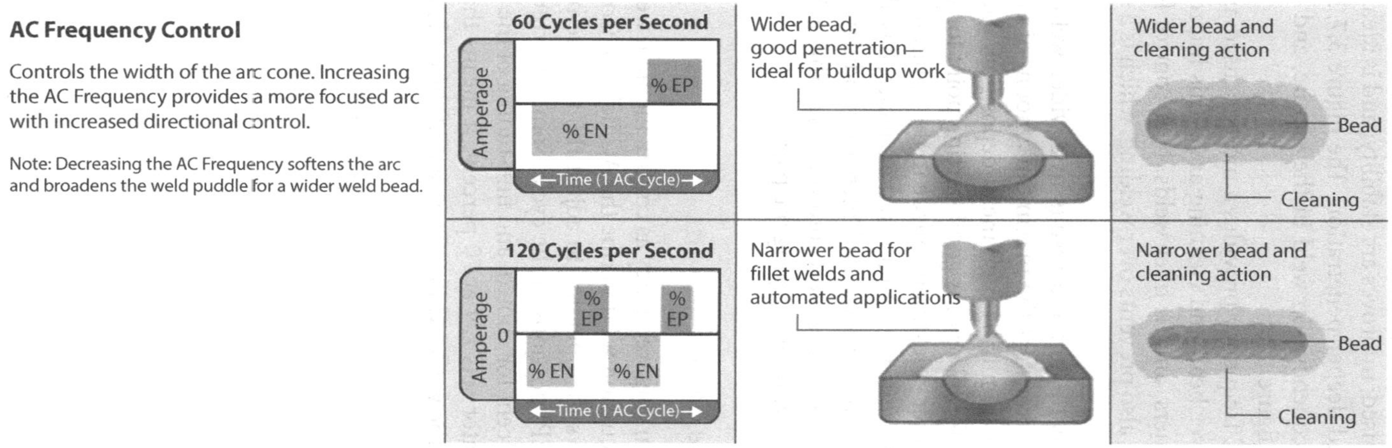

Figure 3.7.2.3 Effect of variation in AC frequency on the weld profile and penetration.

Frequencies between 80 and 120 Hz are deemed most suitable to weld with, it provides increased control of the arc, and boost travel speeds. Setting the frequency between 120Hz to 200 Hz provides an ideal frequency for most aluminum welding. An arc cone at 400 Hz is even tighter and more focused; improves arc stability; and is ideal for fillet welds or other fit-ups requiring precise, and deeper penetration. The Figure 3.7.2.4 below compares two welds done using 40 Hz, frequency, top weld in the picture, and 120 Hz frequency weld picture at the bottom of the figure.

In general, increasing AC frequency provides a more focused arc with increased directional control, and a narrower bead and cleaning area. This improves performance when welding root passes, in corners, and in fillet welds, heat control for achieving toughness properties. A narrower bead also prevents over welding, which is a significant waste of time as well as the filler metal.

A lower frequency softens the arc and results in a wider weld puddle and bead. This removes impurities well and transfers the maximum amount of energy to the weld piece, which speeds up applications requiring high metal deposition, such as cladding, building up a worn part making, and for fill pass. A good starting point for such applications would be to start with 60 Hz with finer adjustments made to the optimum from the start point.

Turning up the frequency while turning down the balance, has allowed some fabricators and manufacturers to reduce distortions, and increase productivity by achieving deeper penetration without putting too much heat into the part. This achieves the desired bead profile and production speed without warping the part.

3.7.2.4 Extended Balance Control

In the above discussion of varying frequency as control method, we have mentioned a term balance, the subsequent discussion we describe the AC Balance Control.

AC balance control allows the operator to adjust the balance between the penetration (EN) and cleaning action (EP) portions of the cycle. Some inverters have adjustable EN as great as 30 percent to 99 percent for control and fine-tuning of the cleaning action.

For instance, if the operator sets EN at 70 percent, it means that 70 percent of the AC cycle is putting energy into the work, while 30 percent of the cycle is used for cleaning.

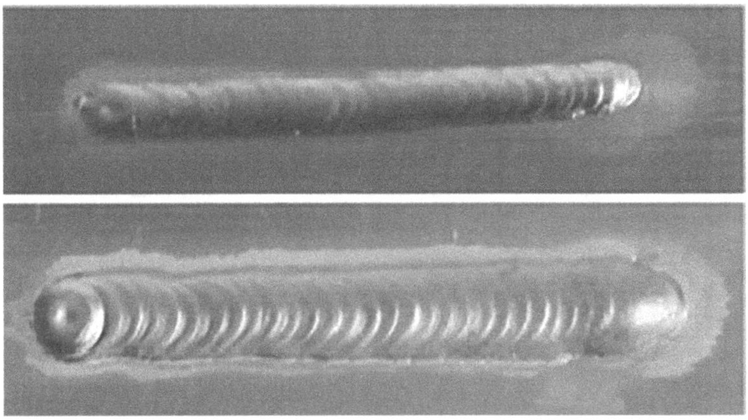

Figure 3.7.2.4 Provides an example of a weld done at 150 Hz and 40 Hz.

A good starting point on clean aluminum welding is between 60 percent and 75 percent, and fine adjust as required from that point. Some fabricators have even experimented with AC GTAW on ferrous metal, where a few extra percentage points of cleaning action have proved beneficial.

Extending the EN portion of the cycle narrows the weld bead, achieves greater penetration that is good for thick welds, and may permit increased travel speeds. It also reduces the size of the etched zone along the length of weld toes, for improved cosmetics. It also reduces balling action of the electrode, thus extends the tungsten electrode life, and may permit the use of a smaller diameter electrode to more precisely direct the heat into the weld. The result of extending the EN portion of the cycle can be seen in the Figure 3.7.2.5 below.

Reducing the EN or welding with EP portion of the cycle widens the weld bead and decreases its penetration, which may be beneficial in catching both sides of the joint. This effect is shown in the Figure 3.7.2.6 below, it is best to compare the weld-pictures

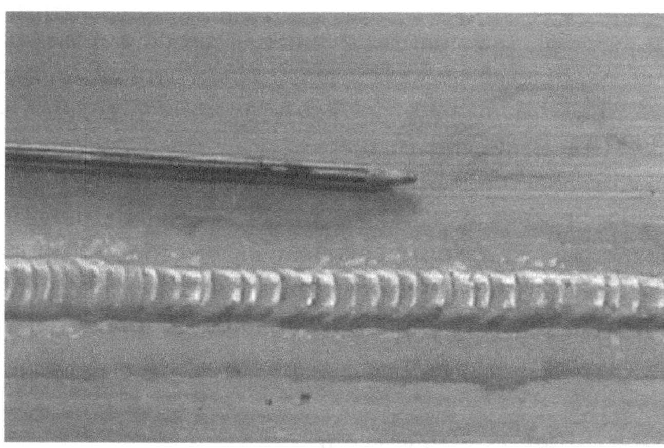

Figure 3.7.2.5 Weld profile as a result of extended EN of the cycle.

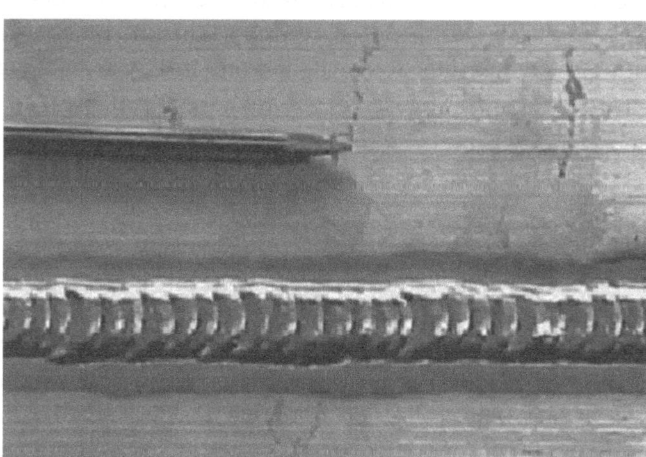

Figure 3.7.2.6 Weld profile as a result of reduced EN cycle.

in Figures 3.7.2.5 and 3.7.2.6 for best understanding of the effect of variation of current in EN cycle.

The reduced EN cycle produces a greater cleaning action to remove heavy oxidation and minimizes penetration, which may help prevent burn-through on thin materials. Reducing the EN cycle, however, decreases tungsten electrode life and increases balling action because more heat is being directed into the electrode. This creates a large ball at the end of the tungsten and causes the arc to lose stability, making it hard to direct the arc weld puddle.

In terms of bead appearance, too much penetration with higher EN, can result in a scummy-looking weld puddle that may still contain oxides and inclusions. Increasing cleaning action will blast away those contaminants. Ultimately, operators should practice adjusting balance control on scrap material and find which settings work best for them.

3.7.3 Diode

Diode is an electronic device that controls the directional flow of the current, it is like a one-way valve, a one-way flow device. The current can flow through the diode in one direction only, it does not allow the current to flow in the opposite direction. Diodes are used to convert alternating current to direct current.

Since the flow of current is controlled in one direction, the diodes are useful in converting AC in to DC. As the current leaves the transformer, through the diode in the circuit. The alternative current entering into the rectifier is changing direction 120 times per second, the current exiting the rectifier through the diode is in one direction only, and it has changed to direct current. In the Figure 2.2.6 in chapter 2 of this book the top portion depicts the diode in the circuit, while the lower portion depicts SCRs in circuit.

3.7.4 Silicon-Controlled Rectifiers (SCRs)

So, what are these Silicon-controlled rectifiers or SCRs?

Silicon-controlled rectifiers or SCRs, are also the devices that control the directional flow of current similar to the diodes, with one very important difference. Unlike diodes, the SCRs get in to action only when they are "switched on". If they are not activated, they are not working as a direction control device. The SCRs have a switch called Gate, when it is turned on it allows the current to flow in the desired direction. Once the SCR is turned on the only way it can be turned off is either by reversing the current direction, or by stopping the current flow. As the reversing action is activated, the SCR stops current flow from the unintended direction, as a result of this current flow from both directions is stopped. The current flow is restored only when the current starts to flow in the correct direction, and the Gate receives a signal to allow the current to flow.

SCRs are placed in the secondary circuit of arc welding power source. By turning on the SCR early in each half-cycle, more current can be obtained, conversely if the turn-on is programmed at the latter in each half-cycle, then less current is derived. The method of turning the SCR at different times allows the control of secondary current. This approach eliminates the need to move primary or the secondary coils, as in manual and mechanical approaches to current control. The method is very effectively used for controlling the transformer output. The way the electrical circuit is built, it allows precise control over SCRs, which in turn allows accuracy in the welding current control. Use of SCRs allows the control of current accuracy required for the applications,

(i) Pulse welding current,
(ii) High initial current to start arc, followed by a study and lower current for the continued welding,
(iii) Programmed increase and decrease in current in out of position welding, where control over weld pool is extremely important.

SCRs are used as current control device in constant current, or constant voltage devices and they are also used in inverter machines. A schematic diagram in a welding circuit is shown in Figure 3.7.4.1 below.

3.7.5 Transistors

Another solid-state device used for current control is a transistor. The application of transistors is mainly in very high-frequency-controlled over 10,000 Hz, inverters and other solid-state controlled power sources.

A transistor is similar to the SCR since it also controls current flow in one direction only, similar to SCRs it can turn the current on at different times. Then what is the difference?

The difference is that a transistor can turn the current off without the need to reverse the current direction as is required for SCR. It can also allow different amounts of current to flow. The current amount flow is related to how much signal is applied to the transistor. More current would flow if more signal is applied, low current will allow low current to flow, and no signal to transistor will result in no current flow.

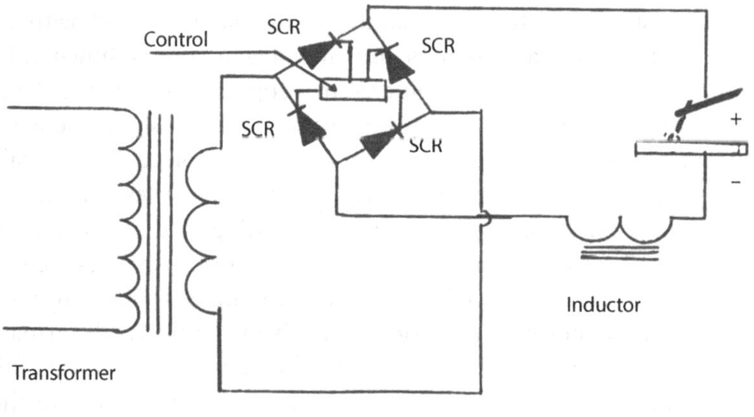

Figure 3.7.4.1 A schematic drawing of single-phase DC power source with SCR bridge control.

3.7.6 A Direct Current Power Source for GTAW

As discussed above, one of the attributes of GTAW power source is its ability to control amperage, or heat input, with a high degree of precision. This attribute is especially of great use if the process is used in the mechanized applications.

The transformer-rectifiers and the generator types are the direct current power source machines. In the transformer rectifier type machine, there are two distinct sections, one is the transformer and the second is the rectifier, together these two make the DC power source.

The transformer section receives the supplied line voltage and current in either 220 volts, 440 volts, and at 60 Hz cycles. The transformer converts the AC line voltage and current to the welding current and voltage of 60 volts to 80 volts, also called open circuit voltage, and the welding current varies to the design of the equipment, and it could be several hundred amperes.

As this low-voltage and high-amperes current exits the transformer it enters in the rectifier section of the machine.

In the rectifier changes the AC to DC. A direct current constant-current output transformer rectifier may be single phase or a three-phase power source.

The rectifiers use devices that are called diodes to convert alternative current to direct current. In diodes the current can flow in one direction only, it does not allow the current to flow in the opposite direction.

Since the flow of current is controlled in one direction, through the diode in the circuit. The alternative current entering into the rectifier is changing direction 120 times per second, the current exiting the rectifier through the diode is in one direction only, and it has changed to direct current. Figure 2.3.2 in the previous chapter describes how a three-phase bridge type rectifier circuit is designed.

The direct current is produced through a generator. Another option is to use an AC alternator with a rectifier. The construction of a generator is relatively simple as compared to an AC alternator with a rectifier.

3.7.6.1 Generator

The generator consists of a rotating armature shaft, that is wound with several independently wound coils. A typical schematic drawing of generator is shown in Figure 3.7.6.1 below. With ends of each armature coil soldered to a copper terminal called a commutator, the commutators kind of half shells wrap on to the armature shaft, the externals of the half-shell commutator are in contacts with carbon or copper piece, that are called Brushes. The two brushes are positive (+) and negative (-) terminals. These brushes are in touch contact with commutator. Encasing the armature windings are stationary (not rotating) wire wound magnets, these are called field windings. The field windings are wires wrapped around an iron core. In this encloser of magnets, the armature is rotated by means of external power, like an electric motor, or a gas or diesel driven engine. As the armature winding rotates it cuts through the magnetic field of the field winding, this cutting of the magnetic field induces current in the armature coil. As the armature wire comes in the horizontal position is enters a position that it does not cut the magnetic field, creating a holiday of current induction, this occurs twice in one full rotation. This induction of current only when

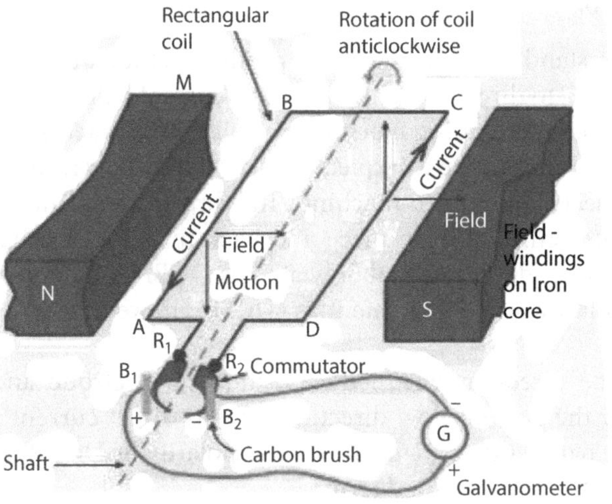

Figure 3.7.6.1 Schematic diagram of a DC generator.

the armature is in vertical position in its rotation, induces current in only one direction, making it the direct current. The induced current is picked up from the commutator by the brushes. More the number of armature coils, field windings and the brushes, better is the flow of direct current. The voltage of a generator is varied by making changes in the current and the field windings. Welding generators are designed to produce low voltage and higher current output.

The study flow of current is essential not only for the efficiency of a welding generator but it is the very essential of the welding system. This consistency of output is ensured by introducing a small motor connected to the field winding, to ensure constant voltage on the main fields and prevent reversal of polarity. These independently powered motors are called exciters. The circuitry shown in the Figure 3.7.6.2 below describes the DC excitation system.

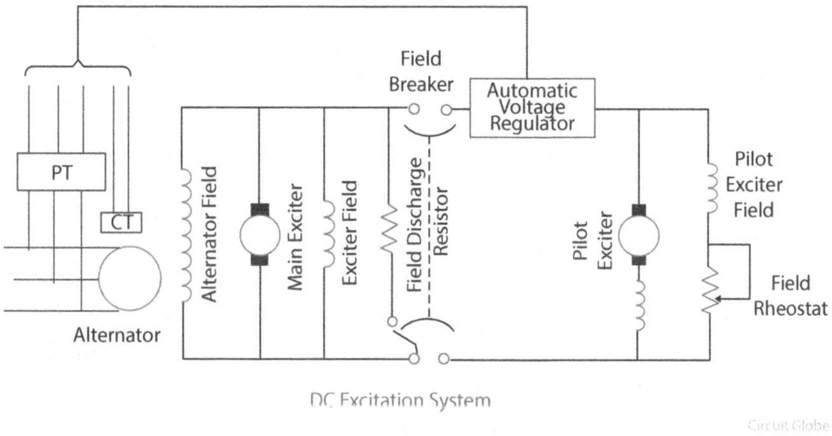

Figure 3.7.6.2 DC excitation circuit.

3.7.6.2 Alternator

Now that we understand that the alternator produces alternative current, so a rectifier is used to change the AC to direct current. The direct current will flow in one of the two direction. It may flow from the welding machine to the electrode, across the arc gap and return back to the machine through the workpiece lead, or it may flow from machine to workpiece and through the electrode back to machine. The direction of flow of direct current from machine to electrode through workpiece and back machine is called electrode negative or direct current electrode negative abbreviated as DCEN. Some people also call it direct current straight polarity and abbreviate it as (DCSP), however DCEN universally is more acceptable term.

Alternatively, the direction where the position of electrode and the work place is reversed, will flow the current flow direction. This is direct current electrode positive or DCEP. Alternative name is direct current reverse polarity or DCRP. Here too, the DCEP is universally accepted and understood term.

As stated above, the polarity can be changed by changing the work lead and electrode positions, however some welding machines have switch, to change the polarity as desired for specific welding task.

Further developments in welding machines have given options of selecting either an alternative or a direct current output from the same source. An AC transformer fitted with a rectifier in series in the circuit allows the selection of the desired current type for the specific welding task.

There are other machines that are called invertor arc welding power sources, these alternative current output machines are an improvement over the transformer-rectifier machines. These are more current in technology and can produce high frequency current in the rage of 1 kHz to 50 kHz across the main transformer. Whereas normal output frequency is 50Hz to 60 Hz alternative current. This improvement in frequency increase allows the welding machines to be reduced in size by about 60% to 70%.

The step-by-step description of how an inverter works starts with first step where input bridge rectifier is installed its function is to convert alternative current to direct current. The input 50Hz or 60 Hz AC may be one or three-phase. This direct current is passed through the inverter switcher that consists of series of SCRs or transistors that are turned on quickly. They chop the DC in to very high high-frequency square-wave alternating current. Single or multiple number of SCRs may be used to direct the one direction flow of the DC current. The chopping of DC current very rapidly by successive SCRs produces high frequency AC current. The SCRs can develop frequencies up to 10m kHz, when even higher frequencies are desired the SCRs are replaced with transistors, they can produce frequencies up to 50 kHz.

This higher frequency wave AC is now used as input to the step-down n transformer. This input is high-voltage, low-current AC power. The output of this transformer is low-voltage high current AC power. At this point an output bridge rectifier is used to change the AC into DC for welding. In the final the device called inductor is used to smooth out the DC output to make is more adoptable for high quality welding processes.

One of the highlights of this welding machine is the feed-back control, this eliminates the need for two welding machines one with constant voltage output, and another for constant current output. Feed-back control allows the one power source to perform multiple welding and cutting functions, like GTAW, GMAW and SMAW, or plasma arc cutting.

Due to electronic circuit and use of SCRs and Transistors, and high frequency power output, these machines are small in size, and light weight, easy to move and mange on work sites. The reduction in the size of the machine is related to the reduction of the size of the transformer. The product of the number of the transformer coils and cross-sectional area of the transformer core is the product of invertor voltage divided by the product of the flux density of core material and the operating frequency of the invertor.

If the number of turns in the transformer coil (N), and invertor voltage (V), and the flux density of the core are held constant, then the cross-sectional area of the transformer (A) can be reduced, while the operating frequency (*f*) is increased. Transformer being the largest and the heaviest part of the system the reduced size of transformer results in machine being smaller and lighter. An advantage that is highly appreciated by the welders, especially more appreciated for site work. A dual source with inverter switching is described in the Alternating current power source.

A key element of GTAW is the power source, which generates the arc. Choosing the right power source can be a confusing decision. Several factors should be considered before purchasing a GTAW power source. The following are few key factors related to the machine's application, apart from the investment cost.

(i) Material(s) to be welded
(ii) Power requirements
(iii) Usage
(iv) Time constraints
(v) Additional capacity

The most important consideration when deciding to purchase a power source is the intended application of the process. One size does not fit all, and if wrong choices are made in purchasing the machine for the job, then nothing but problems and frustration will result.

When choosing a GTAW power source, following two factors must be considered,

a. Output current, and
b. Duty cycle.

3.7.6.3 *The Output Current*

The output current of the power source is one of the primary factors to consider, because the current output determines the thickness of the materials that can be welded using that machine. This is a complicated measurement that requires the understanding of the machine's application in terms of material type, materials' thermal conductivity, and part size. These are all contributing factors that affect amperage requirements.

3.7.6.4 *Duty Cycle*

Duty cycle is a rating of the amount of power a machine can produce in a given time. The duty cycle rating is broken down on a 10-minute scale. For example, a machine with a duty cycle rating of 150 amps at 60 percent duty cycle means that the machine can produce 150 amps for six minutes and then must cool for four minutes. A machine with a rating of 300 amps at 100 percent duty cycle means that the machine can produce 300 amps of current continuously.

The combination of output current and duty cycle is crucial. For instance, the choice of a power source with a lot of output current and a poor duty cycle, would allow welding at higher amperages, but for very short periods of time. For casual welding as hobby, this may not be a problem, but in an industrial setting, where time is critical, waiting for a machine to cool isn't very cool.

The type of input power is an important factor, in selection of suitable welding power source. Industrial application machines are often available to operate at multiple-line input voltages, for example, 208 V, 220 V, and 440 V. As a rule, the higher the input voltage, the less current (amperage) the machine will draw through your electrical system. In a shop where several power sources operate, this is a critical factor.

3.7.7 The Inverter Machines

The modern inverter-based machines offer additional flexibility. This is especially apparent in the use of AC for welding nonferrous materials. Some manufacturers offer power sources with a maximum control over the arc waveform. Some of these newer machines also are considered to be multi-process power sources, these can be used for welding various other electric arc processes. The inverters significantly differ from conventional power sources limited either by constant –current or constant-voltage mode. The inverters however are not limited to either a constant-current or constant-voltage mode.

Other advantage of an inverter type machine is its efficiency in terms of power input versus output. Inverters are energy efficient, and are able to do more work with less energy, thus reducing cost and space. These machines are powerful for their size, but they are also expensive.

Once the application is defined, and power requirements are established, choosing a power source becomes easier. When the output current, duty cycle, and special requirements, are known then matching them to power source specifications becomes a matter comparison, and selection based on individual preferences.

Manufacturers also offer many different equipment packages, which can be customized based on need. Also consider used equipment may be considered, a well-maintained power source that was not abused, can perform well and for far less money.

3.8 Shielding Gases

The primary gases that are used for GTAW process are noble and inert gases like argon (Ar) or Helium (He). These gases are used to envelop the molten pool of weld metal and surrounding hot metal from the reaction with atmospheric gases present in the air to prevent contamination of weld. Argon gas is the most common gas in use, its atomic weight is 39.729 and density is 1.784 g/l, thus it is much heavier than air and able to envelop the hot metal, weld metal and hot tungsten electrode effectively to shield from atmospheric contamination. Helium however is relatively lighter gas, at density 0.0164 g/ml and atomic weight of 6.94, but it is also heavier than air to effectively shield the weld metal and surrounding hot metal from atmospheric contamination. Helium is also relatively expansive gas, and its use is limited, to specific applications where deeper penetration is demanded by the weld design.

The gases other that Argon and Helium used in welding are active gases and they are not used for shielding. Care should be taken to keep two type of gases separate to avoid contamination and serious damage to the equipment and or the weld.

Some very specific welds may use a mix of gasses, these are pre mixed at very precise ratio.

Often the choice of using a mix gas is a deliberate action. The choices are determined due to the process and material demands, process automation, material, and use of higher current also factor in choice of gases, for example a mix of 95% Ar and 5% hydrogen gases is used for certain GTAW process to help increase depth of the penetration, while using higher current, in combination with larger diameter electrode. In any major project, this decision is more of a welding engineering, and welding machine specification call, and is given in the qualified welding procedure specification (WPS) for welders to follow.

Other gases that are used in GTAW process are for the shielding of back-side of the weld from the reaction with atmospheric gases present in the air to prevent contamination of weld. These gases may vary from two inert gases argon (Ar), helium (He), to active gases like nitrogen (N) or hydrogen (H). The selection of these gases is dictated by the material being welded, and the degree of protection desired.

3.9 Gas Regulators and Flowmeters

The gas flow meter is a compact, instruments, used to measure the amount of gas flowing through the point of use. In welding that point of use is at the welding point.

High-accuracy mass flow meter may be electronic and equipped with μF (Micro Flow) sensor chip. It accurately measures the mass flow rate; they are often used to calibrate the mechanical flow meters that fit on the welding gas cylinders.

Theoretically, the fluid flow is defined as the amount of fluid that travels past a given location, this is a straightforward calculation, and uses the following variables.

$$Q = A \times v$$

where;

Q = flow rate, at given space of time,
A = the cross-sectional area of the pipe, and
v = the average fluid velocity in the pipe.

Putting this equation into action, the flow of a fluid traveling at an average velocity of a 1 meter per second through a pipe with a 1 square meter cross-sectional area is 1 cubic meter per second. Note that Q is a volume per unit time. This is the basic principle of measuring volumetric flow of gas.

But all this calculation is redundant, and generally not required in practical welding application, because welding we have a gas that has fixed velocity as flowing from the gas cylinder filled at a specific pressure, and the gas flows through a tube of fixed diameter.

The manufacturers have standardized the dimensions, and further more they calibrate their flow meters to a standard instrument, often an electronic measuring device of very high accuracy.

The flow of gas is measured by a flow-meter, how much gas to flow is regulated by the gas regulator, a calibrated glass or poly carbon tube, or a second gauge, as is shown the Figures 3.9.1A and B below. The second dial measures the pressure at which the gas is flowing from the cylinder.

In welding specific gas flow meters are available and are used, the key point should be to check their calibration before installation, and periodically re-certify their calibrations for

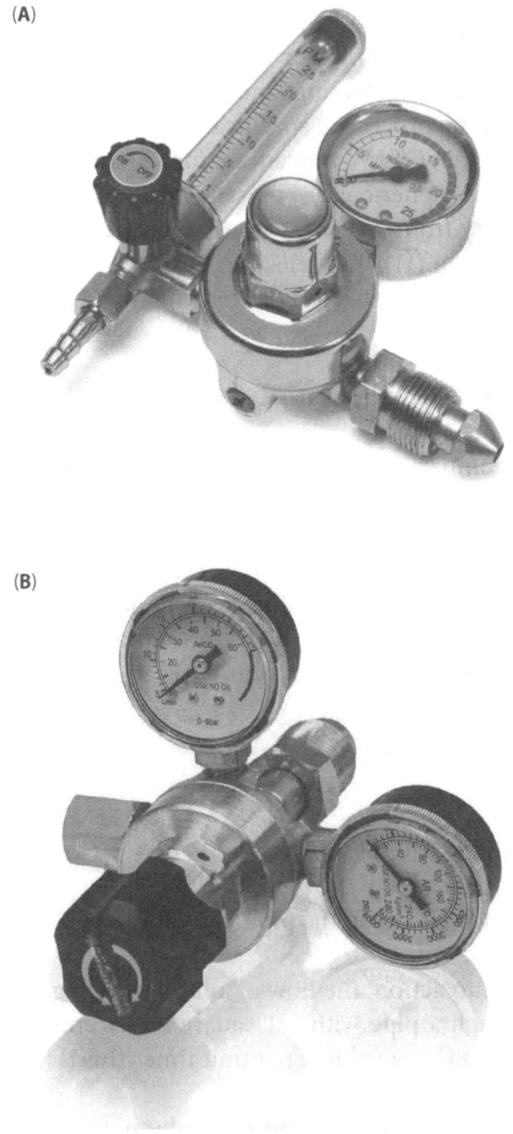

Figure 3.9.1 Gas flow meters (A) shows the tube type flow meter, and the bottom (B) has a gauge type flow meter both calibrated in L/min.

accuracy. Often the argon gas for welding is measured in volume, and rate of flow is measured in volume per hour, example is lpm, (liter per minute) or CF/H or simply CFH, (cubic feet per hour). These are volumetric flow meters.

3.10 GTAW Torches, Nozzles, Collets, and Gas Lenses

GTAW welding torches are the welding heads that hold the tungsten electrode. Torches are the tools used to start the arc, and responsible for the continuance of welding. The manual torches may be air cooled, or equipped with cooling systems water. The manual torch has a handle, while the automatic torch normally comes with a mounting rack, and the tungsten electrode fitted in the nozzle faces vertical down as it travels on the rack.

For the manual torches, the angle between the centerline of the handle and the centerline of the tungsten electrode, known as the head angle, can be varied to the preference of the operator. Air cooling systems are most often used for low-current operations generally up to about 200 amperes, while water cooling is essential for high-current welding up to about 600 amperes. The torches are connected with cables to the power supply and with hoses to the shielding gas source and where used, the water supply. The Figure 3.10.1 below shows the cut out of a typical hand-held welding torch, in the handle portion of the torch we can see the positioning of power cable, water inlet, and outlet hose, the gas hose and in the welding head the positing of the tungsten electrode, gas outlet orifice, gas nozzle, and collet for electrode holding. An assortment of these torch components and tungsten electrode is shown in the Figures 3.10.1 and 3.10.4 below.

The internal metal parts of a torch are made of hard alloys of copper, so they can transmit current and heat effectively, without getting damaged due to heat. The tungsten electrode must be held firmly in the center of the torch with an appropriately sized collet, and ports around the electrode provide a constant flow of shielding gas. Collets are sized according to the diameter of the tungsten electrode they hold. The body of the torch is made

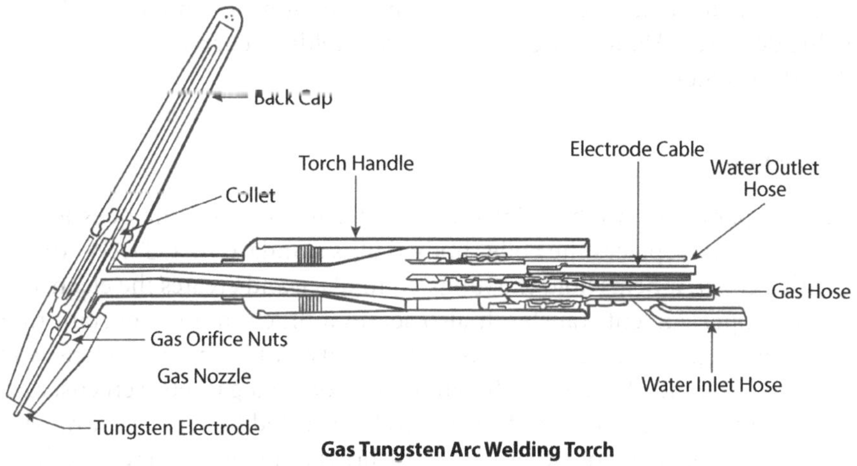

Figure 3.10.1 A typical manual welding torch, note the water cooling, gas supply and tungsten electrode assembly.

Figure 3.10.2 Various nozzles types and sizes.

of heat-resistant, insulating plastics covering the metal components, providing insulation from heat and electricity to protect the welder.

The size of the welding torch nozzle depends on the amount of shielded area desired. Various sizes of nozzles are shown in the Figure 3.10.2 below, note that these nozzles are marked with numbers ranging from 6 to 15, these numbers represent the orifice size 6mm to 15mm. The selection of size of the gas nozzle depends upon the diameter of the electrode, the joint configuration, and the availability of access to the joint by the welder. The inside diameter of the nozzle is preferably at least three times the diameter of the electrode, but there are no hard rules. The welder judges the effectiveness of the shielding and increases the nozzle size to increase the area protected by the external gas shield as needed. The nozzle directly faces the heat from the welding, so it must be able to withstand extreme heat, as a result they are made from heat resistant alloys, ceramic material like alumina, or fused quartz for longevity, and also visibility. Devices can be inserted into the nozzle for special applications, such as gas lenses or valves to improve the control shielding gas flow to reduce turbulence and introduction of contaminated atmosphere into the shielded area. Hand switches to control welding current can be added to the manual GTAW torches.

3.10.1 Gas Lens

In the previous paragraph we have introduced a term Gas Lens, the question is what is a gas lens? A gas lens replaces the collet body that is standard in a GTAW torch. In conjunction with a collet, it helps hold the tungsten in place and creates the electrical contact necessary for proper current transfer. It also serves a much more important function: it improves shielding gas coverage and joint accessibility. A picture of gas lens is shown in the Figure 3.10.3 the metallic lens is shown with associated gaskets, screens, and fittings that allow control of the gas flow, and the gas to be directed in a precise flow pattern.

A typical gas lens is made of either copper or brass body with layered mesh screens of stainless steel since the metal offers greater durability and resistance to rust and corrosion, that helps evenly distribute the shielding gas around the tungsten and along the weld

Figure 3.10.3 A gas lens, with mesh, and holding circlip.

puddle and arc. Gas lenses can be used with all shielding gases and are available for both air- and water-cooled GTAW torches.

The most durable-but also the more expensive gas lenses-feature an engineered porous filter media that improves laminar flow compared to conventional designs. Other gas lenses, such as those composed entirely of brass and with multiple screens, can serve the welding

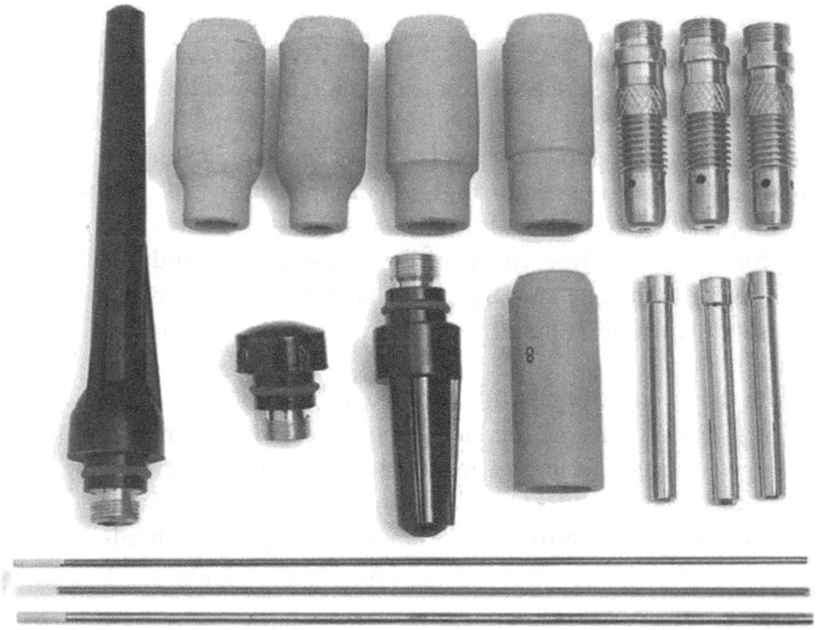

Figure 3.10.4 An assortment of manual welding GTAW torch components.

purposes of less demanding applications, but they can be less conductive and could hinder gas flow after multiple uses.

Gas Lenses offer distinct benefits by reducing the shielding gas turbulence and provide longer, undisturbed laminar flow of the gas to the weld pool. The gas lens also allows the welder to move the nozzle further away from the joint and extend the tungsten electrode past the nozzle by one inch or more. This extension helps minimize tungsten inclusions and improves visibility of the arc and the weld puddle without sacrificing shielding gas coverage, especially on joints that offer very limited access.

Gas lenses are specifically helpful when GTAW welding on alloys that are highly reactive to atmospheric contaminants or on materials used in high temperature applications, as poor gas coverage on these alloys can lead to porosity and material degradation, which may negatively impact the weld's strength. The gas coverage provided by gas lenses also helps prevent oxygen contamination on materials such as stainless steel, titanium, and aluminum by minimizing weld discontinuities. In more basic GTAW applications, such as on steel, gas lenses simply improve shielding gas coverage to ensure more consistent welding performance.

The ability to stick the electrode out further, as the lens ensures the laminar flow of the gas lens, is a major benefit of gas lenses. This additional electrode extension provides better visibility of the acute angle joint and arc, and can improve a welder's ability to lay a proper weld in critical applications such as "T", "K" and "Y" joints in critical structures, and hard-to-reach areas, other configurations.

In some applications, a gas lens can reduce shielding gas consumption, but welders should not use gas lenses solely for this purpose. For a comparison with standard GTAW torch without gas lens, where variables like amperage, torch-to-work distance, and tungsten extension are held constant, the laminar gas flow provided by the screens in gas lenses creates a wider, more even gas coverage with three to five cubic feet per hour less gas.

Gas lenses come in a number of sizes for both air- and water-cooled torches. The type of gas lens a welder chooses for a given application will depend first and foremost on the type of front-end parts 10N or 13N, required for a given torch. A welder also needs to assess the amperage requirements therefore, tungsten size, for the material being welded, along with the joint configuration and amount of joint access. For example, lower amperage applications on non-reactive materials require a different type of gas lens than higher amperage applications on specialty materials and reactive materials.

Gas lenses for 10N series front-end parts, common styles include 17, 18 and 26 series torches, are available in large, extra-large and stubby types. Gas lenses for 13N front-end parts common styles include 9 and 20 series torches, are available in standard, extra-large and stubby types.

A standard gas lens for a torch with 13N front-end parts is sufficient in basic, lower amperage GTAW applications, as is a large type gas lens for a torch with 10N front-end parts. Extra-large gas lenses for both 13N and 10N front-end parts provide improved gas coverage on specialty metals and materials that tend to react to atmospheric contaminants. These larger gas lenses also improve gas coverage on complex and hard-to-reach joints by allowing greater tungsten stick-out for extra visibility of the weld puddle and increased access to the joint.

Stubby gas lenses for both 13N and 10N series feature the same physical orifice, screen size and diameter as the extra-large type lenses, but are noticeably shorter. The smaller torch profile/length increases operator comfort by reducing the overall weight of the torch and allows better access to confined joints. Welders can also gain better torch balance and control if they combine a stubby gas lens with a short back cap.

The Table 3.10.1 below shows the tungsten electrode size and suggested nozzle size, the selection is a guide and should be taken as the starting point, for further improvement as experience is gained.

The nozzle size is certainly going to be different, if gas lens is used, and that should be decided according the selection of N10 or N 13 torch nozzle, discussion in the previous paragraphs.

3.11 Tungsten Electrodes

The function of the tungsten electrode is to serve as one of the electrical terminals for the arc that supplies the heat for GTAW welding process.

Tungsten has a melting point of 3 420°C (6 170°F). At high temperatures, it is thermionic and emits electrons. It is the cooling effect of the electrons boiling from its tip that prevents the tungsten electrode from melting.

The electrodes are classified according to the purity and alloying elements, the classification is indicated by the color coding on the tips of electrodes.

AWS Specification A5.12 classifies tungsten electrodes. EWP is pure tungsten. EWTh-2 is alloyed with the ThO_2 to improve arc stability. EWZr-1 is alloyed with ZrO_2. Normally, straight polarity is used to provide cooling of the electrode.

The electrode tip configuration influences weld penetration and the weld bead. Figure 3.11.1 below shows some of the typical electrode tips used in welding.

For DC welding, the tip should be ground to a specific angle with a truncated end. With AC welding, a hemispherical tip is used. Contamination of the tungsten electrode can occur

Table 3.10.1 Basic matching guide for electrode size and nozzle.

Electrode size (diameter)		Suggested nozzle size (in)
Inch	mm	
0.010	0.25	0.25
0.020	0.50	0.25
0.040	1.00	0.375
0.166	1.6	0.375
0.0938	2.4	0.50
0.125	3.2	0.50
0.156	4.0	0.50
0.188	4.8	0.625
0.25	6.4	0.625

Table 3.11.1 Tungsten electrode tips.

Sharper electrode (included angle at the tip)	Blunt end electrode (included angle at the tip)
Wider weld bead	Narrower weld bead
Low penetration	Better penetration
Easy to initiate arc	Harder to imitate arc
Low current requirement (to initiate arc)	Higher current requirement
Stable arc	Increased potential for arc wander
Shorter electrode life	Longer electrode life

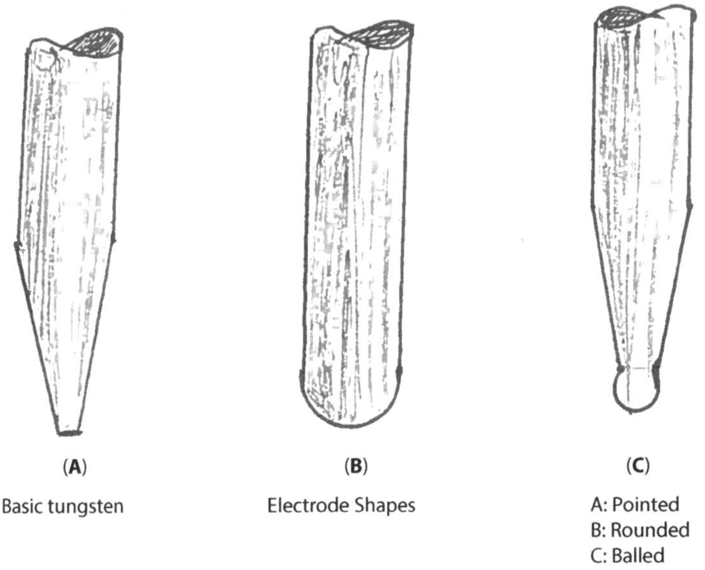

Figure 3.11.1 Electrode tips.

when it touches either the molten pool or the heated filler metal. Improper gas shielding can cause oxidation.

The Table 3.11.3 below lists the types of tungsten electrodes, as classified by AWS specifications their typical compositions and identification color codes.

3.11.1 Grinding of Tungsten Electrode Tips

Grinding a tip angle on the tungsten electrode is a precision job. Typically, the angle can vary from 20 to 90 degrees. The variation in angle of the tip has significant impact on the arc current capacity, and weld penetration. The effect of varying angle is discussed further in the chapter.

The grinding grooves should run in the longitudinal (axial) direction of the tungsten electrode and never around the tungsten electrode (radial), this allows the flow of electrons

Table 3.11.2 Tungsten electrode tips.

Sharper electrode (included angle at the tip)	Blunt end electrode (included angle at the tip)
Wider weld bead	Narrower weld bead
Low penetration	Better penetration
Easy to initiate arc	Harder to imitate arc
Low current requirement (to initiate arc)	Higher current requirement
Stable arc	Increased potential for arc wander
Shorter electrode life	Longer electrode life

Table 3.11.3 Types of Tungsten electrode and their identification.

AWS class	Tungsten composition	Tip color
EWP	Pure tungsten	Green
EWZr	0.25% to 0.5% Zirconium added to tungsten	Brown
EWTh-1	1% Thorium added to tungsten	Yellow
EWTh-2	2% Thorium added to tungsten	Red
EWCe-2	2% Cerium added to tungsten	Orange
EWLa-1	1% Lanthanum added to tungsten	Black
EWG	Unspecified alloying	No color code

in the direction of the arc increasing arc efficiency, and also helps extend the life of the electrode.

A tungsten electrode with the grinding grooves around the tip will hardly ever have a stable arc. The arc searches for the places with the lowest resistance on the grinding grooves and will therefore rotate around the tip of the tungsten electrode.

It is important that the grinding grooves are as small as possible. Deep grinding grooves cause energy loss and unstable arc behavior. Use of belt sanders, sanding sic, flap-disc are not to be used for grinding tungsten tip. The poor quality of the tip would make it difficult to initiate and also maintain the arc required for welding. Furthermore, the adhesive residue from the sanders imbedded in the tungsten groves would contaminate the electrode tip making the arc initiation and maintenance more difficult. Additionally, some of these contaminants and their burned-out oxides will mix in the weld puddle to make the weld defective, and it may not pass the tests. Use of dedicated tungsten tip grinders is strongly recommended.

When sharpening the tungsten electrode tip, care should be taken to produce smother surface with very fine grinding grooves, use of very fine diamond grinding wheels is strongly recommend. This gives a very fine surface to the tip and arc stability and tool life will increase significantly. Overheating of ground surface of electrode, visible through the discoloration should be totally avoided. The discoloration indicates oxidation and formation of oxides at the tipoff the tungsten tip. The oxides would cause difficulty in starting the arc, and contaminate the weld as well.

Centralizing the tip in the middle of the electrode is also an important part of the good welding practices. Electrode tip that is not in the middle can also cause arc deviations, its effect is significantly prominent when mechanized, or automated GTAW process is used.

3.11.2 Tungsten Grind Angles and How They Affect Weld Penetration

Does Tungsten Grind Angle make that much difference?

Grinding a top angle on the tungsten electrode is a precision job. Typically, the tungsten electrode tip angle can vary from 20 to 90 degrees. The variation in angle of the tip is based on the weld design, type and thickness of material being welded, and it has significant impact on the arc current capacity, and weld penetration.

Three pictures below show very distinctly different arc cones from the three different electrode angles ground, in this example we see electrode tips ground to 60°, 30°, and 15° angles.

3.11.2.1 The Impact of Tungsten Tip Angles on Weld

Sharp angle of the tungsten tip is very helpful in easy arc initiation. There are situations where this attribute is very essential in welding. Especially where metal being welded is thin and depth of the penetration is not of much significance, either because of the thin metal or because of the design of the weld. Some examples of this could be the welding of thin tubes in a small heat exchanger, or welding near the cooling fin. Or making a corner or edge weld on thin material.

From the Figure 3.11.2.1 to 3.11.2.3 and also refer to the Table 3.11.2 above, the 15° needle type point is more suited for these type of welding demands.

Welders have used sharper electrode tips, to weld very thing material, they have often used this technique to improvise when they did not have very ideal diameter tungsten for

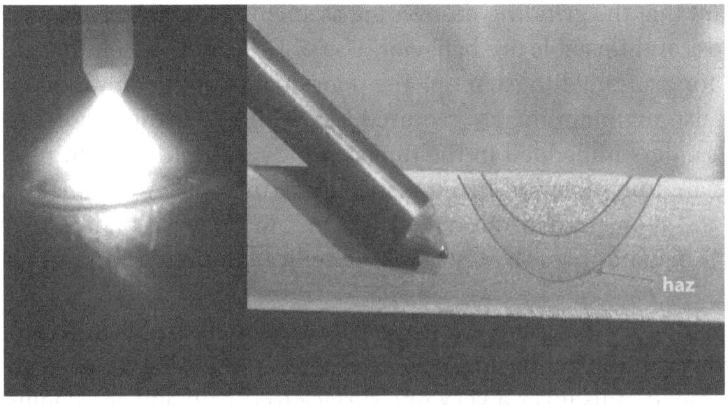

Figure 3.11.2.1 The tip angle 60°, note the depth of the deeper penetration and the shape and depth of the HAZ.

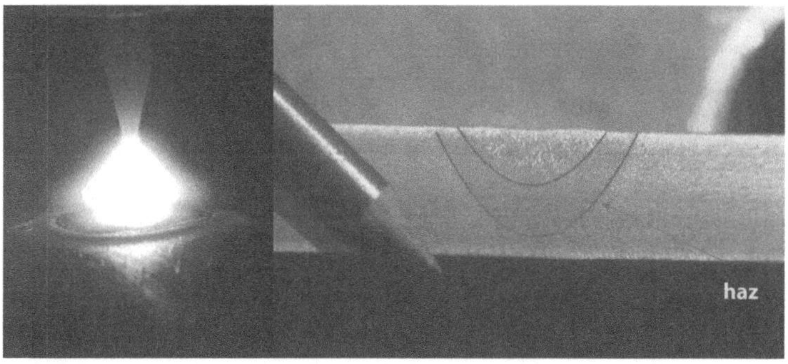

Figure 3.11.2.2 The tip angle 30°, note the depth of the shallower penetration and the shape of the HAZ.

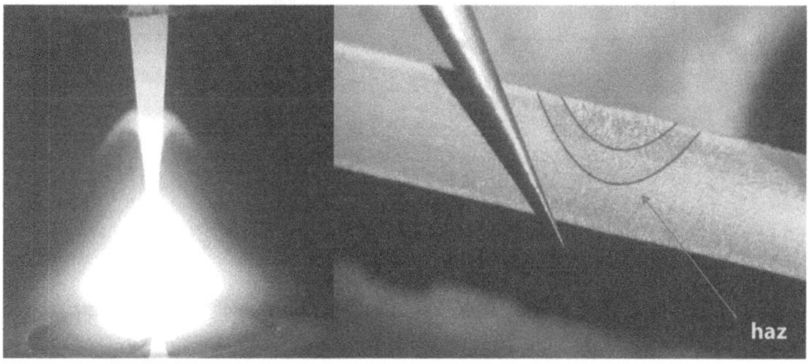

Figure 3.11.2.3 The tip angle 15°, note the depth of the shallowest penetration and the shape of the HAZ.

the job but used what they had with especially thin pin like tip, and very smoothly ground tip to pull of the demands of the work.

In contrast to the thin pin like tips, blunter electrode tips are required to carry lager current densities. Higher amperage to weld thicker metal demands more blunt tapers to penetrate the thick section and it also reduces the risk of tungsten inclusions in the weld.

Welders that see lot of variation the type of work that comes their way keep a set of such electrodes in their boxes and purpose grind for each work. With the experience good welders will know which angle is most suited to their specific work.

3.12 Joint Design

The significance of weld design on the quality of weld cannot be over emphasized. The weld can be designed to carry certain loads, in specific stress conditions, or they may be designed to provide esthetic value to the work, a properly designed weld can reduce the cost of the fabrication and also impart necessary mechanical strength to the component.

The five basic joints shown in Figure 3.12.1 and their variations may be used for welding most of the metals. Figure 2.6.2 in the previous chapter gives more detailed pictures of additional weld designs.

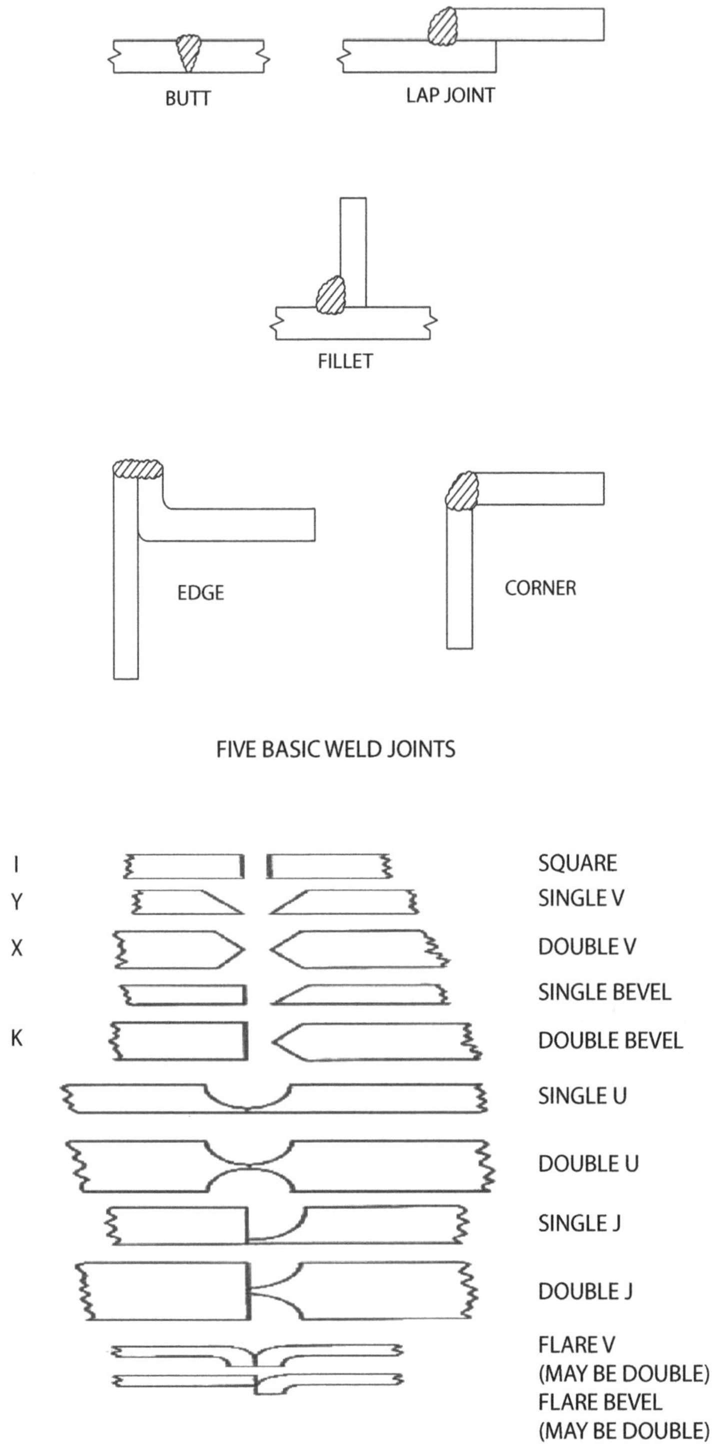

Figure 3.12.1 Five basic weld designs (Courtesy of Indian Air force training manual "Basic Welding Technology").

While designing a weld joint, care must be taken to ensure that there is enough room for proper accessibility for the welder to allow manipulation of the electrode holder in order to obtain adequate fusion of the groove face and addition of filler metal.

The cleanliness of tools used by fabricators and welders for joint preparation, are important. Any contamination with abrasive particles or cutting fluids can cause weld defects. Both the filler metal and the base metal must be cleaned to remove all traces of oil, grease, shop dirt, paint, marking crayon, and rust or corrosion.

3.13 Power Source Remote Control

The remote control on power source gives the GTAW welders more ability to improve weld quality in diversity in terms of materials, and positions of welds are invariable, welders require more controls in their hands to control the arc, and weld pool for a good quality weld, in fact not just good quality welds but even just to weld. Arc behavior varies with number of factors that may include, the current delivery, the position of the weld, and finally and most important factor is the distance from the machine to the location of the weld and the welder. All these challenges demand that the welder is provided with controls that the welder is able to adjust the current, the arc length, and subsequent weld pool for successful welding.

Most welding machines supplied these days include a small, portable remote-control panel that helps distance current adjustment. This remote control allows the welder to turn on or off the machine, adjust the current demand for specific type of welding conditions.

3.14 Installation of Welding Machines

The welding power sources are often connected to the one phase or three phase power supply points. The power factor that these welding machines develop, disturb the power supply if these machines are connected to the same circuit. This requires that other machines connected to the same supply circuit are provided with some power factor correction devices of their own. This is done by connecting a capacitor to give these machines a boost of power to improve their power factor. This arrangement requires the careful planning and consideration and should be done with the expert analysis and to comply with the local electrical code.

3.15 Power Source Cooling System

In the GTAW process cooling occurs ate two different points. (1) at the welding torch, this we have discussed in section 3.10 earlier, and shown in Figure 3.10.1. (2) the second cooling is of the welding power source, the welding machines get hot in their operation, not due to the welding heat, but due to the operation within the machine itself.

Most machines are naturally air cooled, smaller machines are cooled through the gravity fed air to flow through the machine to prevent them getting too hot.

The lager machines require some additional cooling system to keep them from overheating. They use forced air circulation though the system. This is achieved through an electrical motor and fan system. The motor runs the fan that provides the air to cool the machine.

For effective cooling the air passageway is designed for free flow of the air from the entry pint to the exit points. Periodic maintenance, inspection and cleaning of the duct, and openings must be conducted, to remove dirt and dust from the system.

3.16 Welding Connections – Welding Cable and Welding Torch Connections

Welding cables used to carry current from the welding machine to the work and back are also called welding leads, these leads are super flexible large diameter electrical cables.

The lead that is connected to the electrode-holder and often carries current from the machine to the electrode are called welding torch lead. The cable that is connected to the work place often by a clamp, is called **ground cable** or **workpiece lead**. Cables and grounding connection clamps are shown in the Figure 3.16.1 below.

The need to be very flexible to meet the demands of welding activities, they also need to be very well protected since they are carriers of heavy current. The flexibility and insulation are provided by thick rubber covering which is often supported by a layer of reinforcement, by woven fabric layer to provide some rigidity and protection from damage.

Welding leads are produced to be flexible so as to reduce the strain on welders' wrist and hands. The flexibility is achieved by use of about 800 to 2500 fine copper or aluminum wires, wrapped in one single bundle as a cable. Copper leads are more suitable for carrying higher currents, this attribute of copper cable also reduces the diameter cables. But copper cables are heavier that aluminum cables. Aluminum cables are lager in diameter as compared to the copper cable for carrying same current capacity, this is because aluminum can carry only up to 61% current. But the advantage of aluminum is that these cables are lighter in weight.

The electrical capacity is the most key difference between application of copper and aluminum cable. The alloy mix is also determined by the intended use of the welding cables. Copper is considered a better conductor with a higher capacity per volume. However, aluminum has higher capability per weight. The weight difference also is determined by the specified material used.

Figure 3.16.1 Copper and Aluminum welding leads: note the number of fine wires that compose a cable, and the rubber sheathing that covers them.

The changes caused by the metals thermal cycle is more prominent in aluminum than that is for copper. These changes are significant in aluminum due to its thermal growth coefficient, compared to copper. Figure 3.16.1 shows both aluminum and copper cables.

The leads are produced in various sizes, and identified by universal numbering, the number indicates the diameter of the lead, which in turn indicates the current carrying capacity

Table 3.16.1 Welding cable current carrying capacity.

Significance of cable (welding lead) diameter and length and current carrying capacity					
	Lead diameter		Cable length	Cable length	Cable length
			0 to 15 meters	15 to 30 meters	30 to 76 meters
Lead no.	Inch	MM	Amperes	Amperes	Amperes
4/0	0.959	24.4	600	600	400
3/0	0.827	21.0	500	400	300
2/0	0.754	19.2	400	350	300
1/0	0.720	18.3	300	300	200
1	0.644	16.4	250	200	175
2	0.604	15.3	200	195	150
3	0.568	14.4	150	150	100
4	0.531	13.5	125	100	75

Note the drop in current as the length of the lead increases.

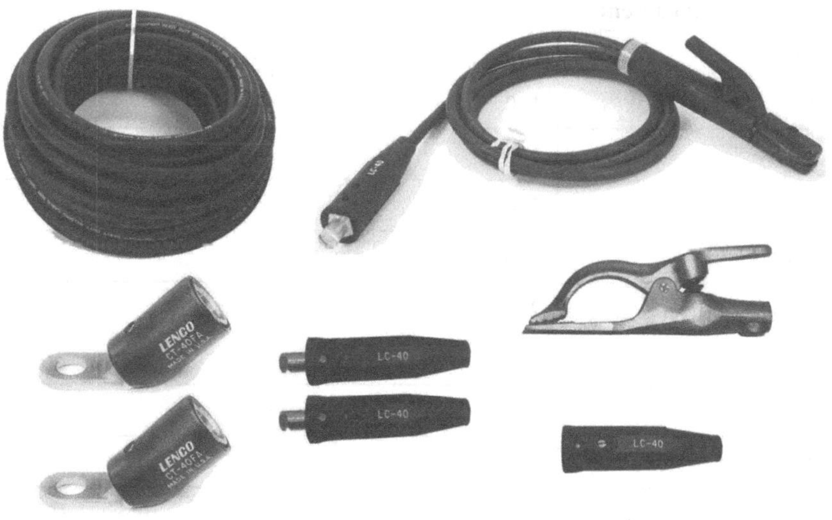

Figure 3.16.2 Various types of cable connectors, and ground clamp. Pictures Courtesy of LENCO catalogue.

of the cable. Larger the number thicker the diameter and lower the current carrying capacity. This is clearly brought out in the table below. The table below also brings out another factor, which is the drop in current as the length of the cable is increased. The length shown in the table includes the length of electrode lead and the workpiece lead. So, the point to note here is that, the reduction in lead diameter, and the increase in the lead length reduces the current capacity of the welding lead. Table 3.16.1 describes the cable identification numbers and corresponding the current capacity at various lengths of the cable.

The corresponding drop in voltage is very low, if all connections are tight and secured, the drop in a copper cable is about 4 volts.

3.17 Welding Power Source Classification by NEMA

Welding machines are electrical machines, so they covered under the guidelines of the National Electrical Manufacturers Association (NEMA). NEMA id a trade association of electrical machine manufacturers.

NEMA classifies the welding machines primarily on the basis of their rated duty cycle output. The NEMA classifications are given below.

1. NEMA Class I: Machines that deliver 60%, 80%, or 100% duty cycles are classified in this group.
2. NEMA Class II: Machines that deliver output at 30%, 40%, and 50% are classified in this group.
3. NEMA Class III: Machines that deliver output at 20% duty cycle are grouped in this class.

Table 3.17.1 Details the NEMA rating and corresponding current output capacity.

Rated output current		
Class I	**Class II**	**Class III**
200	150	18-230
250	175	235-295
300	200	
400	225	
500	250	
600	300	
800	350	
1000		
1200		
1500		

The arc welding machines are described by their following three attributes.

1. Rated current output,
   This is the amount of current measured in amperes that a welding machine is rated to supply at a given voltage. NEMA rated output current for different NEMA class rating described above is in the following table.

3.18 Welding Personal Protecting Equipment

The most important PPE for welder and operators can be easily said to be the Shield and Helmet. The arc welding helmet is used to protect the face and eyes from sparks, spatters, the heat, and ultraviolet rays emanating from the electric arc of the welding. In the GTAW process the welder has his both hands occupied with the welding torch, and filler wire they cannot hold a welding shield, so the welder can only use the welding helmet worn on the head.

The welding helmet consists of a head mount that is supported by a head cover either full or partial, and a head band that wraps around the head on the forehead. The side knobs on the head band allow adjustment to fit different sizes of head. The head and side coving protects the face in general, while the front of the helmet has a window where a dark filter glass of suitable rating is fitted between two plain glass to filter UV rays, and heat to enter in the eye and protects the eye and face from any damage to the welders' eyes.

While welding activity is very safe if proper precaution is taken, the welding arc contains some very damaging rays, they can burn the skin, and damage the eye. The exposure to UV rays in the welding arc can cause eye pain, eye watering, and swelling with irritation as if sand is in the eye, and pain and discomfort can last for about 10 to 20 hours after arc exposure. However, exposure to infrared rays, can injure eyesight. Hence protection is extremely necessary.

The filter glass or lens is an important a part of the shield or helmet, it is required to protect eye of the welder from damaging UV rays, and heat of the welding, while the welder is able to see the arc and progress of his work. Due to its very critical role much study has been done and determined that various shades of lens are required to protect differing intensities of welding arcs.

The density of filter shades is such that the welder cannot see through it until arc is struck.

Helmets have been developed with some "automatic" features, battery is sued to introduce photoelectric cells inbuilt in the helmet, and this allows the lens to be clear until the arc is struck. This is a great advantage to the welders precisely locate their electrode tip on the weld location and the trike the welding arc. This helps produce cleaner welds, reduce damage to the parent metal near welds, less of arc strike outside the weld zone.

For further eye protection from back flashes, some welders also wear a pair of ordinary welding glasses that has # 1 or # 2 lenses on them, this allows them to inspect the weld before or after welding, and cleaning operations, etc., between welding actions.

Various types of welding shields and helmets are described in the chapter 2 of this book, the Miller® welding helmet technology optimizes contrast and clarity in welding and light states. 1/1/1/2 optical clarity rating allows for a lighter light state while not welding — keeping the helmet down thus maximizing safety and productivity. The helmet has four arc sensors and four modes: weld, cut, grind, and X-Mode. X-Mode electromagnetically senses the weld to eliminate sunlight interference and continuously detects the arc even if sensors are blocked. Gen 3.5 headgear with comfort cushion has an ergonomic design that provides extensive adjustability, settings, and enhanced support. Digital controls easily allow welder to adjust shade, delay and sensitivity. AutoSense™ eliminates issues related to setting helmet sensitivity by allowing the welder to push and hold the AutoSense™ button to automatically set the helmet sensitivity for their environment. Auto-on/off power control triggers lens at the strike of an arc.

More complex jobs require more complex safety equipment and PPE. Helmets have been developed and are used where air quality around the welding work is improved by introducing fresh air into the helmet, and some are fitted with air filters that provides clean air for the welder to breath.

The following table gives the recommended safe shade numbers for various welding operations. The user must experiment with the most suitable and safe shade number that suits, the following table gives an indicator and may work as the safe trials tart point in selecting most suitable shade for individual safety in welding.

3.19 Other Essential Clothing for Welders

In welding operation, the molten weld metal often in the form of spatter fly all over and can easily land on the person of the welder. This can cause sever burn and injuries. Welders need protection from such injuries. Most of these accessories are made from leathers, hence they are often collectively referred as *leathers.*

- Welding gloves,
- Welding Gauntlet sleeves,
- Aprons,
- Leggings,
- A Jacket, especially if welding in overhead position
- A cape, especially if welding in overhead position
- Heat protected gloves, these are insulated gloves in addition to the leather welding mentioned above, and used when welding on hot surfaces and for longer time.

3.20 Filler Wires Used in GTAW Process

Unlike Shielded metal arc welding process uses electrode that have solid steel core, the steel core is covered by a coating that is called flux. The filler wire electrodes are identified by the steel core wire diameter and by a series of letters and numbers. The significance and the meaning of these number and letters is explained further in the description.

American welding society (AWS) and several other agencies around the world have developed methods to designate identifier system for electrodes, by far the AWS system is most commonly used and understood.

AWS classifies electrodes for the group of metals that these welding electrodes are developed. Some of these are listed below. They are given a specific AWS specification number as indicated with the material group.

3.21 Classification and Identification of Welding Wires

Welding electrodes are identified by an alfa-numeric numbering code system. Both American (AWS) practice and European practices use similar basis, however their units and sometimes even the sequencing differ. Hence it is important to learn what system is being used to designate the electrode, but in most cases the American practice is prevalent, and offers very less chances of confusion. The description of AWS classification (identification) system for carbon steel, and alloys is given in the following sketches.

ER -XX-S-X

Where ER = Electrode-Rod (It is written as Electrode–Rod, because most of the GTAW wires are also used as electrode in the GMAW process, we will discuss the GMAW process in the next chapter, specific GTAW wires, cut to about 500 mm length, or a meter long, are also available, but they too have similar designations).

Set of 2 XX = Strength of the filler wire weld metal (x1000) psi
S = Solid wire (as opposed to tubular)
X = Chemical composition of the weld metal

For example, the description of a welding wire for Carbon steel ER 70 S-3 is as following;
The wire is Electrode-Rod, it has the weld metal tensile of 70 x 1000 = 70,000 psi, the wire is solid (not hollow). And last letter indicates the chemical composition of the weld metal.

C = 0.06 – 0.15 Ni = 0.15 max Mn = 0.90 –1.40 Cr = 0.15 max Si = 0.45 – 0.75
Mo = 0.15 max P = 0.025 max V = 0.03 max S = 0.035 max Cu = 0.50 max.

A low alloy steel, welding wire designated as ER-90S-B9 is suitable for welding high strength low ally steel including P9 grade of steel. The electrode is designated to give higher that 90 x 1000 = 90,000 psi strength, in actuality the obtained strength is around 120,000 psi. It contains about 9% chromium, nah has very tightly controlled Manganese and nickel combination. Welding such alloys requires very specific heat control and properly developed welding procedure.

Other metals and alloys follow similar pattern of welding wire designation, they either grade them by strength of as in the case of corrosion resistant materials they grade is by the material designation for example the wires for basic 18/8 stainless steel is ER 308 followed by a letter to indicate if the wire is suitable for low carbon content or any other such designations.

3.21.1 Designation of Aluminum Welding Wires

For aluminum the designations system differs from the steel welding wire system, for example, a commonly used welding wire ER 4043 has somewhat different description. The designation simply replicates the grade of aluminum the specific welding wire is used for welding, Table 3.21.1 below lists the aluminum welding alloy designation system. A near matching welding wire may be used for welding an aluminum alloy, however a welding wire may be used to weld number of different aluminum alloys.

ER 4043 welding wire contains approximately 5% silicon. The presence of silicon gives good wetting, properties, and better fluidity to the weld metal, resulting in better weld penetration, and arc control. The ER 4043 wire is frequently used to weld a wide spectrum of aluminum grades such as, alloys 3003, 3004, 5052, 6061, 6063 and Aluminum casting alloys 43, 355, 356, and 214.

Aluminum as a metal is graded in different number system, below is a brief description aluminum and its alloys. The numbering system predominantly indicates the primary alloying element in the alloy. For the alloys with the suffixes like 'T' for example alloy 5052T, indicates that the alloy is in heat treated condition.

3.21.2 Aluminum Alloys and Their Characteristics

There are seven series of wrought aluminum alloys, and as a welder, or a welding engineer, or manager of aluminum fabrication, it is imperative to know these alloys and their differences and understand their applications and characteristics.

Table 3.21.1 Aluminum alloy designation system.

Alloy designation	Grade, primary alloying element	Secondary alloying element	Weldability and heat treatment
1xxx	99.000% Pure Aluminum		Weldable and requires no PWHT
2xxx	Copper	Magnesium may be added.	Weldable and PWHT required
3xxx	Silicon, Copper and Manganese		Weldable and PWHT not required
4xxx	Silicon		Weldable and PWHT not required
5xxx	Magnesium		Weldable and PWHT not required
6xxx	Magnesium	Silicon	Weldable and PWHT is required
7xxx	Zinc	Copper and magnesium may be added.	Weldable and PWHT is required
8xxx	Tin		
9xxx	All other alloys		Weldability and heat treatment to be determined.

1xxx Series Alloys

These are non-heat treatable alloys. In non-heat-treated condition the ultimate tensile strength of these alloy have range from 10ksi, to 27 ksi. The 1xxx series is often referred to as the pure aluminum series, this is because they are required to have a minimum of 99.0% aluminum.

These grades are weldable. However, because of their purity level of aluminum, they have very narrow melting range. This demands both welders' ability, engineering considerations to develop an acceptable welding procedure. When considered for fabrication, these alloys are selected primarily for their superior corrosion resistance such as in specialized chemical tanks and piping, or for their excellent electrical conductivity as in electrical bus-bar applications. These alloys have relatively poor mechanical properties, and would seldom be considered for general structural applications. These base alloys are often welded with matching filler material or with 4xxx filler alloys dependent on application and performance requirements.

2xxx Series Alloys

2xxx series alloys are heat treatable alloys. Their ultimate tensile strength ranges from 27 ksi, to 62 ksi. This increase of strength is attributed to the alloying element, like copper. These are aluminum and copper alloys where copper additions vary from 0.7% to 6.8%. The high strength, and better temperature resistance, these high-performance alloys are often used for aerospace and aircraft structural applications. They have excellent strength over a wide range of temperature.

Some alloy grades with in this series are considered non-weldable by the arc welding processes because of their susceptibility to hot cracking and stress corrosion cracking, both due to the presence of copper in them. However, there are others that can be successfully welded with the correct welding procedures. These base materials are often welded with high strength 2xxx series filler alloys, designed to match their performance, but can sometimes be welded with the 4xxx series fillers containing silicon or silicon and copper, dependent on the application and service requirements.

3xxx Series Alloys

The 3xxx Series alloys are an alloy of aluminum and manganese. 3xxx series alloys are non-heat treatable. These alloys have relatively low tensile strength, where their ultimate tensile strength ranges between 16 ksi and 41 ksi. These are the aluminum and manganese alloys contain from 0.05% to 1.8% manganese. With their moderate strength, they have good corrosion resistance, good formability and are suited for use at elevated temperatures. Among their various uses are the industrial component like heat exchangers in vehicles and power plants, and of course their suitability for manufacturing cooking utensils cannot be ignored. Their moderate strength, however, often precludes their consideration for structural applications. These base alloys are welded with 1xxx, 4xxx and 5xxx series filler alloys, dependent on their specific chemistry and particular application and service requirements.

4xxx Series Alloys

4xxx series alloys are alloys of aluminum and silicon, silicon content varies from 0.6% to 21.5% in various alloys in this series. This series contains both heat treatable as well as non-heat treatable alloys. Their ultimate tensile strength varies from 25 ksi, to 55 ksi.

The effect of silicon added to aluminum reduces its melting point of the alloy, and improves the fluidity of the molten metal, making is very suitable for welding. These characteristics are desirable for filler materials used for both fusion welding and brazing. Consequently, this series of alloys is predominantly found as filler material. Silicon, independently in aluminum, is non-heat treatable; however, when paired with magnesium or copper the alloys can be heat treated for improved strength. Number of these silicon alloys have been designed to have additions of magnesium or copper, which provides them with the ability to respond favorably to solution heat treatment. Typically, these heat treatable filler alloys are used only when a welded component is to be subjected to post weld thermal treatments.

5xxx Series Alloys

5xxx series alloys are non-heat treatable alloys. These are alloy of aluminum and magnesium, where magnesium is from 0.2% to 6.2%. The alloys have very high strength in all non-heat treatable aluminum alloys. They have ultimate tensile strength from 18 ksi to 51 ksi.

Alloys in this series are readily weldable, combined with good strength, and good weldability they are used for variety of fabrication and construction. They are used for a wide variety of applications such as Tankers, Boats, Shipbuilding, Transportation, pressure vessels, bridges and buildings. The magnesium base alloys are often welded with filler alloys, which are selected after consideration of the magnesium content of the base material, and the application and service conditions of the welded component. Alloys in this series with more than 3.0% magnesium are not recommended for elevated temperature service above 65°C (150° F) because of their potential for sensitization and subsequent susceptibility to stress corrosion cracking. Base alloys with less than approximately 2.5% magnesium are often welded successfully with the 5xxx or 4xxx series filler alloys. The base alloy 5052 is generally recognized as the maximum magnesium content base alloy that can be welded with a 4xxx series filler alloy. Because of problems associated with eutectic melting and associated poor as-welded mechanical properties, it is not recommended to weld material in this alloy series, which contain higher amounts of magnesium with the 4xxx series fillers. The higher magnesium base materials are only welded with 5xxx filler alloys, which generally match the base alloy composition.

6XXX Series Alloys

Alloys in 6xxx series are heat treatable, and have the ultimate tensile strength from 18 ksi to 58 ksi. These are the aluminum, magnesium and silicon alloys, where magnesium and silicon additions for each element is around 1.0%, generally less than 1%. These alloys are the most welding and fabrication friendly alloys, they are predominantly used in the form of aluminum extrusions, for use in strength bearing structural components.

The addition of magnesium and silicon to aluminum produces a compound of magnesium-silicide, which provides this material its ability to become solution heat treated for improved strength. These alloys are naturally solidification crack sensitive, and for this reason, autogenous welding of these is not recommended, the addition of adequate amounts of filler material during the arc welding process is essential in order to provide dilution of the base material, to preventing the hot cracking problem. They are welded with both 4xxx and 5xxx filler materials, dependent on the application and service requirements.

7XXX Series Alloys

7XXX series alloys are heat treatable. These are the aluminum and zinc alloys, where zinc additions range from 0.8% to 12.0%. This gives the alloy group the highest strength aluminum alloys with ultimate tensile strength ranging from 32 ksi, to 88 ksi. These alloys are often used in high strength applications such as aircraft, aerospace, and competitive sporting equipment. Like the 2xxx series of alloys, this series incorporates alloys which are considered unsuitable candidates for arc welding, and others, which are often arc welded successfully. The commonly welded alloys in this series, such as 7005, are predominantly welded with the 5xxx series filler alloys.

3.22 The Aluminum Alloy Temper and Designation System

Since we are discussing about welding of aluminum and its various alloys, it is important to learn about the commercial availability of the material, in that the first step would be to know how the material is designated and identified. The Aluminum Association Inc. https://www.aluminum.org registers, maintains, and allocates designations to aluminum alloys, and it updates the register with any new developments. Currently, it is estimated that there are over 400 wrought aluminum, and wrought aluminum alloys, and over 200 aluminum alloys in the form of castings and ingots registered with the Aluminum Association.

The chemical composition limits for all alloys are that are registered are included in the Aluminum Association's two books. As described below, these books are named by the colors.

1. **Teal Book** entitled "International Alloy Designations and Chemical Composition Limits for Wrought Aluminum and Wrought Aluminum Alloys" and
2. The **Pink Book** entitled "Designations and Chemical Composition Limits for Aluminum Alloys in the Form of Castings and Ingot".

These two publications can be extremely useful to the welding engineer, when developing aluminum welding procedures, and when the consideration of chemistry and its association with crack sensitivity is of importance. They can be very useful in selection of material grade for specific service conditions.

Aluminum alloys can be categorized into a number of groups based on the particular material's specific characteristics, theses could be its ability to respond to thermal and mechanical treatment, or the primary alloying element added to the aluminum to make a specific alloy. When we consider the numbering and identification system used for aluminum alloys, the characteristics are identified through these numbers. The wrought and cast aluminums have different identification systems. The wrought alloys having a 4-digit system, and the castings have a combination of a 3-digit and 1-decimal place system.

3.22.1 Wrought Alloy Designation System

The 4-digit (XXXX) wrought aluminum alloy identification system is explained as the following.

The first digit (Xxxx) indicates the principal alloying element, which has been added to the aluminum alloy and is often used to describe the aluminum alloy series, i.e., 1000 series, 2000 series, 3000 series, up to 8000 series as shown in Table 3.21.1.

The second single digit (xXxx), if different from 0, indicates a modification of the specific alloy, and the third and fourth digits (xxXX) are arbitrary numbers given to identify a specific alloy in the series.

For the example of the above explanation let us consider alloy 5183, the number 5 indicates that it is of the magnesium alloy series, the 1 indicates that it is the 1st modification to the original alloy 5083, and the 83 identifies it in the 5xxx series.

The only exception to this alloy numbering system is with the 1xxx series aluminum alloys, known as pure aluminums grades, in which case, the last 2 digits provide the minimum aluminum percentage beyond the basic 99% purity of aluminum. So, in the alloy 1350 would mean that the alloy contains 0.50% more purity above the 99% aluminum that means that alloy 1350 has aluminum purity of 99.50%.

3.22.2 Cast Alloy Designation

The cast alloy designation system is based on a 3 digit-plus decimal designation xxx.x, for example, 356.0. The first digit (Xxx.x) indicates the principal alloying element, which has been added to the aluminum alloy as shown in Table 3.22.1 below.

The second and third digits (xXX.x) are arbitrary numbers given to identify a specific alloy in the series. The number following the decimal point is a system of identification where .0 is casting and 0.1 or 0.2 are ingots. If a capital letter prefix is used it indicates a modification to a specific alloy.

The example of the above description, Alloy - A356.0 the capital A (Axxx.x) indicates that the original alloy 356.0 has had one modification. The number 3 (A3xx.x) indicates that it is of the silicon plus copper and, or magnesium series. The 56 (Ax56.0) identifies the alloy within the 3xx.x series, and the .0 (Axxx.0) indicates that it is a final shape casting and not an ingot.

3.22.3 The Aluminum Temper Designation System

From the description and table presented above, one can see that there are different series of aluminum alloys, what is important to note is the considerable differences in their characteristics. The first point to recognize, after understanding the identification system, is that there are two distinctly different types of aluminum within the series described above. These are the Heat Treatable Aluminum alloys that can be imparted strength through the treatment of heat and cooling cycle. The other group consist of the non-heat Treatable

Table 3.22.1 Cast aluminum designation and numbering system.

Alloy series	Principal alloying element
1xx.x	99.000% minimum Aluminum
2xx.x	Copper
3xx.x	Silicon Plus Copper and/or Magnesium
4xx.x	Silicon
5xx.x	Magnesium
6xx.x	Unused Series
7xx.x	Zinc
8xx.x	Tin
9xx.x	Other Elements

Aluminum alloys. This distinction is particularly important when considering the effects of arc welding on these two types of materials.

- The 1xxx, 3xxx, and 5xxx series wrought aluminum alloys are non-heat treatable, but they are strain hardenable only.
- The 2xxx, 6xxx, and 7xxx series wrought aluminum alloys are heat treatable and,
- The 4xxx series consist of both heat treatable and non-heat treatable alloys.
- The Cast alloys 2xx.x, 3xx.x, 4xx.x and 7xx.x series are heat treatable. Strain hardening is not generally applied to castings.

The heat treatable alloys acquire their optimum mechanical properties through a process of heat treatment, the most common heat treatments are the Solution Heat Treatment, and the Artificial Aging.

Solution Heat Treatment is the process of heating the alloy to an elevated temperature to around 482°C (about 990°F) in order to put the alloying elements in the metal into solution. This is followed by quenching, usually in water, to produce a supersaturated solution at room temperature. Solution heat treatment is usually followed by a process called aging. Aging is the precipitation of a portion of the elements or compounds obtained from a supersaturated solution in order to yield desirable properties. The aging process is divided into two types: aging at room temperature, which is termed natural aging, and aging at elevated temperatures termed artificial aging. Artificial aging temperatures are typically about 160°C (about 320°F). Many heat treatable aluminum alloys are used for welding fabrication in their solution heat treated and artificially aged condition. So, unless stated otherwise, it is assumed that the post weld properties may be required to match that of the base metal welded.

The non-heat treatable alloys acquire their optimum mechanical properties through Strain Hardening. Strain hardening is the method of increasing strength through the application of cold working. The Temper Designation System addresses the material conditions called tempers. The

Temper Designation System is an extension of the alloy numbering system and consists of a series of letters and numbers which follow the alloy designation number and are connected by a hyphen. Examples: 6061-T6, 6063-T4, 5052-H32, 5083-H112. Table 3.21.3 below details the letters used for temper designations, and meaning of those letters.

Further to the basic temper designation, there are two subdivision categories.

(i) addressing the "H" Temper – Strain Hardening, and
(ii) The other addressing the "T" Temper – Thermally Treated designation.
(iii) H Temper – Strain Hardened

- The first digit after the H indicates a basic operation:

H1 – Strain Hardened Only.
H2 – Strain Hardened and Partially Annealed.
H3 – Strain Hardened and Stabilized.
H4 – Strain Hardened and Lacquered or Painted.

- The second digit after the H indicates the degree of strain hardening:

HX2 – Quarter Hard,

HX4 – Half Hard
HX6 – Three-Quarters Hard
HX8 – Full Hard
HX9 – Extra Hard

- Subdivisions of T Temper – Thermally Treated

T1 - Naturally aged after cooling from an elevated temperature shaping process, such as extruding.
T2 - Cold worked after cooling from an elevated temperature shaping process and then naturally aged.
T3 - Solution heat treated, cold worked and naturally aged.
T4 - Solution heat treated and naturally aged.
T5 - Artificially aged after cooling from an elevated temperature shaping process.
T6 - Solution heat treated and artificially aged.
T7 - Solution heat treated and stabilized (overaged).
T8 - Solution heat treated, cold worked and artificially aged.
T9 - Solution heat treated, artificially aged and cold worked.
T10 - Cold worked after cooling from an elevated temperature shaping process and then artificially aged.

Additional digits indicate stress relief.
Examples:

TX51 or TXX51 – Stress relieved by stretching.
TX52 or TXX52 – Stress relieved by compressing.

3.23 Welding Metals Other Than Carbon and Alloy Steels

After the welder has mastered the essential welding techniques, and understood the arc behavior and gained confidence in handling the variance of welding variables, they now be ready to move if so desired, to weld other metals. However, they need to get some basic knowledge of material's physical behavior, effect of heat on its expansion and contraction, melting point, especially under welding heat.

Welding other metals includes the developing skill to read and follow the instructions given in the welding procedures. Welding procedures can vary significantly with information and demands of the job. For example, they may require weld be done in a specific position, they may require tighter control on heat input, the weld may be tested for its hardness, ductility and toughness, among other mechanical properties.

Alloy steel welding is very demanding, as they require very tight control on welding parameters, they may also require post weld heat treatment. In such cases the welder's job is not done until all inspections and all tests are not completed and weld is finally accepted as sound to the specified requirements.

Knowledge of these specifics of each metal is the essence of welding. In welding the knowledge is literally experienced by the welder under the arc, and welder has to manage the changes that occur, to bring out the best of the weld joint.

So, it is important for new welders to develop skills and habits that will make them most desired welder for welding expensive material with minimum possibility of defects and repairs.

3.24 GTAW Welding of Aluminum

We have been introduced with aluminum as material, and its various grades (Series) that are weldable. Some of these grades may require post weld heat treatment or aging to restore the material's strength, after welding. Those specifics should be the part of the developed welding procedure given to the welder to follow.

The following is about welding the aluminum.

It is a very well-known fact that the steel is often the 'default' metal to for learning to weld, and also as the material for the structural construction and welding. We have learned that during the welding, steel gives stage wise temperature indication of what is happening to it by the application of heat, by change of color. Derived from that experience the brain expects somewhat similar temperature indications from the aluminum, when heated during welding aluminum. Therein lies the fundamental challenge of welding aluminum.

Welding aluminum is different than welding steel because of one very basic property, when heated aluminum does not change color as does steel, this lack of color change does not give heat perspective to the welder, specifically to a new welder, while the welder is awaiting some heat related changes to show up, the aluminum material under the heat, collapses.

This is because the aluminum is much greater conductor of heat as compared to steel, and that changes the behavior of heat dwell on the aluminum parts to be welded.

Further to complicate the matter is the fact that aluminum is an active metal and it forms oxides while being heated for welding, it is harder to create a welding filler suitable for welding aluminum. When combined with the metal's high heat conductivity and low melting point, it is very easy for a new welder to completely melt the aluminum pieces involved in the process. As a result, two very important steps need to be taken.

1. First step to arc welding aluminum is to clean the base metal of any oxides or solvent oils, and prevent oxide formation during welding. And
2. The second step is to be mindful of aluminum's behavior under heat,
   (i) Aluminum does not change color,
   (ii) Aluminum does not show molten pool,
   (iii) Aluminum has high conductivity of heat,
   (iv) Aluminum has very low melting point as compared to the steel.

On the positive side, if the welding technique is mastered, the welding of aluminum is less energy intensive, and therefore easier to weld than welding steel.

Another important point to note is that most of the welding machines are tailored to weld steel, they are mostly calibrated and set with those parameters. So, it is important to know the features of the welding machine, and if required reprogram the machine to suite aluminum welding.

The power source, AC with high-frequency stabilized current is most desired source for welding aluminum, however DC reverse polarity is also used with remarkable success for thicknesses up to 2.5 mm (about 0.1-inch).

Table 3.24.1 Aluminum welding procedures using AC high frequency stabilized arc.

Manual GTAW aluminum welding procedures - basic application										
Thickness of the metal at the point of welding	**Weld type**	**Tungsten electrode diameter**	**Filler wire/ Rod diameter**	**Current for in 1 G (flat) position**		**Gas**		**Travel speed**	**Remarks-1**	**Remarks-2**
mm - inch		**mm - inch**	**mm - inch**	**Current type**	**Amperes**	**Type**	**Flow rate (L/min -- cf/h)**	**cm/min -- in/ min**		
1.6 - 1/16	Butt	1.6 - 1/16	1.6 - 1/16	AC	60 - 80	Argon	7 -- 15	305 -- 12	ceramic cup	
1.6 - 1/16	Lap	1.6 - 1/16	1.6 - 1/16	AC	70 - 90	Argon	7 -- 15	254 -- 10	ceramic cup	
1.6 - 1/16	Corner	1.6 - 1/16	1.6 - 1/16	AC	60 - 80	Argon	7 -- 15	305 -- 12	ceramic cup	
1.6 - 1/16	Fillet	1.6 - 1/16	1.6 - 1/16	AC	70 - 90	Argon	7 -- 15	254 -- 10	ceramic cup	
3.2 - 1/8	Butt	2.4 - 3/32	2.4, 3.2 - 3/32, 1/8	AC	125 - 145	Argon	8 -- 17	305 -- 12	Use 2.4 mm (3/32") filler wire for vertical and overhead welding	
3.2 - 1/8	Lap	2.4 - 3/32	2.4 - 3/32	AC	140 - 160	Argon	8 -- 17	254 -- 10	As above	
3.2 - 1/8	Corner	2.4 - 3/32		AC	125 - 145	Argon	8 -- 17	305 -- 12	As above	

(Continued)

Table 3.24.1 Aluminum welding procedures using AC high frequency stabilized arc. (*Continued*)

Manual GTAW aluminum welding procedures - basic application										
Thickness of the metal at the point of welding	**Weld type**	**Tungsten electrode diameter**	**Filler wire/ Rod diameter**	**Current for in 1 G (flat) position**		**Gas**		**Travel speed**	**Remarks-1**	**Remarks-2**
3.2 - 1/8	Fillet	2.4 - 3/32	2.4, 1.6 - 3/32, 1/16	AC	140 - 160	Argon	8 -- 17	254 -- 10	As above	
5.0 - 3/16	Butt	3.2 - 1/8	3.2 - 1/8	AC	190 - 200	Argon	10 -- 21	279 -- 11		
5.0 - 3/16	Lap	3.2 - 1/8	3.2 - 1/8	AC	210 - 240	Argon	10 -- 21	229 -- 9		
5.0 - 3/16	Corner	3.2 - 1/8	3.2 - 1/8	AC	190 - 220	Argon	10 -- 21	279 -- 11		
5.0 - 3/16	Fillet	3.2 - 1/8	3.2 - 1/8	AC	210 - 240	Argon	10 -- 21	229 -- 9		
6.0 - 1/4	Butt	5.0 - 3/16	2.4 - 3/32 or 5mm - 3/16	AC	260 - 300	Argon	12 -- 25	254 -- 10	Use water cooled cups	Use 3.2 mm or 1/8 inch wire for two passes vertical and over head

(*Continued*)

Table 3.24.1 Aluminum welding procedures using AC high frequency stabilized arc. (*Continued*)

Manual GTAW aluminum welding procedures - basic application										
Thickness of the metal at the point of welding	**Weld type**	**Tungsten electrode diameter**	**Filler wire/ Rod diameter**	**Current for in 1 G (flat) position**		**Gas**		**Travel speed**	**Remarks-1**	**Remarks-2**
6.0 - 1/4	Lap	5.0 - 3/16	As above	AC	290 - 340	Argon	12 -- 25	203 --8	Use water cooled cups	Use 3.2 mm or 1/8 inch wire for two passes vertical and over head
6.0 - 1/4	Corner	5.0 - 3/16	As above	AC	280 - 320	Argon	12 -- 25	254 --10	Use water cooled cups	Use 3.2 mm or 1/8 inch wire for two passes vertical and over head
6.0 - 1/4	Fillet	5.0 - 3/16	As above	AC	280 -320	Argon	12 -- 25	203 -- 8	Use water cooled cups	Use 3.2 mm or 1/8 inch wire for two passes vertical and over head

(*Continued*)

Table 3.24.1 Aluminum welding procedures using AC high frequency stabilized arc. (*Continued*)

Manual GTAW aluminum welding procedures - basic application										
Thickness of the metal at the point of welding	**Weld type**	**Tungsten electrode diameter**	**Filler wire/ Rod diameter**	**Current for in 1 G (flat) position**		**Gas**		**Travel speed**	**Remarks-1**	**Remarks-2**
9.5 - 3/8	Butt	5.0 - 3/16 or 6 mm - 1/4 inch	5.0 - 3/16 or 6 mm - 1/4 inch	AC	330 -- 380	Argon	14 -- 30	Travel speed is variable for multi pass welds.	Two passes are generally required.	Use water cooled cups
9.5 - 3/8	Lap	5.0 - 3/16 or 6 mm - 1/4 inch	5.0 - 3/16 or 6 mm - 1/4 inch	AC	330 -- 380	Argon		Travel speed is variable for multi pass welds.	Two passes are generally required.	Use water cooled cups
9.5 - 3/8	Corner	5.0 - 3/16 or 6 mm - 1/4 inch	5.0 - 3/16 or 6 mm - 1/4 inch	AC	350 -- 400	Argon		Travel speed is variable for multi pass welds.	Two passes are generally required.	Use water cooled cups
9.5 - 3/8	Fillet	5.0 - 3/16 or 6 mm - 1/4 inch	5.0 - 3/16 or 6 mm - 1/4 inch	AC	330 -- 380	Argon		Travel speed is variable for multi pass welds.	Two passes are generally required.	Use water cooled cups

(*Continued*)

Table 3.24.1 Aluminum welding procedures using AC high frequency stabilized arc. *(Continued)*

Manual GTAW aluminum welding procedures - basic application										
Thickness of the metal at the point of welding	Weld type	Tungsten electrode diameter	Filler wire/ Rod diameter	Current for in 1 G (flat) position		Gas		Travel speed	Remarks-1	Remarks-2
13.0 - 1/2	Butt	5.0 - 3/16 or 6 mm - 1/4 inch	5.0 - 3/16 or 6 mm - 1/4 inch	AC	440 -- 450	Argon		Travel speed is variable for multi pass welds.	Two to three passes are generally required.	Use water cooled cups
13.0 - 1/2	Lap	5.0 - 3/16 or 6 mm - 1/4 inch	5.0 - 3/16 or 6 mm - 1/4 inch	AC	400 -- 450	Argon		Travel speed is variable for multi pass welds.	Two to three passes are generally required.	Use water cooled cups
13.0 - 1/2	Corner	5.0 - 3/16 or 6 mm - 1/4 inch	5.0 - 3/16 or 6 mm - 1/4 inch	AC	420 -- 470	Argon		Travel speed is variable for multi pass welds.	Two to three passes are generally required.	Use water cooled cups
13.0 - 1/2	Fillet	5.0 - 3/16 or 6 mm - 1/4 inch	5.0 - 3/16 or 6 mm - 1/4 inch	AC	400 -- 450	Argon		Travel speed is variable for multi pass welds.	Two to three passes are generally required.	Use water cooled cups

Table 3.24.2 GTAW stainless steel welding procedures.

Manual GTAW Stainless Steels welding procedures - basic application										
Thickness of the metal at the point of welding	**Weld type**	**Tungsten electrode diameter**	**Filler wire/ Rod diameter**	**Current for in 1 G (flat) position**		**Gas**		**Travel speed**	**Remarks-1**	**Remarks-2**
mm - inch		mm - inch	mm - inch	Current type	Amperes	Type	Flow rate (L/min -- cf/h)	cm/min -- in/min		
1.6 - 1/16	Butt	1.6 - 1/16	1.6 - 1/16	DC	80 -100	Argon	5 -- 10	305 -- 12	ceramic cup	
1.6 - 1/16	Lap	1.6 - 1/16	1.6 - 1/16	DC	100 - 120	Argon	5 -- 10	254 -- 10	ceramic cup	
1.6 - 1/16	Corner	1.6 - 1/16	1.6 - 1/16	DC	80 - 100	Argon	5 -- 10	305 -- 12	ceramic cup	
1.6 - 1/16	Fillet	1.6 - 1/16	1.6 - 1/16	DC	90 - 100	Argon	5 -- 10	254 -- 10	ceramic cup	
2.5 - 3/32	Butt	1.6 - 1/16	1.6 - 1/16	DC	100 - 120	Argon	5 -- 10	305 -- 12	ceramic cup	
2.5 - 3/32	Lap	1.6 - 1/16	1.6 - 1/16	DC	110 - 130	Argon	5 -- 10	254 -- 10	ceramic cup	

(*Continued*)

Table 3.24.2 GTAW stainless steel welding procedures. (*Continued*)

Manual GTAW Stainless Steels welding procedures - basic application										
Thickness of the metal at the point of welding	**Weld type**	**Tungsten electrode diameter**	**Filler wire/ Rod diameter**	**Current for in 1 G (flat) position**		**Gas**		**Travel speed**	**Remarks-1**	**Remarks-2**
2.5 - 3/32	Corner	1.6 - 1/16	1.6 - 1/16	DC	100 - 120	Argon	5 -- 10	305 -- 12	ceramic cup	
2.5 - 3/32	Fillet	1.6 - 1/16	1.6 - 1/16	DC	110 - 130	Argon	5 -- 10	254 -- 10	ceramic cup	
3.2 - 1/8	Butt	1.6 - 1/16	2.4 - 3/32	DC	120 - 140	Argon	5 -- 10	305 -- 12	ceramic cup	
3.2 - 1/8	Lap	1.6 - 1/16	2.4 - 3/32	DC	130 - 150	Argon	5 -- 10	254 -- 10	ceramic cup	
3.2 - 1/8	Corner	1.6 - 1/16	2.4 - 3/32	DC	120 - 140	Argon	5 -- 10	305 -- 12	ceramic cup	
3.2 - 1/8	Fillet	1.6 - 1/16	2.4 - 3/32	DC	130 - 150	Argon	5 -- 10	254 -- 10	ceramic cup	

(*Continued*)

Table 3.24.2 GTAW stainless steel welding procedures. (*Continued*)

Manual GTAW Stainless Steels welding procedures - basic application										
Thickness of the metal at the point of welding	**Weld type**	**Tungsten electrode diameter**	**Filler wire/ Rod diameter**	**Current for in 1 G (flat) position**		**Gas**		**Travel speed**	**Remarks-1**	**Remarks-2**
5.0 - 3/16	Butt	2.4 - 3/32	3.2 - 1/8	DC	200 - 250	Argon	6 -- 13	254 -- 10	Water cooled cup	
5.0 - 3/16	Lap	2.4, 3.2 - 3/32, 1/8	3.2 - 1/8	DC	225 - 275	Argon	6 -- 13	203 -- 8	Water cooled cup	
5.0 - 3/16	Corner	2.4 - 3/32	3.2 - 1/8	DC	200 - 250	Argon	6 -- 13	254 -- 10	Water cooled cup	
5.0 - 3/16	Fillet	2.4, 3.2 - 3/32, 1/8	3.2 - 1/8	DC	225 - 275	Argon	6 -- 13	203 -- 8	Water cooled cup	
6.0 - 1/4	Butt	3.2 - 1/8	4.0 - 5/32	DC	275 - 350	Argon	6 -- 13	127 -- 5	Water cooled cup	More than one passes would be required.

(*Continued*)

Table 3.24.2 GTAW stainless steel welding procedures. *(Continued)*

Manual GTAW Stainless Steels welding procedures - basic application										
Thickness of the metal at the point of welding	**Weld type**	**Tungsten electrode diameter**	**Filler wire/ Rod diameter**	**Current for in 1 G (flat) position**		**Gas**		**Travel speed**	**Remarks-1**	**Remarks-2**
6.0 - 1/4	Lap	3.2 - 1/8	4.0 - 5/32	DC	300 - 375	Argon	6 -- 13	127 -- 5	Water cooled cup	More than one passes would be required.
6.0 - 1/4	Corner	3.2 - 1/8	4.0 - 5/32	DC	275 - 350	Argon	6 -- 13	127 -- 5	Water cooled cup	More than one passes would be required.
6.0 - 1/4	Fillet	3.2 - 1/8	4.0 - 5/32	DC	300 - 375	Argon	6 -- 13	127 -- 5	Water cooled cup	More than one passes would be required.
13.0 - 1/2	Butt	3.2, 4.0 - 1/8, 5/32	6.0 - 1/4	DC	350 - 450	Argon	7 -- 15	Variable, record while qualifying the WPS	Water cooled cup	Two to three passes may be required.

(Continued)

Table 3.24.2 GTAW stainless steel welding procedures. (*Continued*)

Manual GTAW Stainless Steels welding procedures - basic application										
Thickness of the metal at the point of welding	**Weld type**	**Tungsten electrode diameter**	**Filler wire/ Rod diameter**	**Current for in 1 G (flat) position**		**Gas**		**Travel speed**	**Remarks-1**	**Remarks-2**
13.0 - 1/2	Lap	3.2, 4.0 - 1/8, 5/32	6.0 - 1/4	DC	375 - 475	Argon	7 -- 15	Variable, record while qualifying the WPS	Water cooled cup	Three passes
13.0 - 1/2	Corner	3.2, 4.0 - 1/8, 5/32	6.0 - 1/4	DC	375 - 475	Argon	7 -- 15	Variable, record while qualifying the WPS	Water cooled cup	Three passes
13.0 - 1/2	Fillet	3.2, 4.0 - 1/8, 5/32	6.0 - 1/4	DC	375 - 475	Argon	7 -- 15	Variable, record while qualifying the WPS	Water cooled cup	Three passes

The choice of filler wire/rod is dependent on the material grade being welded. Most of the wire and their properties are discussed in the subsequent paragraphs. The selected wire can be combined with the following generic welding procedure to make a specific welding procedure.

The following table details the basic welding procedures using AC high frequency stabilized current.

In GTAW process, the welding process uses shielding gas to keep the weld environment free from atmospheric oxidants to enter in the weld pool or contaminate the weld metal. If the welding area and the surrounding surface in the proximity of the weld zone are cleaned the probability oxidation formation is eliminated.

Alloy 4043 is one of the oldest and most widely used welding and brazing alloys. ER 4043 filler wires can be classed as the general-purpose welding wire suitable of general type of most aluminum alloy grades, like, 3003, 3004, 5052, 6061-T4, 6061-T6, 6063-T6 and 2014-T6.

This specific grade of alloy has about 4% to 6 % silicon in it, silicon is a wetting agent, and it allows the weld metal fluidity and this property makes the electrode use very welder-friendly. The silicon additions result in improved fluidity (wetting action) to make the alloy a preferred choice by welders. The weld metal is less sensitive to weld cracking, and produces brighter, almost smut free welds.

3.25 GTAW Welding of Stainless Steel

Welding stainless using GTAW process somewhat similar to welding basic carbon steel, however there is significant metallurgical issues that are different from, the carbon steel and those need to be understood to make a sound stainless steel weld.

The term stainless steel is very generic, and somewhat misleading. It is essential to understand some fundamental metallurgical aspects of "Stainless-steel".

There are number of groups, and within those groups may be different grades of steels that may be generally classified stainless steels. The following basic discussions are included for those who wants to be more educated about the material they intend to weld.

A basic welding procedure is placed in the Table 3.25.1 below, this procedure should be matched with the filler wire Table 3.25.8 to select right filler material for the specific grade to weld.

3.25.1 Introduction to Stainless-Steels

Stainless steels are iron base alloys that contain a minimum of approximately 11% chromium (Cr), this is an important number that is needed to create a passivating layer of chromium-rich oxide to prevent rusting on the surface. Several stainless-steel grades are produced to address specific demands of the environment that they are expected to protect the material, for this purpose other elements are also added to the steel. Nickel, Copper, Titanium, Aluminum, Silicon, Molybdenum, Niobium, Nitrogen Sulfur and Selenium are some of the commonly used elements that are alloyed to impart the required properties to the specific steel grade.

Stainless steels are marketed in various shapes and sizes and in various finishes. However, for industrial application and ease of understanding we limit this to the following.

Table 3.25.1 Nominal compositions of some of duplex steels.

Type	UNS no.	Nominal composition (wt%)							F_{PREN}
		Fe	Cr	Ni	Mo	N	Cu	W	
Lean	S32101	Bal	21	1.5	0.5	0.16	0.5	-	25
	S32304	Bal	23	4	0.3	0.16	0.3	-	26
	S32003	Bal	20	3	1.7	0.16	-		>30
Standard	S31803	Bal	22	5	3	0.16	-	-	35
	S32205	Bal	22	5	3.2	0.16	-	-	35
25 Cr	S32550	Bal	25	6	3	0.2	2	-	37
Superduplex	S32750	Bal	25	7	3.5	0.27	0.2	-	>40
	S32760	Bal	25	7	3.5	0.25	0.7	0.7	>40
	S32520	Bal	25	7	3.5	0.25	1.5	-	>40
	S39274	Bal	25	7	3	0.26	0.5	2	>40

Bars are available in all grades and come in rounds, squares, octagons, or hexagons of 0.25 inch (6 millimeter) in size.

Wire is usually available up to 0.5 inch (13 millimeter) in diameter or size.

Plate is defined as rectangular shapes of more than 0.1875 inch (5 millimeter) thick and over 10 inches (250 millimeter) wide.

Strip are defined as rectangular shapes of less than 0.185 inch (5 millimeter) thick and less than 24 inches (610 millimeter) wide.

Sheet are defined as rectangular shapes of less than 0.1875 (5 millimeter) thick and more than 24 (610 millimeter) wide.

Further processing is done to produce specific shapes like pipes tubes and structural shapes.

3.25.1.1 Cutting Stainless Steel for Fabrication

Cutting operation is usually necessary to obtain the desired blank shape or size. This is done to trim the part to final size. Mechanical cutting is accomplished by a variety of methods, including straight shearing by guillotine knives, circle shearing by circular knives horizontally and vertically positioned. Blanking by metal punches and dies to punch out the shape by shearing. Nibbling is a process used for cutting by blanking out a series of overlapping holes and is ideally suited for irregular shapes, only some stainless steels can be saw cut by high-speed steel blades.

Normally stainless steel cannot be cut using flame cutting. Another cutting method that is used is the Plasma jet cutting, to make a cut this process uses an ionized gas column in conjunction with an electric arc passing through a small orifice the force of the gas and high heat generated by the gas plasma, melts the metal and makes the cut.

3.25.1.2 Finishing

Surface finish is an important requirement for stainless steel products, depending on the end application. The surface finish is a very important property and it is specified on the fabrication drawing. The main reasons to consider for specifying the surface finish could include one or all of the following.

1. The appearance.
2. Process convenience.
3. Corrosion protection.
4. To facilitate lubrication – often rougher surface is specified.
5. Surface condition specific to facilitate further manufacturing steps.

Stainless steels are produced in variety of different metallurgical class, and in different grades. Each of these classes, and grades have very specific properties to offer, and may behave differently during fabrication and welding. Broadly speaking Stainless steels are of following types,

- Austenitic Stainless steels
- Martensitic Stainless steels
- Ferritic Stainless steels

There are number of grades within these groups.

There are also stainless steels, that are classified as super austenitic and an entirely different class that is termed as Duplex steel that has number of different grades within.

Knowing each class and grade is an important part of being a good "Stainless Steel" welder.

3.25.2 Fabrication of Stainless Steel

After the stainless steel in its various forms are packed and shipped to the fabricator or end user, a variety of secondary processes are needed to make it useful for specific service. Further shaping is accomplished using a variety of secondary processing that may include rolling forming, press forming, forging, press drawing, and extrusion, welding cutting, additional heat treating, machining, and cleaning processes.

3.25.3 Why Stainless Steel

Stainless steel is chosen for a project based on any single or a combination of following specific properties.

1. Resistance to corrosion
2. Resistance to oxidation at higher temperatures.
3. Good mechanical properties at room temperature.
4. Good mechanical properties at low temperature.
5. Good mechanical properties at high temperature.
6. Aesthetic values - Good Appearance.

Stainless steels are corrosion resistant material which relay on surface passivity for resistance to corrosion attack. Use of these materials is governed by the oxidizing characteristics of the environment. For more oxidizing conditions, stainless steel is superior in performance to several more-noble metals and alloys, available for fabrication by welding.

3.25.4 General Welding Characteristics

All the Chromium-nickel (300 Series) austenitic stainless steels with the exception of high Sulphur or selenium added free-machining grade (AISI 303) are easily welded. The welded joints are tough and ductile in "as welded" condition. These welds if used in non-corrosive or mildly corrosive services do not require any post weld heat treatment.

In welding a temperature gradient is achieved ranging from room temperature to molten steel. The area that is heated in this process rage from 425°C to 900°C (800°F to 1 650°F) becomes sensitized as carbides are precipitated. This carbide precipitation may affect the life of equipment under severe corrosive conditions, therefore annealing the welded parts is recommended to restore optimum corrosion resistance. This annealing process is called solution annealing. The process of solution annealing consists of heating the material up to a temperature above sensitizing temperature, generally 1 100°C (or about 1 960°F) and holding it long enough for the carbon to go into solution. After this, the material is quickly cooled to prevent the carbon falling out of the solution. Solution annealed material is in its most corrosion-resistant and ductile condition.

It is not always possible to solution anneal the weldments. This could be for various reasons like its size or other post fabrication process etc. If for any reason the welded part cannot be annealed then extreme care should be taken in welding stainless steel and either a low carbon grade of stainless steels that have less than 0.03% carbon or AISI 321 or 347 grade of steel should be selected.

The grades 321 and 347 are stabilized alloys. They contain Titanium and Columbium respectively. The ratio of these elements is dictated by the percentage of carbon in these steels for example the minimum amount of Titanium in Grade 321 is about 5 times that of carbon in the steel. Similarly, Columbium in Grade 347 is about 10 times that of carbon content of the steel. But accurate ratio of carbon to Titanium or Columbium has to be designed in the steel, based on the specific requirement of the service environment including welding requirements of the project. When these steels are heated during welding and material reaches in the sensitizing range carbide precipitation occurs, as in any other grades of stainless steel, except that due to high affinity of carbon to these elements, the carbide of titanium and Columbium is precipitated, thus leaving Chromium free from Intergranular corrosion. In some very specialized conditions Grade 321 may be further heat treated by heating to 815°C – 900°C range for 2 to 4 hours and air cooled to secure complete carbide precipitation as stable titanium carbides. This heat treatment is some time called stress relief treatment.

In low carbon (less than 0.03%) grades like 304L and 316L the carbon is so low that during welding the heat does not precipitate carbides. The use of these grades of steel is limited to service temperature below 425°C to 870°C (800°F to 1 600°F).

Welds in other corrosion resistant steels like, ferritic and martensitic stainless steels are not as ductile and tough as in austenitic steels discussed above. Ferritic alloy type 405, 430, 442, and 446 are more readily weldable. The martensitic grades like 403 and 410 are more weldable than types 420 and 440.

3.25.5 Protection Against Oxidation

A welding process must protect the molten weld metal from the atmosphere during arc transfer and solidification. Fluxing may be required to remove the chromium and other oxides from the surface and the molten weld metal. Gas shielded processes do not require fluxing since the shielding gas prevents oxidation.

3.25.6 Welding and Joining

The practical welding techniques make it easy to weld in the flat and horizontal position. However, welding in vertical uphill position is tougher to master. This is because stainless steels have lower conductivity, as a result the electrode heats up quickly and then maintaining the arc is difficult. Even if the weld is completed the appearance of the weld bead is not uniform, it looks peaked to the crown. Experienced welders recognize the situation beforehand, and control the arc by lowering the amperage. This specific situation is caused due the lower thermal conductivity of the stainless-steel property.

During the welding, the weld deposition rate changes very soon from the rate at the start, this is because of the hot electrode starts to melt faster than when it was colder, at the start of the weld.

Some welders use weave technique to spread out the weld metal, but not all engineering specifications accept this method due to the metallurgical changes that happens due to the longer arc-dwell time.

Good welding technique is to start at the minimum possible level of amperage, and keep control of the arc during the welding by adjusting the amperage.

A new welder training to weld stainless steel, should practice on the same machine that is actually to be used for the test welding. This will allow familiarization with the accuracy of the output amperage, and also the adjustments response of the machine. In other words, the welder should familiarize with the machine before actual welding.

3.25.7 Importance of Cleaning Before and After Welding

The high chromium content of stainless steels promotes the formation of tenacious oxides that must be removed for good welding results. Surface contaminants affect stainless steel welds to a greater extent than they affect carbon steel welds.

The surfaces to be joined must be cleaned prior to welding. An area surrounding the weld joint for at least 12 mm (0.5 inch) is cleaned, far wider area if thicker plates are being welded. As a general rule of thumb cleaning a band of metal about 1.5 times the plate thickness will be considered good practice, it would avoid contaminations.

Special care in surface cleaning is required for gas shielded welding because of the absence of fluxing.

Carbon contamination can adversely affect the metallurgical characteristics and corrosion resistance of stainless steel. Pickup of carbon contaminants or embedded particles must be prevented.

Suitable solvents are used to remove hydrocarbon and other contaminants such as cutting fluids, grease, oil, waxes, and primers. Light oxide films can be removed by pickling or by carefully selected mechanical means of cleaning. Acceptable pre-weld cleaning techniques include:

1. Stainless steel wire brushes that are used only for stainless.
2. Blasting with clean sand or grit.
3. Machining and grinding with chloride-free cutting fluid.
4. Pickling with 10% to 20% nitric acid solution.

Table 3.25.8 Stainless steel welding wire rod and heat treatments.

AISI type	Recommended heat treatment		Common recommended electrode for welding
	Pre-weld	**Post weld**	
301, 302	Not required if steel temp is above 15°C	Rapid cooling from temperature between 1065°C to 1150°C (1950°F to 2100°F), if service condition is moderate to severe corrosive.	308
304	As above	Rapid cooling from temperature between 1010°C to 1095°C (1850°F to 2000°F), if service condition is severe corrosive.	308
304L	As above	Not required for corrosion resistance.	308L or 347
309, 310	As above	Not required for corrosion resistance, because steel is usually at higher temperature in service.	309, 310
316	As above	Rapid cooling from temperature between 1065°C to 1150°C (1950°F to 2100°F), if service condition is severe corrosive.	316
316L	As above	Not required for corrosion resistance.	316L
317	As above	Rapid cooling from temperature between 1065°C to 1150°C (1950°F to 2100°F), if service condition is severe corrosive.	317
317L	As above	Not required for corrosion resistance.	317L
321, 347	As above	Not required for corrosion resistance.	347
Ferritic and Martensitic Steels			
403, 405	150 to 300°F Light gauge sheet need no pre heat	Air Cool from 1200°F/1400 °F (650°C/760 °C)	410

(Continued)

Table 3.25.8 Stainless steel welding wire rod and heat treatments. (*Continued*)

AISI type	Recommended heat treatment		Common recommended electrode for welding
	Pre-weld	Post weld	
410	As above	Air Cool from 1200°F/1400 °F (650°C/760 °C)	410
430	As above	Air Cool from 1400°F/1450 °F (760°C/785 °C)	430 Can be welded with 308, 309 or 310 without pre heat.
442	As above	Air Cool from 1450°F/1550 °F (785°C/840 °C)	446
446	300°F to 500°F	Rapid cooling from temperature between 840°C to 900°C (1550°F to 1650°F)	446
501	300°F to 500°F	Air Cool from 1325°F/1375°F (715°C/745°C)	502
502	300°F to 500°F Light gauge sheet need no pre heat	Air Cool from 1325°F/1375°F (715°C/745°C)	502 Can be welded with 308, 309 or 310 without pre heat.

Thorough post weld cleaning is required to remove welding slag. The surface discoloration is best removed by wire brushing or mechanical polishing.

3.25.8 Filler Metals

Covered electrodes and bare solid and cored wire are available to weld most of the grades. The chemical composition of allweld metal deposits vary slightly from the corresponding stainless-steel metal composition to ensure that the weld metal will have the desired microstructure and be free from cracks.

3.25.9 Austenitic Stainless Steels

3.25.9.1 Metallurgical Concerns Associated with Welding Austenitic Stainless Steels

The properties of austenitic stainless steel as value to industry are, high ductility, excellent toughness, strength, corrosion resistance, weldability, and excellent formability and castability. Because of these properties, austenitic stainless steels are the most commonly used material from the family of stainless steels. There is a virtual continuum of austenitic alloys containing Fe, Cr, Ni, and Mo.

The distinction between highly alloyed stainless steels and lower-alloyed nickel-base alloys is somewhat arbitrary. Nickel alloys must satisfy either (a) Cr >19; Ni>29.5; Mo >2.5, or (b) C r>14.5; Ni >52; Mo >12 over their entire composition range, those that do not meet this criteria e.g., alloy 20, UNS N08020 are classified as stainless steels.

Unified Numbering System (UNS) alloy numbers starting with a prefix "S" are grouped with the austenitic stainless steels discussed in this section whilst super austenitic stainless steels, defined in this report as alloys with *F*PREN greater than 30.0 are discussed in the section entitled "Super austenitic Stainless Steels. The alloys that begin with prefix "N" are grouped with the nickel-based alloy.

3.25.9.2 Mechanical Properties of Stainless Steels

The lower-alloyed austenitic stainless steels, such as type 304 and 316 (UNS S30400 and S31600), possess yield strengths around 30 to 40 ksi (210 to 280 MPa) in the annealed condition. Some higher-alloyed austenitic stainless steels with nitrogen have higher yield strengths.

Cold working often increases strength, especially in higher-alloyed austenitic stainless steels.

Cold deformation during fabrication, although less severe than that applied during temper rolling, can produce martensite in some austenitic stainless steels, thereby increasing their susceptibility to hydrogen embrittlement. Fabrication processes can also induce residual stresses that may help increased prospects of stress corrosion cracking (SCC).

Many of the common austenitic stainless steels can be readily welded using matching filler metals. Higher alloy grades are normally weldable, but non-matching, over-alloyed nickel-base filler metals are used.

Generally, these alloys are readily weldable whether for longitudinal seam welded pipe or girth welds, etc., via a range of processes (SAW, GTAW, GMAW, SMAW, and PAW etc.). They are usually welded with matching composition filler metal. For some of the molybdenum-containing grades, over-alloyed filler with an extra 1-3% molybdenum and higher nickel content are specified. Normally, argon is used for both shielding and backing gases. Austenitic stainless steels typically require care in welding and adherence to good stainless-steel welding practice.

Since these alloys are in austenitic phase and they do not have phase transformation on cooling they do not require Preheat or Post weld heat treatment (PWHT). Except in some specific cases where solution annealing may be specified after welding and hot working.

Welding technology for the typical austenitic stainless steels is common practice using standard consumables like, ER308L, etc. Welding of higher-strength (650-690 MPa UTS), 200-series austenitic stainless steels can be done with standard 308 L-type fillers if matching the strength of the base metal is not critical. Use of duplex ER2209 filler metal is one way of matching or exceeding the strength of the base metal, but toughness and embrittlement concerns restrict use of this approach to service temperatures of about - 40 to 315 °C (-40 to 600 °F). For cryogenic applications, such as liquefied natural gas (LNG) equipment, use of less-standard fillers such as E16-8-2 or 316 L Mn, or use of nickel-base fillers such as UNS N06022, are often of selected to take advantage of necessary strength and cryogenic toughness.

3.25.9.3 Welding of Austenitic Stainless Steels

At the beginning of this chapter, we discussed general requirements of welding and addressed fundamental essentials like sensitization control, difference in welding stainless steel and carbon steel and importance of weld hygiene. We take those discussions further in the subsequent paragraphs.

The austenitic steels have high coefficients of thermal expansion and low thermal conductivity and are particularly susceptible to distortion during welding. They have better ductility and toughness than carbon steels and excellent notch toughness even at cryogenic temperatures. They are stronger than carbon steels above 500°C (1 000°F) and have good oxidation resistance.

When austenitic stainless steels are joined to carbon steel, construction codes often mandate PWHT in the temperature range of about 550-675°C (1 025-1 250°F) for relief of residual stresses. These heat treatments can adversely affect intergranular corrosion and stress corrosion cracking (SCC) resistance of the stainless steel. In these situations, use of a low-carbon grade type 304-L or stabilized grade like type 347-L is recommended. It may however be noted that the service temperature range of 304-L and 347-L is limited to –40°C to 315°C. If PWHT is one of the limiting factors imposed on the design, then other alternatives must be thought, one of them is to butter the carbon steel as described below.

A buttering layer of austenitic stainless-steel electrode/filler wire is deposited to the carbon steel. Often the selection is based on the available chromium in the as deposited weld metal, after compensating for the dilution, if the resulting weld metal is close to the austenitic level then the buttering is completed, the most common interface electrode for welding and buttering austenitic steel and carbon steel is E 309 grade of consumable. Once the buttering is completed, the buttered carbon steel is heat treated as required. After the post weld heat treatment (PWHT) is carried out to relive stresses in carbon steel, then the stainless-steel member is welded on to the PWHT buttered section of carbon steel.

Heat input ranges and inter-pass temperatures are not especially important for the austenitic stainless steels. Interpass temperatures up to 150 °C (300 °F) is usually permissible.

After welding, there is usually a heat tint in the weld/HAZ area, and it is usual to remove this in some applications where feasible. The heat tint is often removed by manual (but not mechanical) brushing, by mechanical abrasives, such as a flapper wheel, or by a suitable pickling paste or gel. The inside of small-bore pipe welds, flowlines, and clad line pipe are difficult to clean, this requires that welding procedure uses inter gas as backing to keep the inside surface oxide free.

Weld deposit microstructures are very different from wrought metals with the same composition. 100% austenitic structure in welds is prone to cracking. Some amount of

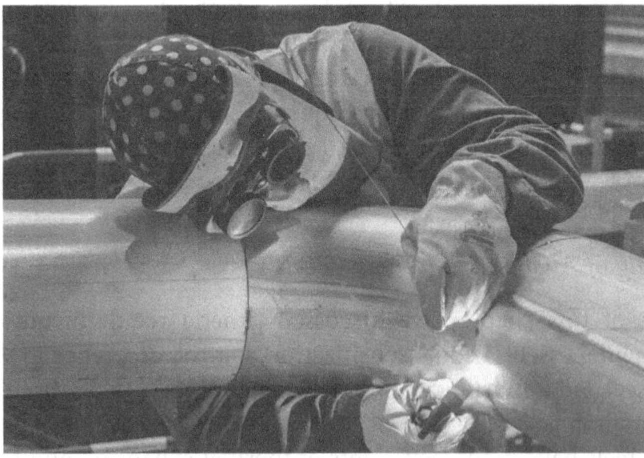

Figure 3.25.9.3.1 Welder is tacking a pipe prior to welding.

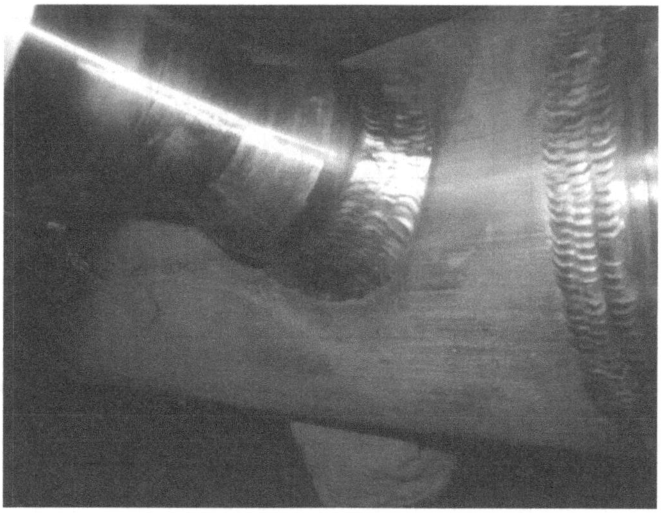

Figure 3.25.9.3.2 A nozzle is welded on a pipe header.

ferrite is essential to control cracking of these welds. In austenitic welds small pools of delta ferrite often form and carbides may also be present. The weld metal ferrite control is essential; Schaeffler and DeLong diagrams see Figure 3.25.10.1 and Figure 3.25.10.2, are used to predict as-welded microstructures. These diagrams are also useful in selecting the electrode for keeping control on the ferrite in austenitic steel weld metal.

3.25.10 Welding Super-Austenitic Stainless Steels

3.25.10.1 Material Properties and Applications

Like the austenitic stainless steels, the super-austenitic stainless steels are highly ductile; they have excellent toughness, high strength, outstanding corrosion resistance, good weldability, and excellent formability. The super-austenitic stainless steels are normally used where greater resistance to corrosion, especially protection from chloride pitting and crevice corrosion, is needed. Super-austenitic stainless steels are defined as austenitic, iron-based alloys that have PREN greater than 40.

The higher PREN values are achieved primarily by adding nitrogen (N) to these alloys, and upper working temperature limits of 400°C (750°F) are generally imposed by industry codes to prevent Σ (sigma) or X (chi) phase embrittlement.

Many of the super-austenitic stainless steels, especially those containing nitrogen, possess higher yield strengths in the annealed condition than the standard austenitic stainless steels.

These alloys are generally available in most product forms (bar, wrought plate, castings, pipe, forgings, etc.), and are usually supplied in the solution-annealed condition. Specialized parts (fittings, fasteners, etc.) of these grades are not generally inventoried by stockicst, but often custom manufactured. As an alternative suitable nickel alloys are often selected. Castings are solution annealed to homogenize the as-cast, cored, dendritic structure.

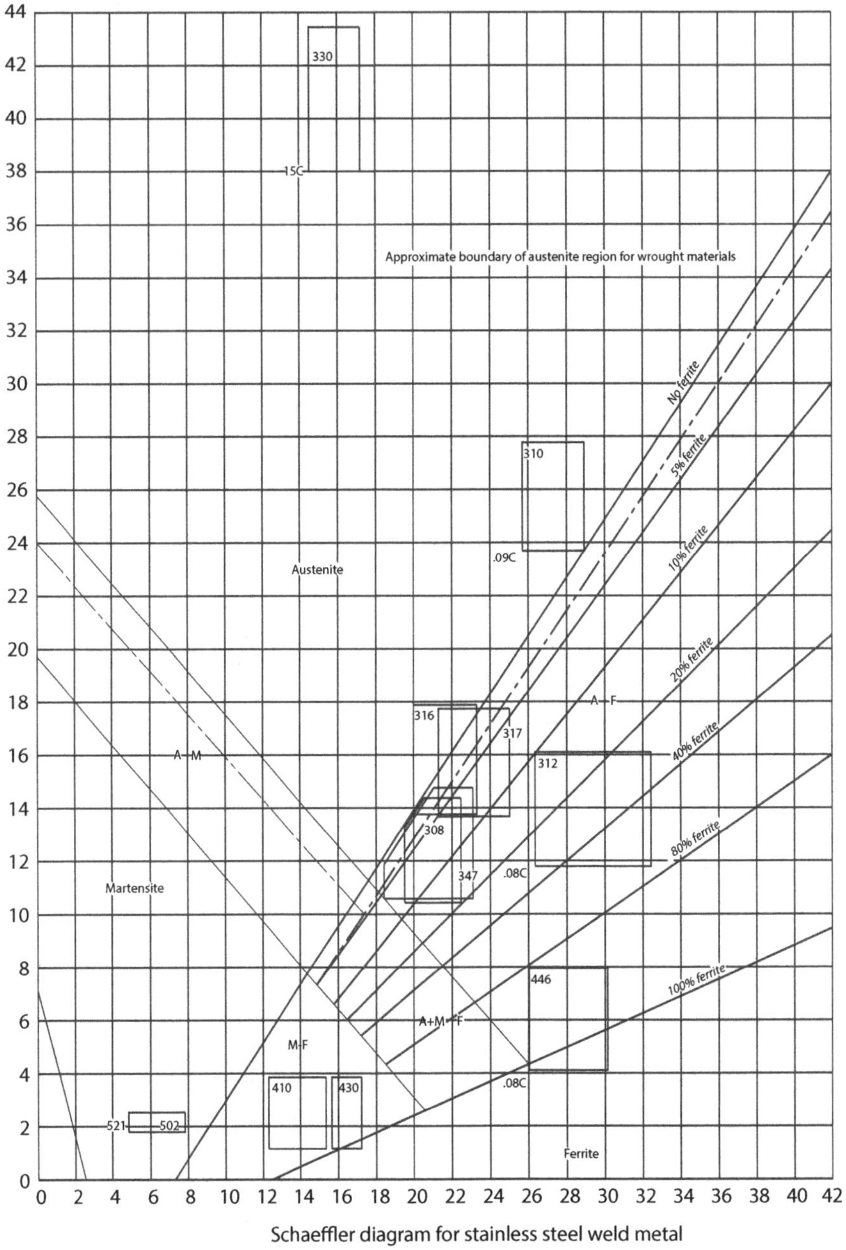

Figure 3.25.10.2.1 Schaeffler diagram.

The super-austenitic stainless steels are generally used in the solution-annealed and rapid-cooled condition. Prolonged heating in the temperature range of about 510 to 1 070 °C (950 to 1 960 °F) can cause precipitation of carbides, nitrides, or intermetallic phases. This precipitation increases susceptibility to intergranular corrosion, IGSCC, and chloride pitting and crevice corrosion. These alloys cannot be strengthened by heat treatment.

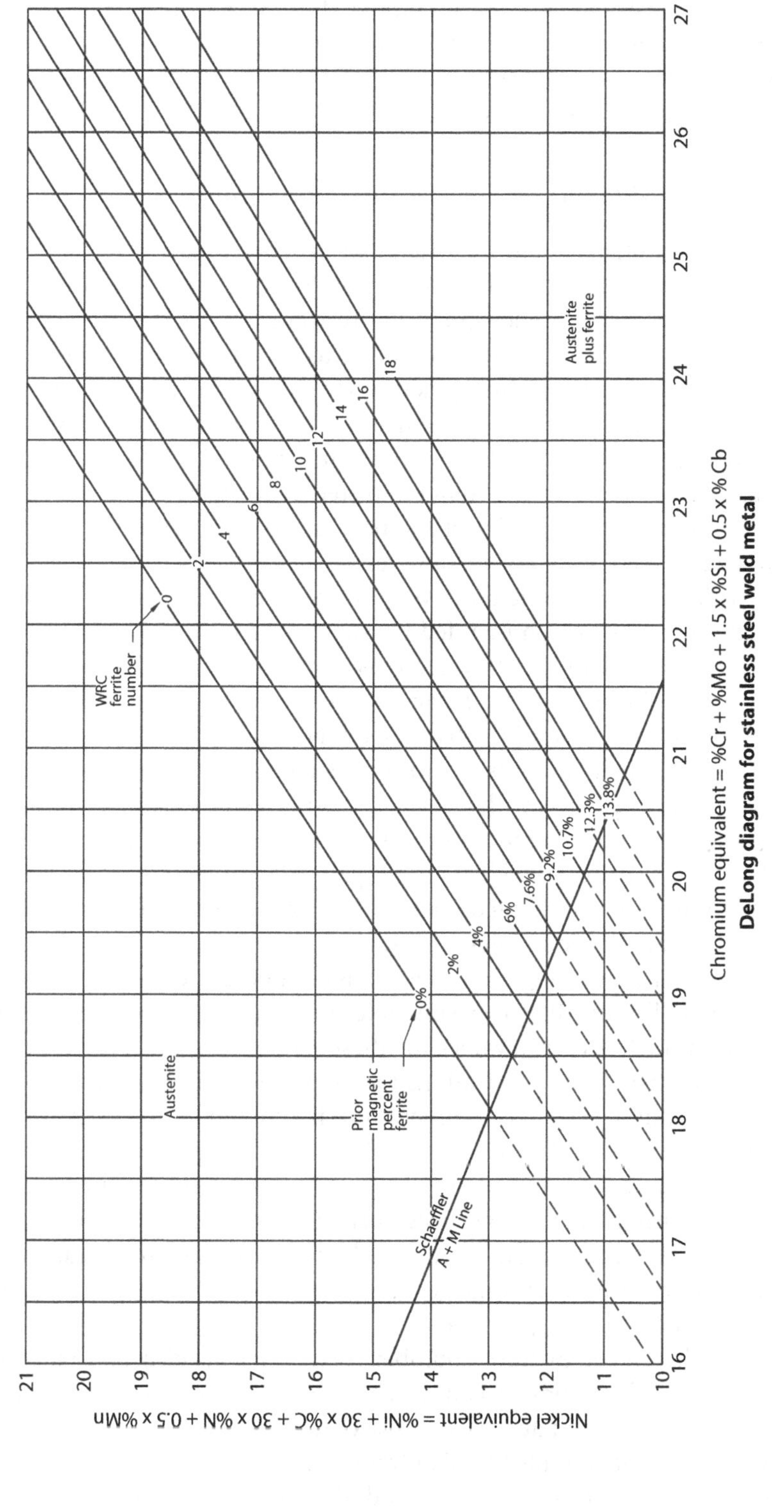

Figure 3.25.10.2 DeLong diagram.

The higher alloy contents of the super-austenitic stainless steels give them greater resistance to the formation of martensite during cold working. Thus, they show reduced susceptibility to hydrogen embrittlement (HE) as compared to the austenitic stainless steels. The higher alloy contents of the super-austenitic stainless steels also give them greater resistance to stress corrosion cracking (SCC) when compared to the austenitic stainless steels. Thus, fabrication induced residual stresses are less likely to cause SCC in these alloys.

3.25.10.2 Welding and Joining of Supper-Austenitic Stainless Steels

These alloys are easily weldable, and by a range of processes like, SAW, GTAW, GMAW, SMAW, etc. All these processes have been discussed in this book. Because in part these alloys rely on molybdenum to provide corrosion resistance properties, the segregation of molybdenum that occurs during solidification of welds can impair the corrosion resistance of welds. To counter this effect the normal practice is to use over-matching composition filler metal. The over-alloyed fillers typically contain about 1.5 times the molybdenum to that of the base metal. To keep these high levels of molybdenum in solid solution, nickel-based fillers are used. This ensures that even the solute-depleted dendrite cores will have local PREN values meeting or exceeding the PREN of the base metal. Examples of such filler metals include UNS N06625, N06022, and N06686 wires.

From the above description of weldability of these molybdenum alloyed steels the autogenous welding is not normally recommended for super-austenitic stainless steels, although it has been performed successfully in thin sections of < 2 mm and with special gases. Some specialized post weld solution annealing can also restore autogenous welds to corrosion resistance levels approaching that of the base metal. Normally, argon is used for shielding gas, but the addition of small amounts of nitrogen is considered more beneficial. Backing gases can be argon or nitrogen. As with high alloy stainless steels, care is typically taken in welding super-austenitic, and adherence to good stainless steel welding practice is generally considered to be good practice. Suitable joint design, inter-pass temperatures and low heat inputs is the path to successful welding of these alloys. Pre and post weld heat treatment are not required for super-austenitic stainless steels.

During welding, heat input ranges and inter-pass temperatures is very carefully monitored and controlled. The maximum permissible heat input and inter-pass temperature increase with section thickness. The values for these parameters generally decrease as the alloy content increases. Specialist publications for suitable values for a specific joint should be consulted. If heat input or inter-pass temperatures is kept too high, the risk of precipitating sigma or chi phases in the HAZ or weld metal increases. These intermetallic phases are rich in chromium and molybdenum, thus leaving chromium depleted area around them, which is responsible for reducing the localized corrosion resistance in these metals.

The austenite stainless steels contain a combined total chromium, nickel, and manganese content of 24% or more, with the chrome generally above 16%. Nickel and manganese stabilize austenite to below room temperature.

Ferrite content is designated by ferrite number (FN). Ferrite is difficult to measure accurately although automated equipment is now available. The importance of ferrite in weld microstructure cannot be understated, as it increases resistance to hot cracking. Ferrite provides sites with good ductility for interstitial or tramp elements to distribute. However, excessive ferrite can also lower corrosion resistance and high temperature properties of material.

Welding parameters and technique have a significant effect on the amount of ferrite formed and retained in a weld and they must be controlled to reduce desired properties in the weld.

3.25.10.3 Difficulties Associated with Welding Stainless Steel

Austenitic stainless-steel welding appears to be similar to the normal carbon steel welding. But little more in-depth observation will reveal that it is anything but similar to the carbon steel welding. This is basically due to the metallurgical difference between the two types of steel, Stainless steel do not undergo the normal phase changes associated with carbon steels. While ferritic steels are austenitic and nonmagnetic at elevated temperatures, they transform to ferrite, pearlite, martensite and other phases as they are cooled through the transformation range.

In contrast when stainless steel is cooled, all or nearly all of the material retains the austenite structure at room temperature. No phase changes occur, and no hardness increase is associated with cooling. This property of austenitic steels reduces the need to pre heat or post heat.

When these steels are welded two very specific points need to be considered.

1. Carbide precipitation or sensitization.
2. Micro fissuring, ferrite content and Sigma phase formation.

The corrosion resistance of austenitic steels depends on the addition of various alloys of which Chromium is of primary importance.

In our earlier discussion on stainless steel welding, we introduced terms like sensitization, sigma phase etc., we discuss them in little more details. When austenitic steel is heated to a temperature range called "sensitization range" that is 425°C to 870°C (800°F to 1 600°F), some of the chromium in the solution can combine with any carbon that is available and form a chromium rich precipitate called chromium carbide, thus reducing the chromium in the steel, now less chromium is available in the alloy to carry its primary duty that is to resist corrosion. This reduction of chromium by forming Chromium carbide is called sensitization.

The steel in such venerable conditions is easily attacked by acidic environment. Under certain conditions, austenitic welds are subject to intergranular corrosion. A narrow band of metal in the heat-affected zone is always heated to the sensitizing range. The amount of precipitation that occurs is a function of the carbon content. Carbon levels higher than 0.03% are particularly susceptible.

Many metallurgical solutions are available to overcome the formation of sensitized area in austenitic welds; we discuss them in subsequent paragraphs.

Since we have discussed the importance of austenite in steel for corrosion resistance, we may make it clear that while pure austenite has excellent mechanical and corrosion resistance properties, its ability to absorb impurities without cracking during solidification is severely limited. During the cooling process the low meting impurities are forced out to the grain boundaries. Excessive amounts of such grain boundary accumulations weaken the material at the grain boundary. They create grain boundary flaws called "microfissures". This condition is of significance in welding because in relation to the parent metal a very small area is associated with welding. One method to reduce such micro fissuring is to disperse these impurities among the disconnected gain boundaries that surround the island of second phase. This can be accomplished by modifying the chemical composition of steel that would allow creation of islands of ferrite in the welds. Ferrite has enormous capacity to absorb impurities, and ferrite islands are dispersed throughout the microstructure. The presence of ferrite has potential to very slightly reduce the corrosion resistance of steel, however it can certainly prevent micro fissuring, which is more serious and can lead to catastrophic failures. On the other hand, too-much ferrite is also detrimental to the material. It can cause other problems called "Sigma phase" developed within the welding temperature range. This is a very brittle constituent and is caused by very evenly dispersed ferrite, even a small amount of sigma phase will embrittle large areas of stainless steel.

It is clear that both minimum and maximum limitations on ferrite phase are desirable in stainless steel welding to prevent micro-fissuring and sigma phase embrittlement.

Since carbon rapidly decreases the corrosion resistance and changes the properties of austenitic welds, it must be carefully controlled. Filler metals are usually chosen to match the base metal composition.

It is established that the Ferrite content must be appropriate for the weldment's service requirements. Many electrodes are developed to produce deposits containing ferrite limits within the range of 4 to 10 "ferritic number". There are methods that can measure ferrite in the weld, one of them is magnetic ferrite gage as described in AWS A 4.2.

3.25.11 Welding Martensitic Stainless Steels - Properties and Application

Martensitic stainless steels are Fe-Cr-C alloys that are capable of the austenite-martensite transformation under all cooling conditions. Compositions for most of martensitic steel alloys are covered by number of specifications, such as ASTM A 420 or API 13 Cr L80 and 420 M with additional small amounts of Ni and/or Mo. Although 9Cr1Mo is not strictly a martensitic stainless steel; it is often included in this alloy group, especially because of challenges associated with welding of 9-Cr-Mo steel is in many ways similar to this group of materials.

The martensitic stainless steels are generally used in the quenched and tempered, or normalized and tempered condition. For services where hydrogen evolution or presence of Sulphur is expected as in sour gas services in oil and gas industry a maximum hardness of 22 HRC is specified by most of the specifications, and for most of the alloys. Some of the alloys like type 410 and 420 develop quench crack if quenched in water, so they are quenched only in oil, or polymer, or air-cooled before tempering.

Some alloys like type 410, type 415, and J91540 (CA6NM) receive a second temper treatment called "Double tempering" at a temperature lower than the first tempering temperature, to reduce the untampered martensite in type 410, type 415, and J91540 (CA6NM). Double tempering has not been shown to improve resistance to stress corrosion cracking in type 420 tubular products and for 9Cr1Mo tubular or forgings.

The mechanical properties of typical base metal strength (SMYS) are grouped as 414 MPa (60 ksi), 517 MPa (75 ksi), 552 MPa (80 ksi), and 586 MPa (85 ksi), with hardness controlled to the maximum 22 or 23 HRC, and often have specified to maximum yield strength of 95 to 100 ksi (660 to 690 MPa). For sour service applications the tubular products are generally used according to the API Specification 5CT, or L80, strength level; forgings and castings are generally specified with hardness not exceeding 22 on Rockwell C scale. Higher strengths are used in sweet service; however, corrosion resistance and ductility are adversely affected as the strength of steel is increased.

3.25.12 Welding Martensitic Stainless Steels

Weld design strength levels range from 414 MPa (60 ksi) upward, but they can be different than the parent metal; for example, a 552 MPa (80 ksi) mandrel could be welded with a duplex or austenitic stainless filler metal that results in a lower weld joint strength, provided this has been considered to meet the design and operation demands.

The martensitic stainless steels are easy to work with, including welding, the welding processes used include SMAW, GMAW (MIG/MAG), FCAW, GTAW (TIG), SAW, EBW, and laser beam welding (LBW) Typical welding consumables include 410 Ni Mo (matching weld metal), or 2209, 309LSi (overmatching consumables), while some limited application of autogenous welding is also practiced. These alloys are not used in the as-welded condition in more demanding environments like, sour service.

Extreme care is typically required when these alloys are welded, because they are susceptible to high hardness. Tubing and casing are generally not welded.

When welding type 410, high pre-heat temperatures are used. The alloys classified as type 410, type 415 (F6NM), and J91540 (CA6NM) are tempered again as a post weld heat treatment after welding to ensure that they have maximum specified strength and hardness. These alloys have been welded using nominally matching filler metals. The use of non-matching austenitic types consumables can increase the risk of fusion boundary cracking in sour service, this increase in fusion boundary cracking is irrespective of the hardness limits in the weld area.

These alloys are known for moderate corrosion resistance, heat resistance up to 535°C (1 000°F), relatively low cost, and the ability to develop a wide range of properties by heat treatment.

If left in as-welded condition, the intergranular (sensitization) cracking is common occurrence in both sweet (CO_2 containing) and sour conditions. These problems also arise as a result of poor PWHT cycles, where the treatment has been ineffective in refining structure and reducing HAZ hardness.

They are capable of air hardening from temperatures above 815°C (1 500°F) for nearly all section thicknesses. Maximum hardness is achieved by quenching from above 950°C (1 750°F). They lack toughness in the as-hardened condition and are usually tempered.

Martensitic alloys can be welded in any heat treat condition. Hardened materials will lose strength in the portion of the heat affected zone. With a carbon content of 0.08% and 12% Cr (Type 410), the heat affected zone will have a fully martensitic structure after welding. The steep thermal gradients and low thermal conductivity combined with volumetric changes during phase transformation can cause cold cracking.

The hardness of the heat-affected zone depends primarily on the carbon content and can be controlled to some degree by developing an effective welding procedure. As the hardness of the heat affected zone increases, its susceptibility to cold cracking become greater and its toughness decreases.

Weldability is improved when austenitic stainless-steel filler is used because it will have low yield strength and good ductility. This also minimizes the strain imposed on the heat-affected zone.

Martensitic steels are subject to hydrogen-induced cracking like low alloy steels. Covered electrodes used for welding must be low-hydrogen and maintained in dry condition.

Preheating and good inter-pass temperature control is the best means to avoid cracking. Preheating is normally done in the 200°C to 315°C (400°F to 600°F) range. Carbon content, joint thickness, filler metal, welding process, and degree of restraint are all factors in determining the pre-heat, heat input, and post weld heat treatment requirements.

Post weld heat treatment is performed to temper or anneal the weld metal and heat affected zone with aim to decrease hardness and improve toughness, and to decrease the residual stresses associated with welding. Subcritical annealing and annealing are performed. When matching filler metal is used, the weldments can be quench hardened and tempered to produce uniform mechanical properties.

Types 416 and 416Se are free machining grades that must be welded with caution to minimize the hydrogen pickup. ER312 austenitic filler metal is recommended for welding type 416 and 416Se alloys, since it can tolerate the sulfur and selenium additions.

Type 431 stainless can have high enough carbon to cause heat affected zone cracking if proper preheat, preheat maintenance, and slow cooling procedures are not followed.

3.25.13 Welding Ferritic Stainless Steels

Properties and application

Ferritic stainless steels are Fe-Cr-C alloys with ferrite stabilizers such as aluminum (Al), columbium (Cb), molybdenum (Mo), and titanium (Ti) to inhibit the formation of austenite on heating. Therefore, they are non-hardenable. In annealed conditions, Lower-alloy ferritic stainless steels have mechanical properties somewhat similar to the low-alloy austenitic stainless steels like type 304. The typical yield strength is in the rage of 30 to 50 ksi (205 to 345 MPa). Alloys with increased chromium, molybdenum, and nickel content have higher strengths.

In the high-chromium-containing alloys such as UNS S44626, the welding procedure typically developed to minimize interstitial pickup during welding and to retain material toughness. These alloys are predominantly utilized as thinwalled tubing products.

These alloys generally exhibit a loss of toughness with increasing section thickness, and a maximum thickness has been established for each alloy depending on the toughness requirements. In high-chromium-containing alloys, the interstitial contents have been carefully controlled for this purpose.

First generation ferritic alloys (Types 430, 422, 446) are subject to intergranular corrosion after welding and exhibit low toughness.

Second generation ferritic alloys (Types 405 and 409) are lower in chromium and carbon and have powerful ferrite formers and carbide formers to reduce the amount of carbon in solid solution. Although they are largely ferritic, some martensite can form as a result of welding or heat treating. Ferritic alloys are low cost, have useful corrosion resistance with low toughness properties.

Recent improvements in melting practice have resulted in third generation ferritic alloys with very low carbon and addition of nitrogen e.g., Types 444 and 26-1 steel. Stabilizing with powerful carbide formers reduces their susceptibility to intergranular cracking after welding, improves toughness, and reduces susceptibility to pitting corrosion in chloride environments and to stress corrosion cracking.

The most important metallurgical characteristic of the ferritic alloy is the presence of enough chromium and other stabilizers to effectively prevent the formation of austenite at elevated temperature. Most grades do form some small amount of austenite since interstitials are present.

Since austenite does not form and the ferrite is stable at all temperatures up to melting, these steels cannot be hardened by quenching. The small amounts of austenite which may be present and transform to martensite are easily accommodated by the soft ferrite.

Annealing treatment at 760°C to 815°C (1 400°F to 1 500°F) is required to restore optimum corrosion resistance after welding. Ferritic stainless steels cannot be strengthened appreciably by heat treatment. These steels are generally used in the annealed condition. The cooling rate from the annealing temperature chosen depends on the particular alloy. The importance of proper heat treatment is emphasized by the fact that the higher-chromium-containing alloys are subject to embrittlement by sigma or alpha prime phase if not properly heat-treated.

All ferrites if heated above 927°C (1 700°F) are susceptible to severe grain growth, due to this the material toughness is reduced and it can only be restored by cold working and annealing.

3.25.13.1 Welding Ferritic Steel

Types 430, 434, 442, and 446 are susceptible to cold cracking when welds are made under heavy restraint. A 150°C (300°F) preheat can minimize residual stresses that contribute to cracking. These steel grades are also susceptible to intergranular corrosion.

Filler material selection would include any of the three available options.

(1) Matching compositions,
(2) Use of austenitic stainless steels consumables, and
(3) Use of nickel alloy consumables.

Matching fillers are normally used only for Types 409 and 430. Austenitic stainless steels electrode or filler wire matching E 309 or E312 grade or nickel alloys are often selected for dissimilar welds.

The need for preheating is determined by the chemical composition, desired mechanical properties, thickness, and conditions of restraint. High temperatures can cause excessive grain growth and heat affected zone cracking can occur in some grades. Low 150°C (300°F) and inter-pass temperatures are usually recommended.

If post weld heat treatment is deemed necessary, it is done in the 700°C (1 300°F) to 843°C (1 550°F) range to prevent excessive grain growth. Rapid cooling through the 538°C (1 000°F) to 371°C (700°F) range is necessary to prevent embrittlement.

3.25.14 Welding Precipitation Hardening Stainless Steels

Properties and application of Precipitation Hardening steels

Precipitation hardening (PH) stainless steels can develop high strength with simple heat treatments. They have good corrosion and oxidation resistance without the loss of toughness and ductility that is normally associated with high strength materials. Precipitation hardening is promoted by alloying elements such as copper (Cu), titanium (Ti), columbium (Cb), and aluminum (Al). Submicroscopic precipitates formed during the ageing treatment increase hardness and strength.

Martensitic PH steels provide a martensitic structure which is then aged for additional strength. Semi-austenitic precipitation hardened steels are re-heated to form martensite and also aged. Austenitic precipitation hardened steels remain austenitic after cooling and strength is obtained by the ageing treatment.

As a group, the precipitation hardened steels have corrosion resistance comparable to the more common austenitic stainless steels. Corrosion resistance is dependent on the heat treatment and the resulting microstructure. Welding can reduce corrosion resistance by over-aging and sensitization.

Precipitation hardened steels tend to become embrittled after exposure to temperatures above 300°C (580°F), particularly if heated for long periods of time in the range of 370°C to 427°C (700°F to 800°F) temperature. After welding, the maximum mechanical and corrosion resistance properties can be obtained by solution heat treatment followed by ageing. For some applications, only ageing treatment is sufficient.

Martensite precipitation hardened steels are often fabricated in the annealed or overaged condition to minimize restraint cracking. Solution heat treatment and ageing is performed after fabrication.

3.25.14.1 Welding Precipitation Hardened (PH) Steels

Semi-austenitic precipitation hardened steels are welded in all conditions. Austenitic conditioning and ageing are performed after welding for maximum mechanical properties. Austenitic precipitation hardened steels are difficult to weld because of cracking problems. Matching, nickel alloy, or austenitic filler materials are used. The selection of suitable filler metal is dependent on the post-weld heat treatment and final property requirements.

The following are the key points that must be kept in mind for selection of material as well as welding of all stainless steels discussed thus far.

- Thermal expansion, thermal conductivity, and electrical resistivity have significant effects on the weldability of stainless steels.
- The relatively high coefficient of thermal expansion and low thermal conductivity of austenitic stainless steel require better control of distortion during welding.
- Low thermal conductivity for all stainless steels indicate that less heat input is required.
- The weldability of the martensitic stainless steels is affected mainly by hardenability that can lead to cold cracking.
- Welded joints in ferritic stainless steels have low ductility as a result of grain coarsening related to the absence of an allotropic transformation.
- The weldability of the austenitic stainless steels is governed by their susceptibility to hot cracking.
- The precipitation hardening stainless steels have welding difficulties associated with transformation (hardening) reactions.

- Stainless steels which contain aluminum or titanium can only be welded with gas-shielded processes.
- Joint properties of stainless-steel weldments will vary considerably as a result of their dependence on welding process and technique variables.
- Suitability for service conditions such as elevated temperature, pressure, creep, impact, and corrosion resistance must be carefully evaluated. The complex metallurgy of stainless steels must be accounted for.

3.26 Mechanical Properties

The mechanical properties of the different types of duplex stainless steel are shown in Table 3.26.2 below. The mechanical properties of the cast versions of these alloys (e.g., UNS J93380, J92205, etc.) are lower than their wrought counterparts. ASTM A 995 "Standard Specification for Castings, Austenitic-Ferritic (Duplex) Stainless Steel, for Pressure-Containing Parts" details the compositions and mechanical properties of cast duplex alloys that are used for pressure-containing parts.

The duplex stainless steels used by the oil and gas industry have a roughly 50/50 austenite/ferrite, in general the duplex steel of various types would present a phase balance within the range 35% to 65% ferrite. They have adequate toughness at low temperatures, the alloy is commonly used to temperatures as low as minus 60°C (−76 °F).

Super duplex stainless steel (UNS S32760) has been successfully used up to minus 120 °C (−184 °F), but this requires a well-developed welding procedure and closely monitored welding parameters during the production process.

On long exposure to temperatures above 320°C (608°F) and up to about 550°C (1 022°F), the ferrite decomposes to precipitate alpha prime. This phase causes a significant loss of ductility; hence, duplex stainless steels are not normally used above 300°C (572°F).

In oil and gas service applications these alloys have fared very well in both sour and sweet environmental conditions.

3.26.1 Heat Treatment of Duplex Steels

Generally, these alloys are used in the annealed or annealed and cold worked condition. Prolonged heating at temperatures between 260 and 925 °C (500 and 1 700 °F) can cause the precipitation of number of phases, including sigma, which reduces toughness and can

Table 3.26.2 Nominal mechanical properties of duplex stainless steels.

Type	0.2% proof stress (MPa)	Tensile strength (MPa)	Elongation (%)
Lean Duplex	450	620	25
Standard Duplex	450	620	25
25 Cr Duplex	550	760	15
Superduplex	550	750	25

reduce SCC resistance. Any prolonged heating below the minimum solution-heating temperature is to be normally avoided. Low-temperature toughness generally decreases with decreasing cooling rates in annealing.

Cold-worked alloys are usually not welded because the mechanical strength of the weld would be lower than the base metal. Annealed alloys are easily welded. The weld filler metal is chosen to produce a desired volume fraction of ferrite and austenite. Hence, fabrication using autogenous (without filler) metal can result in welds that are poorer in mechanical and corrosion-resistant properties. The welding procedure is typically developed to control and balance the ferrite/austenite phase, this is essential to prevent deleterious phases or intermetallics.

These alloys are readily weldable, by SMAW, SAW, GTAW, GMAW processes. Other processes are also successfully used. Where the weld is to be heat treated after completion it is usual practice to weld with matching composition filler metal.

In as welded application, it is normal to use an over alloyed filler metal with an extra 2 to 2.5% nickel (Ni). This helps in getting the austenite/ferrite phase balance of about 50/50, if the weld is cooled rapidly. The lean duplex grades are welded with the filler metal used for 22% Cr duplex stainless steels. Except for thin sheets of up to 2 mm thickness autogenous welding is normally not recommended for duplex stainless steels.

Normally, argon gas is used for both shielding and backing gases, and welding does not begin until the oxygen level is dropped below 0.1%. As with high alloy stainless steels, care is to be taken in welding duplex alloys, and adherence to good stainless-steel welding practice discussed earlier in this chapter a good practice. Good joint design, control of inter-pass temperatures and keeping low heat inputs are other essential variable for good welding. Preheat and post weld heat treatment are not required for duplex stainless steels.

Maximum permissible heat input and inter-pass temperature increase with section thickness. The values for these parameters generally decrease as the alloy content increases. If heat inputs or inter-pass temperatures are too high, there is a risk of precipitating sigma (Σ) or chi (X) phases in the heat affected zone (HAZ) or weld metal. These are intermetallic phases, rich in chromium and molybdenum that leave a denuded area around them, which reduces the localized corrosion resistance. Sigma and chi phases also reduce impact toughness properties. In many applications especially in some oil and gas applications, the low temperature toughness is compromised for the corrosion resistance properties.

After welding, there is usually a heat tint in the weld and heat affected zone (HAZ), and it is normal to remove this, by manual brushing, by mechanical abrasives, or by a suitable pickling solutions or gels.

While developing welding procedures it is common and advised to include a corrosion test for example testing according to ASTM G 48, (http://www.astm.org/) as part of the weld qualification procedure.

The corrosion test sets important weld parameters; hence it is essential that the qualified parameters of welding are followed very closely during production welding. The experience tells us that sometimes "less experienced" welders have difficulty passing the corrosion test. Although the weld made by these less experienced welders would meet the mechanical requirements, it may not meet the corrosion tests as specified above. This increases the importance of welders/operators' qualification test and production weld parameters monitored by inspectors. In a very limited way this problem is resolved with use of 2% nitrogen gas along with argon as shielding gas. The reasons for the corrosion test failure can be due to the development of third phases, which are the result of poor supervision and control over the heat input and the inter-pass temperature.

Duplex Stainless-steel welds usually have lower impact toughness than their parent metals. The welding process used often affects the level of toughness, the GTAW welds being the toughest and SAW being the poorest. The weld-metal toughness is a function of both the heat input and the type of flux used. The experience suggests that a minimum of 70-Joule Charpy impact toughness in the parent metal ensures adequate toughness in a duplex weld, and it is easily achieved when correctly welded. To improve the low temperatures toughness requirements, especially for very low temperature services, it is worth considering the use of a nickel alloy filler metals, taking into account that other properties are not compromised for example, the nickel alloy weld must have the same strength as the parent duplex stainless steel. A practical example of above would be the selection of C-276 (UNS N10276) filler metal to improve the impact toughness of cast super-duplex (UNS J93380) at minus120 °C (-184 °F) service.

Some specifications for duplex material that are used in subsea environments with cathodic protection (CP) require maximum austenite spacing. In a weld, this cannot be controlled and the result cannot be changed by any heat treatment. However, duplex welds usually have a fine microstructure and meeting a maximum austenite spacing of 30 μm, is usually not difficult. Although the welding in itself does not necessarily degrade the resistance of duplex stainless steel against HISC, the presence of higher stress and stress raisers like weld toe, poses a significant problem when uncoated duplex stainless steels or steels with defective coating are exposed to CP under mechanical stress. Failures have occurred as a result of this effect, and guidance to avoid them can be sought from industrial specifications.

Though nearly everything to know about welding duplex steels have been said in the earlier paragraphs, there is no specific on size fits all welding guide that can be given to meet all grades of duplex steels. Welding any grade of duplex steels is knowing the metallurgy and how that affects the status of the material in welding.

However, the following summary of all the previous discussions may be able to remind what is best practice of welding duplex steels.

3.26.2 How to Weld Duplex Stainless Steel

Question is often asked, about the correct parameters for welding duplex stainless steels. Such question arises from the un-clarity about the metallurgical behavior of the material, resulting in a lot of confusion on which practices and quality checks are best. It is also driven by the way the final product will be put in service.

The common fact is that all Duplex grades require specific care and adherence to he developed procedure with details on shielding gases, amps, volts and other settings, to maintain corrosion and mechanical properties, after welding.

3.26.2.1 Filler Metal

Most duplex stainless steels have an over alloyed filler wire or rod to match these filler metals are nearly identical in chemistry to the base metal, except for an extra 2% nickel is added. The increased nickel content aides in forming the 50/50 mix of austenite and ferrite phases. The benefit of these over alloyed fillers is that it reduces the need for the post weld heat treat the fabrication. Sometimes using a filler wire with higher nickel, chromium, and molybdenum increases the corrosion resistance and makes a sound weld.

3.26.2.2 Heat Input and Interpass Temperatures

One of the most common errors seen during the welding of duplex stainless steel is the heat input and inter-pass temperatures control, if they are not followed correctly the weld will not meet the desired strength and corrosion properties. Duplex steels are welded with relatively high heat input and low inter-pass temperatures. It is not at all difficult to maintain these parameters in tight range for a successful welding. The reason these parameters are important is because duplex stainless rely on a nearly equal balance of ferrite and austenite. Following two situations can arise.

a. If the weld deposit does not see enough time at temperature, it may have high ferrite and not enough austenite.
b. If the weld deposit sees too much time at temperature, the austenite may be too high and there is an increased chance of forming nitrides and Σ (sigma) phase in the base metal. These changes will decrease the corrosion resistance and toughness.

3.26.2.3 Quality Checks

Note that, welding duplex stainless steel is not fool-proof, it is important to perform quality checks. Apart from the normal test and inspections the following very critical checks and tests must be performed during the qualification of the welding procedure.

i. test the corrosion resistance,
ii. the impact toughness, and
iii. the ferrite/austenite mix in the weld qualification.

Once the WPS is written to provide good corrosion, toughness, and ferrite the final production weld should also yield good results. Since it is often impossible to fully test the final product, it is a normal practice to test the production weld with insitu ferrite content. Typically, incorrect ferrite levels will be the first indication of a problem. If a problem is found, further testing should be performed, it may lead to revisiting the WPS parameters.

3.27 Welding Nickel Alloys

Nickel is a very versatile material with excellent weldability. Nickel and nickel alloys are major corrosion resistant materials in use in chemical and petrochemical industries. They are also used in other industries like, marine engineering, aeronautical and automobile making.

Nickel is an element, with the symbol Ni and atomic number 28, as shown in the periodic table below.

In appearance, it is a silvery-white lustrous metal with a slight golden tinge. Nickel belongs to the transition metals and it is hard, and ductile. Nickel is slow to react with air because of passivation, an oxide layer forms on the surface and prevents further corrosion.

Nickel is slow to oxidize in air at room temperature and is considered corrosion resistant. It is used for plating iron and copper alloys like brass, and coating chemistry equipment, and manufacturing alloys that retain a high silvery polish, one such alloy is called German silver, which contains about 60% copper, 20% nickel and 20% zinc.

Nickel like iron, cobalt, and gadolinium is ferromagnetic at room temperature. That property allows its use to make permanent magnets. The metal is valuable in modern times primarily to make various alloys, primary among them are various grades of stainless steels, and cupro-nickel alloys.

Nickel 200 is the purest of nickel commercially available in wrought condition. The metal is weldable. Among the various alloys of nickel most of them are commercially named with number identifiers, some of them are discussed below for introduction purpose.

Nickel and Nickel alloys are welded by number of welding processes including SMAW process. The wrought nickel alloys can be welded under conditions similar to those used to weld austenitic stainless steels. Cast nickel alloys, particularly those with a high silicon content, present difficulties in welding.

The most widely employed processes for welding non-age-hardenable (solid-solution-strengthened) wrought nickel alloys are shielded metal arc welding (SMAW), gas-tungsten arc welding (GTAW), and gas-metal arc welding (GMAW).

Nickel alloys are usually welded in the solution-treated condition. Precipitation-hardenable (PH) alloys should be annealed before welding if they have undergone any operations that introduce high residual stresses.

Most of the time these materials do not require post weld chemical or heat treatment, however in some cases a full solution anneal is desired to improve corrosion resistance. Heat treatment may be necessary to meet specification requirements, such as stress relief of a fabricated structure to avoid age hardening or stress-corrosion cracking (SCC) of the weldment in hydrofluoric acid vapor or caustic soda. If welding induces moderate-to-high residual stresses, then the PH alloys would require a stress-relief after welding and before aging.

Nickel and nickel alloys are susceptible to embrittlement by low-melting-point elements like, lead, sulfur, phosphorus, and other. These materials can exist in contaminants like grease, oil, paint, marking crayons or inks, forming lubricants, cutting fluids, shop dirt, and other processing chemicals. Hence thorough cleaning of parts to be welded is very essential.

Work-pieces should be thoroughly cleaned of all foreign material before they are heated or welded. Shop dirt, oil and grease can be removed by either vapor degreasing or swabbing with acetone or another nontoxic solvent. Paint and other materials that are not soluble in degreasing solvents may require the use of methylene chloride, alkaline cleaners, or special proprietary compounds. If alkaline cleaners that contain sodium carbonate are used, then the cleaners themselves must be removed clean of the material, prior to welding. Spraying or scrubbing with hot water is recommended. Marking ink can usually be removed with alcohol.

Processing material that has become embedded in the work metal can be removed by grinding, abrasive blasting, and swabbing with 10% HCl solution, followed by a thorough water wash. Oxides must also be removed from the area involved in the welding operation, primarily because oxides get imbedded in the weld as inclusions, because the oxide have much higher meting point than the base metal melting points. Oxides are normally removed by grinding, machining, abrasive blasting or pickling.

Nickel alloys, both cast and wrought and either solid-solution-strengthened or precipitation-hardenable, are often welded by the GTAW process, with great success.

3.27.1 Welding of Precipitation Hardenable Nickel Alloy

The PH alloys, alloy 718 is the primary of them are very good weldability, but they also have higher susceptibility of cracking, both in the base-metal, and heat affected zone, hence they require special welding procedures for successful welding.

Cracks also occur post weld and during the operation if the service temperature is greater than the aging temperature, and residual stresses developed during welding or stresses induced during the precipitation. Before welding these alloys, a full-solution anneal is usually performed. After welding, the appropriate aging heat treatment is performed. To further improve alloy properties, a full anneal after welding, followed by a post-weld heat treatment, can be incorporated in the welding procedure.

Any part that has been subjected to severe bending, drawing or other forming operations should be annealed before welding. If possible, heating should be done in a controlled atmosphere furnace to limit oxidation and minimize subsequent surface cleaning.

A generic welding procedure for most commonly used, and weldable nickel and nickel alloys with GTAW wires and rods are listed below. However, the information is not sufficient and more detailed procedure must be developed by the Welding engineer responsible for the project welding.

Heat input during the welding operations should be held to a moderately low level in order to obtain the highest possible joint efficiency and minimize the extent of the HAZ. For multiple-bead or multiple-layer welds, many narrow stringer beads should be used, rather than a few large, heavy beads. Any oxides that form during welding should be removed by abrasive blasting or grinding. If such films are not removed as they accumulate on multiple-pass welds, then they can become thick enough to inhibit weld fusion and produce unacceptable laminar type oxide stringers along the weld axis.

3.27.2 Welding of Cast Nickel Alloy

Cast nickel alloys can be joined by the GTAW processes. For optimum results, casting should be solution annealed before welding to relieve some of the casting stresses and provide some homogenization of the cast structure. Light peening of solidified metal after the first pass will relieve stresses and, thus, reduce cracking at the fusion line, or the interface of the weld metal and the cast metal. The peening of the subsequent passes is of little, if any, benefit. Stress relieving after welding is also recommended.

3.27.3 Nickel - Chromium Alloys

- Alloy 600,
  This alloy has nominally about 72% nickel and 16% Chromium, among other alloying elements. It is resistant to oxidation at high temperatures.
  Weldability: Has good weldability.

Recommended GTAW wire/rod: ER NiCrFe-5, ER NiCr-3.

- Alloy 601,
  This alloy has nominally about 60% nickel and 25% Chromium, 1% Aluminum, Iron among others. It has higher strength and is excellent resistant to oxidation at high temperatures.
  Weldability: Has good weldability.
  Recommended GTAW wire/rod: ER NiCr-3
- Alloy 617,
  This nickel, chromium, molybdenum, and cobalt alloy is metallurgically stable alloy, that serves well as corrosion resistant in wide range of corrosive environment, and in high temperature environment, while maintain it strength.
  Weldability: Has good weldability.
  Recommended GTAW wire/rod: ER NiCrCoMo-1
- Alloy 625,
  This Nickel, chromium, and molybdenum alloy is designed to be an excellent corrosion resistant material by adding niobium, which stabilizes the structure matrix and the strength of the material at high temperature applications. One of the most significant properties is the high resistance to pitting in vary corrosive environment.
  Weldability: Excellent weldability.
  Recommended GTAW wire/rod: ER NiCrMo-3
- Alloy 718,
  This precipitation hardenable, nickel and chromium alloy with iron and niobium, and molybdenum presents excellent creep resistance properties, and resists cracking after welding.
  Weldability: Excellent weldability.
  Recommended GTAW wire/rod: ER NiFeCr-2

3.27.4 Nickel - Copper (Cupro-Nickle Alloys)

- Monel alloy 400,
  This alloy with nominal, about 60% nickel, and 30% copper, is excellent material for service in sea water, and other acids like sulfuric acid, and hydrofluoric acid environment.
  Weldability: Good weldability.
  Recommended GTAW wire/rod: ER NiCu-7
- Monel alloy 401,
  This copper and nickel alloy with nominal, about 45% nickel, and reminder copper, is excellent material for electrical application service.
  Weldability: Not considered.
- Monel alloy 450,
  This cupro-nickel alloy with nominally 70% copper, and 30 % nickel is resistant to biofouling and sea water corrosion.
  Weldability: Superior weldability.
  Recommended GTAW wire/rod: AWS Class, ER CuNi
- Monel alloy K-500,

This precipitation hardenable alloy of nominally 60% nickel, and 30% copper with other alloying elements is a version of Monel 400 discussed above, but it has greater hardness, and strength.
Weldability: Not considered.
Recommended GTAW wire/rod: ER NiCu-7

3.27.5 Nickel – Iron – Chromium Alloys

- Alloy 800,
  This Nickel- Iron- Chromium, with good creep properties is excellent alloy for service in oxidizing and carburizing environment in high temperature atmosphere service.
  Weldability: Good
  Recommended GTAW wire/rod: ER NiCr-3
- Alloy 825,
  The nickel-iron chromium with added molybdenum and copper has excellent resistance to both oxidizing and reducing acids. Resists SCC and has good pitting and crevice resistance.
  Weldability: Very Good,
  Recommended GTAW wire/rod: ER NiFeCr-1
- Alloy 902
  The nickel-iron chromium is designed for precipitation hardening by adding aluminum and titanium.
  Weldability: Not considered.
- Alloy 330
  The nickel-iron chromium with added silicon for increased resistance to oxidation, is good alloy in high temperature service in both oxidizing and carburizing environment.
  Weldability: Good.
  Recommended GTAW wire/rod: ER NiCr-3
- Alloy 020
  The nickel-iron chromium with added copper and molybdenum and stabilizer niobium. The alloy has good resistance to general corrosion, and localized corrosion forms like pitting and crevice corrosion occurring in chlorides and sulfuric, nitric, and phosphoric acids.
  Weldability: Good.
  Recommended GTAW wire/rod: ER Ni-1

3.27.6 Minimizing Discontinuities in Nickel and Alloys Welds

The discontinuities including the metallurgical issues encountered in the arc welding of nickel include can be listed as the following.

1. Porosity
2. Susceptibility to high-temperature embrittlement by sulfur and other contaminants
3. Cracking in the weld bead, caused by high heat input and excessive welding speeds
4. Stress-corrosion cracking in service.

3.27.6.1 Porosity

From welders' point of view, cleaning of the parent metal, and surrounding area can reduce the possibility of porosity. However, from the metallurgical angle more reactions are possible that can cause porosity, gases that are either intentionally present in the weld area, or present due to the service environment of the material being welded can react in welding heat and cause discontinuities. Oxygen causes oxidation, carbon dioxide is a reducing agent, nitrogen allows formation of nitrides, or hydrogen reacts with atmospheric oxygen to form water vapor that cause porosity, all of these gases can cause porosity in welds. In the SMAW processes porosity can be minimized by using electrodes that contain deoxidizing or nitride forming elements, such as aluminum and titanium. These elements have a strong affinity for oxygen and nitrogen and form stable compounds with them. Presence of deoxidizers in either type of electrode serves to reduce porosity. In addition, porosity is much less likely to occur in chromium-bearing nickel alloys than in non-chromium-bearing alloys.

3.27.6.2 Weld Cracking

Hot shortness of welds can result from contamination by sulfur, lead, phosphorus, cadmium, zinc, tin, silver, boron, bismuth, or any other low-melting-point elements, which form intergranular films and cause severe liquid-metal embrittlement at elevated temperatures. Many of these elements are found in soldering and brazing filler metals.

Hot cracking of the weld metal usually results from such contamination. Cracking in heat-affected zone is often caused by intergranular penetration of contaminants from the base-metal surface. Sulfur, which is present in most cutting oils used for machining, is a common cause of cracking in nickel alloys.

Weld metal cracking also can be caused by heat input that is too high, as a result of higher current and slower travel speed. Welding speeds have a large effect on the solidification pattern of the weld. High welding speeds create a tear-drop molten weld pool, which leads to uncompetitive grain solidification at the center of the weld. At the weld centerline, residual elements will collect and cause centerline hot cracking or lower transverse tensile properties.

In addition, cracking may result from undue restraint. When conditions of the high restraint are present, as in circumferential welds that are self-restraining, all bead surfaces should be slightly convex. Although convex beads are virtually immune to centerline splitting, concave beads are particularly susceptible to centerline cracking. In addition, excessive width-to-depth or depth-to-width ratios can result in cracking may be internal (that is subsurface cracking).

3.27.6.3 Stress Corrosion Cracking

Nickel and nickel alloys do not experience any metallurgical changes, either in the weld metal or in the HAZ that affects normal corrosion resistance. When the alloys are intended to contact substances such as concentrated caustic soda, fluorosilicates, and some mercury salts, however, the welds may need to be stress relieved to avoid stress corrosion cracking. Nickel alloys have good resistance to dilute alkali and chloride solutions. Because resistance to stress-corrosion cracking increases with nickel content, the stress relieving of welds in the high-nickel-content alloys is not usually needed.

3.27.6.4 Effect of Inclusions on Weld Metal

GTAW process is very clean process, it is also very unforgiving process meaning that any inclusion will be immediately noticeable to the welder, and will stand out in the radiographs. Mostly the inclusion comes from the tungsten electrode, but that is also an indicator of welders' inability to weld properly, and not maintain the electrode tip.

Since fabricated nickel alloys are ordinarily used in high-temperature service and in aqueous corrosive environments, all inclusions should be removed from finished weldments. If not removed in these types of application, then crevices and accelerated corrosion can result. Inclusions tend to reduce the strength of the weld.

3.28 Later Developments in GTAW Process

From the fundamental welding process that is mostly used in the industry, there have been several developments to suite specific needs and applications of specific job. Variants have been developed to counter the limitations of basic GTAW processes, primary limitations are (a) slow deposition rate (b) the limited ability to penetrate thicker metals. Hot wire, multiple wire feed systems have been in the market for some time now and they find specific use in the industry. Such changes have been mostly proprietary and their specifics are not available in public domain, but some introductory information is available and that is what we can get. These changes are mostly tuned for the mechanized welding applications, such as for cladding or building up surcease of valves, and other such jobs mostly in shop environment.

Other type of applications includes use of welding heads with rotator on a guide, which is put on pipe circumference, to weld girth weld, or on a rack to travers longitudinally to make straight line single plane welds. they are often called by their band names and bugs etc.

Yet another and more recent development with trade name of K-Tig® (k-tig.com) is a very useful development very useful for piping and similar welding application. The process goes beyond the basic GTAW welding by using bigger diameter tungsten electrodes and mix gases to achieve deeper penetration on stainless steel and other CRAs welds. Thus, reducing number of passes, and time to weld. This significant development has cut down some of the specific advantages of Plasma Arc Welding (PAW) process over traditional GTAW process.

3.29 Plasma Arc Welding

The Plasma Arc Welding (PAW) is not part of the GTAW process, it is a process by itself, and should be dealt with independently. However, because it is often noted that people try to find a comparison between the two process it is deemed necessary to introduce to the basic fundamentals of the PAW process.

In Plasma arc welding (PAW) process the heat is produced between an electrode and the work piece by heating them with a constricted arc. Shielding is obtained from hot ionized gas delivered through the torch. A supplementary shielding gas is usually provided. From the process point of view, it is the constricted gas flow that differentiates PAW process from the GTAW process.

The plasma issues from the nozzle at about 16 650°C (30 000°F) this allows for better directional control of the arc and high heat results in deep penetration and very small heat affected zone. The major disadvantage of PAW is high equipment expense.

As stated in the introduction of this process, PAW process is about striking an arc and forcing the hot ionized gas through the nozzle. The PAW process is an extension of the Gas Tungsten Arc Welding (GTAW) process. However, it has much higher energy density and greater velocity of the plasma gas through a constricting nozzle.

The process involves in directing the orifice gas through the constricting nozzle's Plenum chamber. The tungsten electrode, in the center and the walls of the constricting nozzle, forms the plenum chamber. The chamber's exit end is constructed to give a tangential vector this forms a swirl to the exiting plasma gas. The Throat length and the Orifice diameter define the constricting nozzle. The tungsten electrode located in the center of the constricting nozzle is set offset (set at a distance) from the opening of the constricting nozzle. This distance is called electrode setback distance. This is a significant departure from the GTAW process where the electrode is extended out of the nozzle to strike the arc on the work piece.

The offset of the tungsten electrode, allows for the collimation of the arc that is focused on to the relatively very small area of the work piece, since the shape of the arc is cylindrical as contrasted with focused beam, there is no change in the area of contact as the standoff varies during the welding process. Further, as the orifice gas passes through the Plenum chamber, it is heated and expanded, increasing in the volume and pressure; this increases the velocity of the gas that exit from the orifice. This is a very important variable in welding, as too powerful gas jet can cause turbulence in the weld pool. Hence the gas flow rate at the orifice is controlled within 0.25 to 5L/min (0.5 to 10 ft^3 per hour). The additional shielding gas is introduced to protect the weld pool from atmospheric contamination. The gas flow rate of shielding gas is kept in the range of 10 to 30 L/Min (20 to 60 ft^3/hour).

The effect of plasma jet created by constricted flow is following.

1. Improvement in directional stability of the plasma jet, due to its ability to overcome effects of magnetic fields.
2. Higher current density and temperature can be produced, by plasma arc.

It may be noted here that the heat produced by a non-constricted arc like that in the GTAW process is high enough to melt most of the metals that are welded. Hence the objective is not to generate too high heat. The objective is to get more directional stability and focusing ability of the plasma jet and thereby of the plasma arc. This is an efficient use of the energy supplied by the process.

The degree of arc collimation, arc force, and energy density available on the work piece are function of several parameters. These parameters can be altered to a degree to produce very high to very low thermal energies as required by the work piece.

- Plasma Current
- Orifice diameter

Table 3.29.1 Advantages and limitations of PAW process.

	Advantages of PAW process	Limitations PAW
1	Energy concentration is greater, resulting in, • Higher welding speed. • Lower current required to produce given weld. • Lower Shrinkages and distortion. • Adjusting welding variables can control the depth of penetration. • Keyhole technique allows for higher thicknesses being welded with minimum distortions, minimum addition of weld metal, and the wine glass appearance of weld cross section.	Very low tolerance for joint misalignment.
2	Arc stability is improved.	Manual PAW is not very feasible to use, hence automation is the most practical application.
3	Arc Column has greater directional stability.	The constricting nozzle requires, regular inspection and maintenance.
4	Higher depth-to-width ration results in less distortion.	
5	Fixturing cost can be reduced.	
6	This operation is much easier for adding filler metal to the weld pool since torch standoff distance is generous, and electrode cannot touch the filler or the weld pool. This also eliminates tungsten contamination of the weld.	
7	Reasonable variations in torch standoff distance have little effect on bead width or heat concentration at the work, this allows for out of position welding.	

- Type of orifice gas
- Flow rate of orifice gas
- Type of shielding gas

Two different arc modes Transferred Arc and Non-transferred Arc are used in Plasma Arc Welding process.

In the Transferred Arc, the arc is transferred from electrode to the work piece because the work piece is made part of the electrical circuit. Mostly positive polarity is used for welding of all steel and nickel alloys, light alloys like Aluminum and magnesium are welded with DCEP or Ac with continuous frequency stabilization. In the Transferred arc system, the heat is obtained by both the anode spot on the work piece as well as from the plasma jet. Obviously, this system has advantage of greater energy over the Non-transferred arc system, hence used for welding.

In contrast, the arc in the Non-transferred arc system the work piece is not part of the circuit, the arc is established and maintained between the electrode and the orifice of the constricting nozzle. The application is the Plasma jet which is the sole supplier of the required heat in this system. This process with relatively low energy is mainly useful for cutting, especially useful for cutting and joining nonconductive materials.

3.30 Review Your Knowledge

To review the knowledge of GTAW process, take the test on the question set on the content of this chapter in the book. After writing the answers, check the chapter again and where necessary read those chapters again, and correct your answers. After a couple of hours take the test again, till you are able to answer all questions correctly.

1. What does GTAW process stands for?
2. State the effect of Helium gas on the arc behavior and weld quality. What would be difference if argon gas is used in place of Helium gas?
3. Describe the effect of tungsten tip on quality of welding?
4. What types of AC welding machines are discussed in this chapter?
5. What are the electronic devices used to control the current and voltage output from the welding machines?
6. What is the role of HF unit in GTAW process?
7. How does AC help in aluminum welding?
8. What are different types of DC welding machines?
9. What is the normal no-load voltage range in DC welding arc?
10. What is the method to stabilize the AC arc?
11. Explain the DCEN and DCEP, in each circuit what direction the electrons flow?
12. State the importance of having tight connections to the work, and welding torch?
13. What are the dangers of arc flash?
14. Is there any coating on the GTAW filler wires as in SMAW electrodes?
15. State the variables that are considered in the selection specific DC arc polarity?
16. What is the resistance of a welding circuit when the arc voltage is 20 volts and the current is 130 amperes? Use Ohm's law and step by step calculation.
17. How a weld is properly ended?
18. Describe what you understand by the GTAW welding wire designations (i) ER 80S-D2 and (ii) ER 309-15, (iii) ER 310-L?
19. What is the weld position that allows easiest manipulation of welding arc?
20. Justify the choice of smaller diameter electrodes for welding.
21. Give factors responsible for running a good weld bead.
22. Describe the weld profile of a bead that is welded with long arc length.
23. How is the weld restarted in the middle of the plate?
24. What points as a welder you need to remember when welding aluminum?
25. What is porosity, and how to control it? Give all that may cause the porosity.
26. What is the difference between pre-heat and Interpass temperatures?

27. What is the difference in welding carbon steel and aluminum?
28. How is welding carbon steel different from welding stainless steel?
29. What do you understand by Post weld heat treatment, and why is it necessary for PH alloys?
30. What is additional knowledge that a welder should have to successfully weld Nickel alloys?
31. What are special precautions taken for welding alloy steels, as compared to welding carbon steel?

4
Gas Metal Arc Welding

4.1 Synopsis

The chapter discusses in number of sections and paragraphs the GMAW welding process, and its relatively recent variants. The description of the process includes various equipment types, their accuracy in delivering required power for welding, and control methods. The chapter has a section on the important aspects of metal transfer types and their attributes. In the practical application of various metals being welded with GMAW process the detailed description of Aluminum, Stainless steels and Nickel alloys is included with welding guide in the form of WPSs.

The chapter focuses on how to make new welders to improve their efficiency in GMAW welding.

4.2 Keywords

Metal transfer modes, short circuit arc transfer, spray transfer mode, pulsed arc transfer mode, gas tungsten arc welding, gas shielding, GMAW power sources, DC power.

4.3 Introduction to Gas Metal Arc Welding Process

The Gas metal arc welding process, is also termed as GMAW or simply MIG process, it is one of the several electric arc welding processes that is very commonly used to join several types of metals. The process has wide spectrum of use from hobby welding, to small fabrication needs and high quality and heavy industrial applications. The gas metal arc process is dominant today as a joining process among the world's welding fabricators. Despite its sixty years of history, research and development continue to provide improvements to this process, and all those efforts have been rewarded with high quality results, and continues to do so.

4.3.1 Developmental History of GMAW Process

The GMAW process was developed as the need for a faster production, and better weld quality s was felt by the industry, especially during the high production demands of world war era, a process that would have better production rate than the existing SMAW process.

Ramesh Singh. Arc Welding Processes Handbook (209–298) © 2021 Scrivener Publishing LLC

That need led to the development of the present-day gas metal arc welding process (GMAW). The process had its industrial introduction in the late 1940's. Air Reduction Company, researched and developed the first use of a continuously fed aluminum wire electrode, shielded with 100% argon gas. This was the start of the axial spray metal transfer GMAW process. The axial spray transfer for aluminum was the earliest metal transfer mode for the process. This eventually led to the use of argon plus small additions of oxygen as the shielding gas. The introduction of the oxygen gas improved the arc stability, and finally permitted the use of axial spray transfer on ferrous materials.

The process was limited in its application, primarily due to the high energy level of the axial spray transfer to weld thick plate material. In the early 1950's, the work of Lyubavshkii and Novoshilov initiated the development of the GMAW process to include the use of large diameters of steel electrode (wire), that was with the shielded with carbon dioxide, an active gas, in fact CO_2 is a reactive rather than an active gas.

The process development at this stage was high in weld spatter, and the level of heat generated by the arc made the process uninviting to welders.

In the late 1950's improvements in power source technology, and the interface of small diameter electrodes, in the range of 0.035" - 0.062" (0.9 - 1.6 mm) diameter, allowed the introduction of another mode of metal transfer known as the short-circuiting transfer. This development permitted the use of lower heat input welding on thin sections of base material, and it opened up the opportunity for all-position welding.

In the early 1960's, the research and development in welding power sources led to the introduction to the GMAW process the pulsed spray transfer mode. The idea for pulsed spray transfer, GMAW-P, occurred in the 1950's and it conceptually involved the use of a high-speed transition between a high-energy peak current to a low background current. The motivation behind the idea was the need to decrease spatter and eliminate incomplete fusion defects, that plagued the process at the time. The pulsed arc process incorporated the benefits of axial spray transfer, that resulted in the clean and spatter-free welds with excellent fusion properties, with lower heat input. The lower average current provided by GMAW-P allowed for out-of-position welding capability with improved weld quality, when compared with short-circuit transfer.

In next decade of 1970's was introduced the technology, which further enhanced the development of the GMAW process in particular the development was in the GMAW-P process. In this period the introduction and use of the thyristor power sources for pulsed GMAW was initiated. The Welding Institute (TWI) The Welding Institute - Home of the United Kingdom, was the leading source responsible for determining the linear relationship between pulsed frequency and wire feed-speed. The algorithm for this mathematical relationship permitted a fundamental base for subsequent synergic transistor-controlled power sources. In the recent years, the developments of the high-speed electronic controls have improved the interface between the laboratory developed welding sophistications, and the welding shop floor. The new descriptor for this development was the word "Synergic." As it relates, synergy means: one knob control, this manifest itself as the welder changes (increases or decreases) the wire feed speed, a predetermined pulsed energy is automatically applied to the arc. Synergic power sources made it easier to use GMAW-P.

In the 1990's, research and development in welding power source technology continued to evolve. The incorporation of an inverter-based transformer design with a high speed, computerized control circuit is the face of the modern GMAW welding system. Software developed programs provide an expansive array of synergic and non-synergic optimized arc welding programs for the following welding processes:

- GMAW — Gas Metal Arc Welding
- FCAW — Flux-Cored Arc Welding
- GTAW — Gas Tungsten Arc Welding
- SMAW — Shielded Metal Arc Welding
- CAC-A — Carbon Arc Cutting Process

Several proprietary systems have been developed and in market, among the newer advanced Waveform Control Technology™ processes is Surface Tension Transfer™, or STT™. The STT process is a low heat input mode of weld metal transfer, which incorporates a high-speed reactive power source to meet the instantaneous needs of the arc. The power source is a waveform generator, which is therefore neither a constant current nor constant voltage power source. Unique to STT, is the application of applying welding current independent of the wire feed speed. This feature has the benefit of increasing or decreasing the welding current to correspondingly increase or decrease heat input. Fundamentally, STT provides an answer for controlling the welding conditions, that was responsible for incomplete fusion in the earlier versions of GMAW process. The STT welding mode has the dual benefit of increasing productivity, and improving overall weld quality.

The GMAW process is flexible in its ability to provide sound welds for a very wide base material type and thickness range. Central to the application of GMAW is a basic understanding of the interplay between several essential variables:

- The thickness range of the base material to be welded would dictate the electrode diameter, and the useable current range.
- The shielding gas selection would influence the selection of the mode of metal transfer, and will have a definite effect on the finished weld profile.

Now, the GMAW process is defined as an arc welding process which produces the coalescence of metals by heating them with an arc between a continuously fed filler metal electrode and the work. The process uses shielding from an externally supplied gas to protect the molten weld pool, as result the weld is normally of good quality, and if done correctly has low defects and requires low post weld cleaning.

The Figure 4.3.1 below depicts the essential components of a typical GMAW process.

Further, in this chapter we discuss the basic concepts of the gas metal arc welding (GMAW) process, and then provide an examination of more recent process developments. Technical data pertaining to the GMAW welding and equipment is also introduced, so that the readers would be able to optimize the operation of the GMAW process and its several variants.

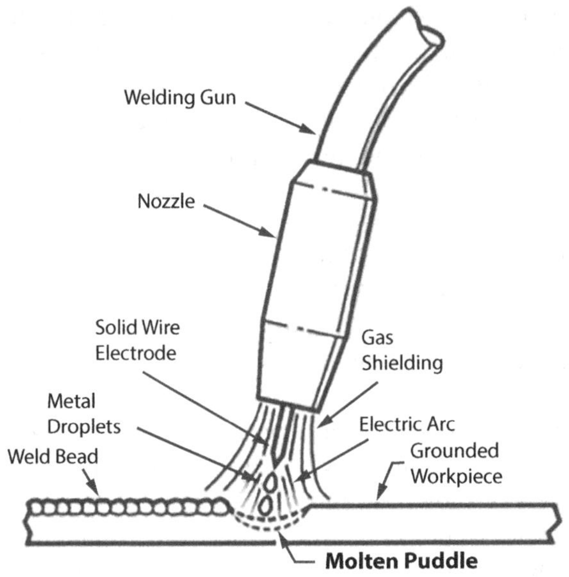

Figure 4.3.1 Typical GMAW welding.

The process uses shielding from an externally supplied gas to protect the molten weld pool. The application of GMAW generally requires DC+ (reverse) polarity to the electrode. The process is also termed as MIG welding process, where the MIG stands for Metal Inert Gas, and its corresponding term is MAG where MAG stands for the Metal Active Gas. The term GMAW is a combination of the two types of gases, as it covers both inert gas and active gas shielding versions, in fact the GMAW not only covers the inert and active gas but more, as it covers various combination of the two types of gases as shielding gas for the process.

The process is applied to weld a wide range of metals using both the solid carbon steel, and tubular metal-cored electrodes, though the use of solid wire electrode is the most common version.

The materials that are commonly welded with GMAW process are carbon steel, stainless steel, aluminum, magnesium, copper, nickel, silicon bronze and tubular metal-cored wires as surfacing alloys. The GMAW process is easily adopted for use in a mechanized and robotic automation.

The GMAW process uses a consumable welding wire that is also an electrode to initiate arc that provides the heat to melt the metal being welded, and this electrode is consumed as filler wire in the weld pool. The weld, molten weld pool and surrounding parent metal, is covered under an envelope of either an active or inert gas, or a combination of active and inert gases, the envelop protects the weld zone from atmospheric contaminations.

The GMAW process, is relatively inexpensive, at the hobby and small fabrication level of its application. As the application level is increased the complexity of the equipment increases and cost rises. The process requires some training to master it, but it is also a very well used process across the industrial spectrum.

The main disadvantage of GMAW is that it its initial cost may be higher than SMAW process. The cost of the equipment is the capital cost, but the requirements of gas, which is a recurring cost can be something to think about in a small fabrication set-up. But the fast production rate can offset such costs.

4.3.2 The Advantages of GMAW

The appeal to use the GMAW process has extended, and is increasing, because of its ability to provide high quality welds, for a wide range of ferrous and non-ferrous alloys, and at a relatively low cost. The process uses shielding from an externally supplied gas to protect the molten weld pool. The list of GMAW processes' advantage is big, some of these are general across the board for all GMAW versions, and some are specific to a particular variant. Some of these variants are proprietary, thus they have some very specific advantages over the other variants. The advantages of the GMAW process would include the following.

1. The ability to join a wide range of material types and thicknesses.
2. GMAW is an All-position welding process.
3. GMAW has higher electrode efficiencies, usually between 93% and 98%, when compared to other welding processes. Less material is lost as weld spatter or stubs.
4. Low slag makes weld clean up fast and easy.
5. Higher welder efficiencies and operator factor, when compared to other open arc welding processes.
6. Simple equipment components are readily available and not too expensive.
7. GMAW is easily adapted for high-speed mechanized welding, robotic, welding applications.
8. Excellent weld bead appearance.
9. The process is a Low-hydrogen weld process, easily producing weld with 4 mL of hydrogen per 100 g of weld metal.
10. Adopted to the Low-heat input welding.
11. Less welding fumes when compared to SMAW (Shielded Metal Arc Welding) and FCAW (Flux-Cored Arc Welding) processes.
12. Generally, lower cost per length of weld metal deposited when compared to other open arc welding processes.
13. Lower cost electrode.
14. Less distortion with GMAW-P (Pulsed Spray Transfer Mode), GMAW-S (Short-Circuit Transfer Mode) and STT™ (Surface Tension Transfer™).
15. Handles poor fit-up with GMAW-S and STT modes.
16. Minimal post-weld cleanup.

4.3.3 Limitations of GMAW

1. The lower heat input characteristic of the short-circuiting mode of metal transfer restricts its use to thin materials.
2. The higher heat input axial spray transfer generally restricts its use to thicker base materials.

3. The higher heat input mode of axial spray is restricted to flat or horizontal welding positions.
4. The use of argon based shielding gas for axial spray and pulsed spray transfer modes is more expensive than 100% carbon dioxide (CO_2).

4.4 Process Description

4.4.1 Gas Metal Arc Welding (GMAW) Process Introduction

Beyond the initial introduction in the previous section the GMAW process is one of the most used, arc welding process the Gas Metal Arc Welding (GMAW), as it is known, and identified by the American Welding Society. The process uses a continuously fed filler wire from a spool of welding filler wire that is either housed inside the power source, or fed from an external wire feeder. This wire or filler material is fed through a welding gun. The filler wire, serves dual purpose

i. it is the consumable filler wire, as the term suggests,
ii. it is also an electrode, since this same wire (or rod) is also used to initiate, and maintain the welding arc.

The power source is used for the GMAW to start the welding arc, and the control systems in the electrical circuit allow the maintenance of the arc as the welding progresses. The filler wire, protruding out of the welding gun's nozzle known as the stick out, also works as the electrode for the arc initiations and maintenance. The arc melts the base-metal as well as the filler wire and brings the coalescence of both molten metal to weld the joint.

The filler wire for the GMAW is primarily a solid wire, mostly it is the same class of spooled wire that is used for the GTAW process. The process does not have any flux in the form of added flux, or as covering of the electrode as in the SMAW process to protect the weld puddle from the atmospheric contaminations. The filler wire/electrode is bare wire, hence the GMAW process requires the use of a shielding gas to protect the weld metal from atmospheric contaminations. The shielding gases are various, ranging from active to inert gases, and both types of gases are used, however a combination of the two is often used, and the exact ration of mixture is dependent of number of factors. The selection of what type of shielding gas to use depends on number of factors, that includes the material being welded, the level of quality demanded of the weld etc.

Since several of the skill required to make a good weld is taken away from the operator and given to the machine, the operator skill for the GMAW process is very much comparable to a machine operator's skill. The operator is required to set the welding parameters and start welding. In semi mechanized versions, the operator has little responsibility, that is to hold the weld head study to maintain the wire stick out length for the study arc operations. Because the machine feeds the wire, the understanding of the machine controls, to maintain the arc length is required. In comparison to GTAW and SMAW welders this skill level is less demanding. This is one of the reasons that the GMAW is the fastest growing welding process. In the mechanized or automated version that skill set is given to the machine and the demand on the operator skill is reduced or moved to set the parameters in the welding machine.

The welding operator holds the gun in one hand, squeezes the trigger, and welds. The use of the suitable shielding gas for the specific metal being welded, makes for a very smooth and stable arc. Since other processes typically require very specific electrode positioning and manipulation, making them a bit difficult to master.

The higher welding speeds, and higher filler metal deposition rates due to the continuously fed electrode, and the operating factor being above 30% and up to 50% the GMAW processes becomes more economical to operate. Couple this GMAW attributes with low demand on welding operators' skill, the process becomes most desired process.

GMAW can be used on all of the major commercial metals. This process also can be used over a wide range of material thickness and operate in all positions. For these reasons, the GMAW is usually the welding processes of choice for most fabrication and production shops.

On the downside, equipment for GMAW is more complex, more costly and traditionally less portable than SMAW process, however as stated above, for the hobby welding, and small fabrication units, some portable models are marketed, and they are not very expansive. Welding is typically done within a 10 to 12 foot distance of the wire feeder, the work is usually brought to the weld stations, and positioned for the welding.

How the Gas Metal-Arc Welding process works

As stated above in the introduction, the GMAW uses a continuous fed solid wire electrode, which is also the filler metal. The weld zone consisting of the arc, the molten pool, and other heat affected zone, is shielded under an externally supplied gas, typically from a high-pressure gas cylinder for protecting from the atmospheric contamination. The wire is usually matched to the material being welded. Figure 4.4.1 shows the typical weld,

Figure 4.4.1 A GMAW operator welding on an offshore pipeline.

filler wire as electrode and as filler metal, molten metal puddle, welding head, and flow of shielding gas.

The selection of the welding wire for example, the carbon and alloy steel welding wire are of the compatible grade, and usually electroplated with a thin layer of copper to protect it from rusting. The copper coating also improves the wire's electrical conductivity, increase contact tip life and generally improves arc performance. Thin copper coating does not contaminate the weld in normal welding situation.

It is easy process to get used to, and if done properly, operator appeal and weld appearance are excellent with GMAW and it is most welders' favorite process to use. Good technique yields good results. The properly made finished weld has no slag and virtually no spatter. A "push" gun angle of about 5° to 15° towards the direction of weld progression is normally used to enhance gas coverage and get the best results. For good weld quality, it is essential that the material being welded is free from dirt and rust or paint. Light grinding to remove these contaminates is recommended. Grinding until the shiny bare metal appears is recommended, but that must not remove material to reduce the designed thickness of the material, especially where material's strength is an essential part of the construction, design.

GMAW process is used to weld most construction metals, including low carbon steel, low alloy steel, and stainless steel, and aluminum with excellent success.

The process predominantly uses DC current with electrode on the positive terminal (DCEP) that is reverse polarity in USA.

The weld appearance is very good, with a thin glass like coating on the bead that is removed easily. The process is adopted for a verity of job conditions and requirements, this is also possible because of number of variables that can be changed for specific welding, that may include variations in electrical parameters like voltage, amperage travel speed, wire feed rate and the metal transfer method.

As stated above in the introduction of the Gas Metal Arc Welding (GMAW) process, the continuous (solid wire) consumable electrode is fed in to the weld by number of deposition modes. In the following paragraphs we describe each mode with their technical details, their deposition rates vary, the Table 4.4.1 below lists the deposition rate for each mode.

The term transfer mode is introduced in the earlier paragraph with some description. The following is the detailed explanation and description of modes of metal transfer in the GMAW process.

Table 4.4.1 Deposition rate of various GMAW metal transfer mode.

Metal transfer mode	**Metal deposition rate**	
GMAW	**Kg/hr**	**lbs/hr**
Short circuit	0.9 – 2.7	2 – 6
Globular	1.8 – 3.2	4 – 7
Spray	2.7 – 5.4	6 – 12
Pulsed spray	0.9 – 2.7	2 – 6

4.4.1.1 *Short Circuiting Transfer (GMAW-S)*

Short-circuiting metal transfer, also described by the acronym GMAW-S, is a mode of metal transfer, where a continuously fed solid or metal-cored wire electrode is deposited during repeated electrical short-circuits. The metal droplets are detached from the wire by a Pinch effect. The transfer of a single molten droplet of electrode occurs during the shorting phase of the transfer cycle. Physical contact of the electrode occurs with the molten weld pool. Number of short-circuiting events can occur up to 200 times per second.

The current delivered by the welding power supply rises, and the rise in current accompanies an increase in the magnetic force applied to the end of the electrode. The electromagnetic field, which surrounds the electrode, provides the force, which squeezes the molten droplet from the end of the electrode. This squeeze and subsequent detachment of the droplet from the wire is commonly known as the Pinch effect. The short-circuiting mode is very suited to the root pass welding applications on heavier plate or pipe groove welds. The shielding gas selection includes 100% CO_2, and binary blends of argon + CO_2 or argon + O_2. Occasionally ternary blends, (three gas mixes), of argon + CO_2 + oxygen is used, to meet the needs of a particular application.

The short-circuiting metal transfer mode is the low heat input mode of metal transfer for GMAW. Because of the low-heat input associated with short-circuiting transfer, it is more commonly applied to sheet metal range of material thickness. However, it has frequently found use for welding the root pass in thicker sections of material in open groove joints. All of the metal transfer occurs when the electrode is electrically shorted (in physical contact) with the base material or molten puddle. Short Circuit transfer occurs in the lowest range of welding currents for that electrode diameter. For short circuit transfer to occur conditions have to be suitable, the diameter of the wire is also relatively very small generally ranging between 0.6 mm 1.1 mm or 0.025" to 0.045", and the electrodes is shielded with either 100% CO_2 or a mixture of 75-80% argon, plus 25-20% CO_2.

As described above the short circuit transfer mode of GMAW produces a small, fast freezing weld suited for joining thin sections, for welding out of position, and for bridging large root openings. The wire electrode actually contacts the weld pool at the rate of 20 to 200 times per second. Inductance is used in the power supply to control the amount of heat available before the short circuit occurs.

Inductance is the property of an electric circuit that slows down the rate of the current change. The inductance is an important aspect of the short circuit weld-metal transfer method. Inductance can be understood by the following example.

If the set welding current is, 125 amps, then at the time of the short-circuiting, the amperage could rise to about as much as 3.33 times that is about 416 amps, this high current could burn the metal instead of welding it. This is controlled by an inbuilt inductance circuit in the machine.

Machines have the inductance coil built into them to activate the inductance by developing a magnetic field, which creates a current in the welding circuit that is in opposition to the welding current. The increasing of inductance in the welding machine will slow down the increase of the welding current. And decreasing the inductance current will increase the welding current. When too little inductance is used, the welding current rises rapidly, often too rapid rise causes the molten metal at the tipoff the electrode to explode, this causes a lot of spatter. Alternatively, if too high inductance is used, the current rise will be too slow and unable to heat the tip off the electrode to melt and make a weld. A balance is created between the two extremes to make the inductance work for the short circuit welding transfer.

What is an inductance and how it is important for short circuit metal transfer mode?

We have referenced inductance in the earlier discussions. Following is the description of inductance and what it does.

The application of an inductance control feature is typical for most GMAW power sources. Inductance affects only in the short-circuit transfer mode. The use of either a fixed or a variable inductance depends on the design of the specific power source. A fixed inductance power source indicates that an optimum level of inductance is built into the power source, and variable inductance indicates that the amount of inductance applied to the arc is adjustable. Inductance controls the rate of current rise following the short-circuit condition. Consequently, its use is beneficial because its adjustment facilitates adding or decreasing energy to the short-circuit condition. Inductance plays a role in the frequency of droplet transfer per unit of time: as the inductance increases, the frequency of short-circuit metal transfer decreases. Each droplet contains more energy and weld-toe wetting improves. As the inductance decreases, the short-circuit events increase, and the size of the molten droplet decreases. The objective for the variable inductance control feature, on any given power source, is to transfer the smallest molten droplet possible with the least amount of spatter, and with sufficient energy to ensure good fusion. Additions of inductance will provide the essential energy to improve toe wetting. Inductance is measured in Henries, and in a variable inductance power source it is the resulting arc performance characteristic that results from the interplay of a combination of electrical components. These components typically include the choke filter, capacitors, and power resistors.

The low welding current gives the process the ability to bridge poorly fitted joint with better penetration. This is a low heat input process suited for toughness-controlled welds where high impact values are desired.

The process in present form can be used for welding thin sheets and/or positional welding where very precise control over weld metal-pool is required. The method is effectively used to weld the root pass in pipe welding, in very critical applications.

The low heat input attribute makes it ideal for sheet metal thickness materials. The useable base material thickness range for short-circuiting transfer is typically considered to be 0.024" – 0.20" (0.6 – 5.0 mm) material.

Short circuit arc transfer mode has some very important advantages, these can be surmised as below,

i. Lower heat input reduces weldment distortion.
ii. It can weld in all All-positions, (FHVOH).
iii. Handles poor fit-up extremely well, and is capable of root pass work on pipe applications.
iv. Ease of use makes it more operator friendly.
v. Higher electrode efficiencies, 93% or more.

Limitations of Short-Circuiting Transfer

i. Restricted to thin section of metals, like sheet metal and open roots of groove joints on heavier sections of base material.

Table 4.4.1.1 WPS for carbon steel and low alloy steels with short circuit transfer mode.

GMAW short circuit transfer welding details								
Joint design: Sq. butt, thin sheets, or for root pass on bevel edges. Position: Flat and horizontal (see note below)								
Single pass material thickness	**Electrode/ wire diameter**	**Electrical parameters current and polarity (DCEP)**		**Wire feed speed**	**Travel speed**	**Shielding gas flow rate**	**Shielding gas**	
mm	**mm**	**Welding current**	**Arc voltage**	**Meter/minutes**		**CFH**	**Carbon steel**	**Low alloy**
0.6	0.6	30 - 50	13 - 15	3.0 – 4.5	0.254 – 0.5	15 - 20	100% CO_2, or 75% Ar with 25% CO_2 or A90 – 95% Ar, with 5% to 10% CO_2	75% Ar with 25% CO_2 or 60 to 70% He with 25 to 35% Ar and remaining CO_2 with
	0.8		15 - 17	2.2 – 2.5	0.3 - 0.5			
0.8	0.8	40 - 60	15 – 17	2.2 – 3.0	0.46 – 0.56	15 – 20		
0.9	0.9	55 - 85	15 – 17	1.8 – 3.0	0.89 – 1.0	15 - 20		
1.3		70 - 100	16 - 19	2.2 - 4.5				
1.6		80 - 110	17 - 20	3.0 – 4.6	0.76 – 1.0	20 - 25		
2.0	0.9	100 - 130	18 - 20	4.1 – 5.6	0.64 – 0.76	20 - 25		
3.2		120 – 160	19 - 22	5.33 – 7.4	0.51 – 0.64			
	1.1	180 - 200	20 – 24	5.33 – 6.1	0.7 – 0.82			
4.7	0.9	140 – 160	19 - 22	5.33 – 7.36	0.36 – 0.48		As above	
	1.1	180 - 205	20 - 24	5.33 – 6.22	0.46 – 5,59			
6.4	0.9	140 - 160	19 - 22	6.1 – 7.37	0.28 – 0.38			
	1.1	180 - 225	20 - 24	5.33 – 7.37	0.305 – 0.46			

Notes: (1) Reduce current by about 10 to 15% for vertical and overhead welding.
(2) Adjust current as required during actual welding.

Table 4.4.1.2 Aluminum WPS for short circuit.

A typical GMAW short circuit transfer aluminum welding guide				
Metal grades	Al type 1060, 1100, 3003, 5052*, 6063, 6061 and castings	Amperes	Voltage	Shielding Gases
Electrode type	AWS Class 5.10			
Electrode diameter	0.8 mm	45 - 75	15 - 19	Argon, Argon + Helium Helium + Argon
	1.2 mm	65 - 175		
Electrode extension	6 to 9 mm			
Position	All positions			
Adjust current based on the actual performance.				

*Use specific alloy suited electrode, for each grade and job requirements. For example; 5000 series aluminum is alloyed with magnesium and it would require specific welding wire.

ii. Poor welding procedure control can result in incomplete fusion. Cold lap and cold shut are additional terms that serve to describe incomplete fusion defects.
iii. Poor procedure control can result in excessive spatter, and will increase weldment cleanup cost.
iv. To prevent the loss of shielding gas to the wind, welding outdoors may require the use of a windscreen(s).

The Figure 4.4.1.1 shows the time sequenced with the arc cycle to describe the short circuit mode of transfer. It shows the Zero Current Voltage, Reignition, Extinction Arcing Period 1, 2, 3 4 and 5 points for the metal transfer. These five points are described below.

1. The welding wire and electrode makes physical contact with the molten puddle. The arc voltage approaches zero, and the current level increases. The rate of rise to the peak current is affected by the amount of applied inductance.
2. This point demonstrates the effect of electromagnetic forces that are applied uniformly around the electrode. The application of this force necks or pinches the electrode. The voltage very slowly begins to climb through the period before detachment, and the current continues to climb to a peak value.
3. This is the point where the molten droplet is forced from the tip of the electrode. The current reaches its maximum peak at this point. Jet forces are applied to the molten puddle and their action prevents the molten puddle from rebounding and reattaching itself to the electrode.
4. This is the tail-out region of the short-circuit waveform, and it is during this downward excursion toward the background current when the molten droplet reforms.

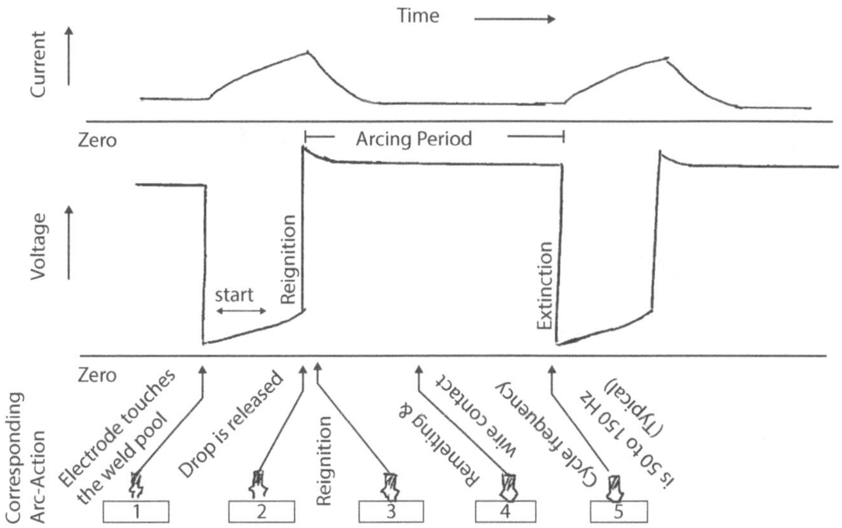

Figure 4.4.1.1 Short circuit transfer (arc-action and cycle).

5. The electrode at this point is, once again, making contact with the molten puddle, preparing for the transfer of another droplet. The frequency of this varies between 20 and 200 times per second. The frequency of the short-circuit events is influenced by the amount of inductance and the type of shielding gas. Additions of argon increase the frequency of short-circuits and it reduces the size of the molten droplet.

4.4.1.2 Globular Transfer

Globular Transfer Globular metal transfer is a GMAW mode of metal transfer, whereby a continuously fed solid or metal-cored wire electrode is deposited in a combination of short-circuits and gravity-assisted large drops. In this transfer mode, the larger droplets are irregularly shaped. During the use of all metal-cored or solid wire electrodes for GMAW, there is a transition where short-circuiting transfer ends and globular transfer begins, see Figure 4.4.1.2. The globular transfer takes place when the current is relatively low but higher than the short circuit mode range.

In this current range the metal transfers occur across the arc as large droplets, often larger than the size of the welding electrode wire, and are irregular in shape.

The drop forms as a globule at the end of the electrode and grows, subsequently due to its own weight, it detaches from the electrode wire and drops in the weld pool, just by gravity.

Globular transfer characteristically gives the appearance of large irregularly shaped molten droplets that are larger than the diameter of the electrode. The irregularly shaped molten droplets do not follow an axial detachment from the electrode, instead they can fall out of the path of the weld or move towards the contact tip. Cathode jet forces, that move upwards from the work-piece, are responsible for the irregular shape and the upward spinning motion of the molten droplets. The process at this current level is difficult to control,

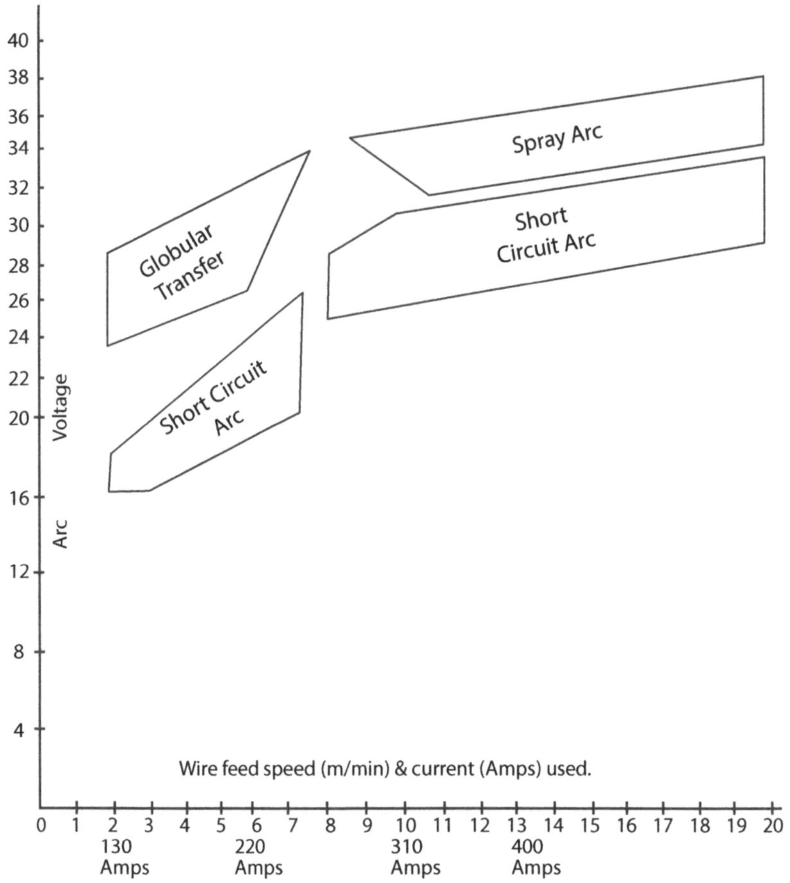

Figure 4.4.1.2 Current voltage range for various transfer mode.

this results in severe spatter. Gravity is instrumental in the transfer of the large molten droplets, with occasional short-circuits. Since, the weld-pool is fed by the gravity the GMAW process with globular transfer is suitable only for flat welding position. The method is not suited for welding overhead, or where the weld position is such that the gravity is not an adding force.

In the past the globular transfer mode was very popular mode of metal transfer for high production sheet metal fabrication. The transfer mode is associated with the use of 100% CO_2 shielding, but it has also seen heavy use with argon/CO_2 mix. For general fabrication on carbon steel, it provides a mode of transfer, just below the transition to axial spray transfer, see the Figure 4.4.1.2 this allows to increase the speed of welding production. The use of globular transfer in high production settings is being replaced with advanced forms of GMAW. The change is being made to GMAW-P, which results in lower fume levels, lower or absent spatter levels, and elimination of incomplete fusion defects.

As sated above, the mode is used with all kinds of shielding gases, but if CO_2 is used as the shielding gas then the dropping globule causes excessive spatter as it travels through the arc. Keeping a short arc helps control the spattering, with CO_2 gas shielding. Keeping the current a bit higher also helps in controlling the spatter.

However, if an inert gas is used the spatter is reduced to minimum.

Advantages of Globular Transfer

i. Can be welded using the inexpensive shielding gas CO_2. However, to improve the quality of welding a mix of argon and CO_2 gas is frequently used.
ii. Fast production rate can be achieved.
iii. Relatively inexpensive welding equipment.
iv. The process can use both the solid or metal-cored electrodes.

Limitations of Globular Transfer

i. Very high levels of spatter levels require extensive cleanup.
ii. Process is fast but results in weld defects such as, excessive cold lap, cold shut, incomplete fusion.
iii. Welders do not like the process very much.
iv. Poor wetting at the weld-toes.
v. Weld bead shape is convex, giving a typical peaking cross section to the weld.
vi. Losses due to high levels of spatter reduces electrode use efficiency to about 87 – 93%.

4.4.1.3 Spray Transfer

Spray transfer mode produces a very stable spatter-free axial spray transfer, when the current level is above the minimum transition current, and above the globular transfer voltage and current levels. This axial spray metal transfer is the higher energy mode of metal transfer, whereby a continuously fed solid or metal-cored wire electrode is deposited at a higher level of energy, this results in a stream of small molten droplets, falling in to the weld pool. The droplets are propelled axially across the arc into the pool.

Spray transfer is supported by either the use of solid wire or metal-cored electrodes. Fine droplets of the molten metal travel through the arc column, and due to the higher current, the speed of the travel of droplets is also accelerated, and they are deposited in the weld pool.

The spray action is possible under an inert gas shielding, for welding carbon and alloy steels that may be 90% argon, plus 10% CO_2, however for nonferrous metals, and stainless-steel welding 100% argon gas is used. Normally, 98% - 2% and 95% - 5% argon and CO_2 mix are commercially available. Though a combination of 75% argon and 25% CO_2 gas mix is also used for carbon steel welding; such gas mix and other variations are commercially available from gas suppliers.

The transitional current setting, as shown in the Figure 4.4.1.2 is so essential for the spray transfer mode to occur, if the transitional current is not exceeded the transfer would remain in the globular transfer mode. Above the transitional current range, the pinch force is much active and it applies the pinch at much smaller droplets to detach itself from the wire and drop through the arc column in to the weld pool, this action is repeated in extremely rapid succession, this makes the metal transfer a spray mode transfer.

Spray transfer may be used with all of the common metals and alloys including: aluminum, magnesium, carbon steel, stainless steel, nickel alloys, and copper alloys. For most of the diameters of filler metal alloys, the change to axial spray transfer takes place at the globular to spray transition current. But the material specific precautions must be included in the welding procedures.

The transitional current itself is a function of material being welded, electrode extension, and the metallurgical composition, and wire diameter of the electrode. For example; the transitional current for steel is much higher than that is required for aluminum.

The spray mode creates deeper weld penetration, and the weld metal direction is easier to control by the welder. The process can be used only in the flat and horizontal positions since the weld pool is large. As stated above, argon or argon rich gases are required for this process. For welding aluminum, titanium and magnesium and their alloys the argon-helium mix gas is often used. For welding ferrous material small amounts of oxygen or carbon dioxide is added to stabilize the arc and eliminate excessive spattering of material.

The spray transfer or the axial spray transfer process as it is also known, can be characterized and described as the following,

A stream of fine metal droplets travels axially from the end of the electrode through the arc to the weld pool. The high puddle fluidity restricts its use to the horizontal and flat welding positions only. For carbon steel, axial spray transfer is applied to heavier section thickness material for fillets and for use in groove type weld joints. The use of argon shielding gas compositions of 95%, with a balance of oxygen, creates a deep finger-like penetration profile, while shielding gas mixes that contain more than 10% CO_2 reduce the deeper (finger-like) penetration profile and provide a more rounded type of penetration. The selection of axial spray metal transfer is dependent upon the thickness of base material and the ability to position the weld joint into the horizontal or flat welding positions. Finished weld bead appearance is excellent, and operator appeal is very high.

Advantages of Axial Spray Transfer

- Electrode's deposition rate is more than 93%.

The process has high deposition rates.

- Can use number of welding electrode wires, of variety of diameters.
- Weld bead appearance is good quality.

The Table 4.4.1.3 below details the transition currents for Spray Transfer, the table further details the spray transfer range shown in the graph, in the Figure 4.4.1.2 above.

4.4.1.4 Pulsed Spray Transfer Mode

This mode of transfer is in fact a variation of the spay transfer mode discussed above. The pulsed spray metal transfer, also known by its GMAW-P, the hyphened letter P is the indicator of Pulsed current. The current level is above the Transition Current level for the spray transfer to occur, as is for the spray transfer mode, however an electrical circuit in the

Table 4.4.1.3 The transition current for spray transfer currents.

GMAW axial spray transition currents for solid and composite carbon steel and stainless-steel solid wire electrodes					
Material type	**Electrode (wire) diameter**		**Shielding gas**		
	mm	**inch**	**% Reactive/ Active gas**	**% Inert gas**	**Approximate current range**
Carbon Steel	0.8	0.03	10 CO_2	90 Argon	155 - 165
	0.9	0.035			175 - 185
	1.2	0.045			215 - 225
	1.3	0.052			265 - 275
	1.6	0.062			280 - 290
Low alloy steel	0.9	0.035	2 O_2	98 Argon	130 - 140
	1.2	0.045			205 - 215
	1.3	0.052			240 - 250
	1.6	0.062			265 - 275
Stainless steel	0.8	0.03	2 O_2	98 Argon	120 - 130
	0.9	0.035			140 - 150
	1.2	0.045			185 - 195
	1.6	0.062			250 - 260
	Composite electrodes				
	0.8	0.03	2 O_2	98 Argon	130 - 140
	0.9	0.035			200 - 210
	1.2	0.045			145 - 155
	1.6	0.062			255 - 265

welding machine is used to create a pulsed current. A low-level current in the globular transfer range, called *background* current, is used to maintain the arc. The current is on for a very brief period in the cycle of transfer so no actual globular mode is achieved, and the current is raised at a regular frequency to above the transitional current, this current is called the *peak current* which is the time in the cycle when the transition current is achieved, and the spray transfer of the weld metal occurs.

The process is a highly controlled variant of axial spray transfer, in which the welding current is cycled between a high peak current level to a low background current level. Metal transfer occurs during the high energy peak level in the form of a single molten droplet. GMAW-P was developed for two demanding reasons

1. Control of weld spatter and
2. The elimination of incomplete fusion defects common to globular and short-circuiting transfer.

The process was developed to weld out of position, high strength steel alloys used for Navy's submarine and ship building projects. The advantages that it brought to the shipbuilding industry included the following,

- Allows the use of higher efficiency electrodes, and is capable of delivering low-hydrogen weld deposits, that improves ductility especially the low temperature ductility.
- The mode employs solid-wire electrode diameters from 0.030" to 0.0625" (0.8 - 1.6 mm) and metal-cored electrodes diameters ranging from 0.045" to 0.078" (1.1 - 2.0 mm) diameter.
- It is used for welding a wide range of material types.
- Argon based shielding gas with a maximum CO_2 not exceeding 18% supports the use of pulsed spray metal transfer with carbon steels.

In the process the welding current alternates between a peak current and a lower background current, and this controlled dynamic of the current results in a lower average current than is found with axial spray transfer. The time, which includes the peak current and the background current, is a period, and the period is known as a cycle (Hz). The high current excursion exceeds the globular to spray transition current, and the low current is reduced to a value lower than is seen with short-circuiting transfer. Ideally, during the peak current, the high point of the period, a single droplet of molten metal is detached and transferred across the arc. The descent to the lower current, known as the background current, provides arc stability and is largely responsible for the overall heat input into the weld. The frequency is the number of times the period occurs per second, or cycles per second. The frequency of the period increases in proportion to the wire feed speed. Taken together they produce an average current, which leverages its use in a wide material thickness range.

The process uses periodical change in the current level to move from low current level, to the higher current in transitional rage to effect spay transfer. This periodical change in quick frequency causes the current to pulse, hence the name pulsed spay transfer mode. The process has advantage in increased rate of transfer, of smaller droplets with increased pulsed frequency. Normally the pulse frequency varies from 60 to 120 pulse per seconds, but higher pulsing machines are also available.

The main advantage of the pulsed spray transfer is that in this mode the welder can use the machine to weld out of position welds. The intermittent lower current level allows the lover average current. Due to this low average current, the pulsed spay mode is also useful for brazing, and welding thin section. Other advantages and limitations are listed below.

Advantages of Pulsed Spray Transfer

i. Very low levels of spatter.
ii. Excellent weld bead appearance.
iii. Ability to weld out-of-position.
iv. Lower hydrogen process weld deposit.
v. The process helps reduce chances of lack-of-fusion defects, most often associated with other modes of GMAW metal transfer, especially the Globular transfer.
vi. Offers an engineered solution for the control of weld fume generation.
vii. Reduced levels of heat induced distortion.
viii. Reduces the tendency for arc blow.
ix. Handles poor fit-up.
x. When compared to FCAW, SMAW, and GMAW-S, pulsed spray transfer provides a low cost high-electrode efficiency of 98%.
xi. Easily atomized for respective and consistent quality weld.
xii. Provides greater weld travel speed, the process can give the arc travel speeds of greater than 50 inches per minute (1.2 M/min.).

Limitations of Pulsed Spray Transfer

- Higher capital cost as the welding equipment for the process are more expensive than traditional systems.
- The Argon based Shielding gas blends not easy to find everywhere, and are expensive too.
- Adds complexity to welding.
- Higher arc energy requires the use of additional safety protection for welders and bystanders.
- Requires the use of windscreens outdoors.

Joint Design

The deep penetration characteristic of spray transfer permits to weld in joint preparation that has smaller included angle, which is an advantage as it allows for reduced filler metal consumption and labor cost to complete the joint. But such weld joint preparation must be carefully evaluated for potential metallurgical problems arising due to the high depth to width ratio, which can lead to cracking of the weld.

The above introduction to and the description is of the most of the basic types of GMAW metal transfer processes and included with them are the references to some of the newer variants.

Most of the new variants revolve around these basic welding and electrical principles, with specific amplification of one or the other electrical attributes to either enhance or suppress electrical reactions to achieve the desired welding objective.

Most of the newer variants are mostly proprietary in nature, and are patent protected and not much details can be easily obtained. In most cases they are also designed to address a specific type of work requirements. The aspiring welding personnel will learn and develop the specific types as they encounter them on their jobs. The knowledge gained here will be the basic in understanding those new systems.

Table 4.4.1.4.1 Carbon steel - Basic training WPS for spray transfer welding.

GMAW spray transfer welding details carbon Steel – WPS for training									
Material thickness		**Electrode/ wire diameter**	**Electrical parameters current and polarity (DCEP)**		**Wire feed speed**	**Travel speed**	**Shielding gas flow rate**	**Shielding gas**	
mm	**Type of joint**	**mm**	**Welding current**	**Arc voltage**	**Meter/minutes**		**CFH**	**Carbon steel**	**Low alloy**
1.3	Fillet	1.1	280	26	8.9	4.83	25	95% to 98% Ar with 2% to 5 % O_2	98% Ar with 2% O_2
	Sq. Groove	1.1	270	25	8.64	4.57			
1.6	Fillet	1.1	325	26	9.14	3.8	35		
	Sq. Groove	1.1	300	28	8.89	3.56			
2.0	Fillet	1.1	325	27	9.14	3.3	35		
	Sq. Groove		325	29	9.14	2.79			
	Sq. Groove		300	29	8.89	2.67			
3.2	Fillet	1.6	380	28	5.33	2.16	35		
	Sq. Groove	1.1	350	29	9.65	2.83			
4.8	Fillet	1.6	425	31	6.60	1.91	35		
	Sq. Groove	1.6	425	30	8.13	1.93			

Notes: (1) Reduce current by about 10 to 15% for vertical and overhead welding.
(2) For thicker metals subsequent passes must be filled with current and speed as can be controlled by the welder.
(3) Adjust current as required during actual welding.

Table 4.4.1.4.2 Aluminum - Basic training WPS for spray transfer welding.

A typical GMAW spray transfer aluminum welding guide training WPS				
Metal grades	Al type 1060, 1100, 3003, 5052*, 6063, 6061 and castings	Amperes	Voltage	Shielding Gases
Electrode type	AWS Class 5.10			
Electrode diameter	0.8 mm	85 - 95	23 - 32	Argon, Argon + Helium Helium + Argon
	1.6 mm	90- 120- 450		
Electrode extension	16 to 20 mm			
Position	All positions, difficulty in OH position welding			
Adjust current based on the actual performance.				

*Use specific alloy suited electrode, for each grade and job requirements. For example; 5000 series aluminum is alloyed with magnesium and it would require specific welding wire.

4.4.2 Gas Metal Arc Welding: Newer Variants

Most of the new GMAW developments come with new names given by the developers and manufacturers. These new variants and their specific methods of variations are often proprietary, patented and the specific to the task they intend to accomplice and the details are guarded with utmost secrecy.

These new developments are possible due to the modern electronic devices installed in the welding machines that use innovative digital control and regulation concepts, including the clock-pulse controls.

As discussed above, changes have occurred in both Short Circuit Arc transfer, Spray Arc, as well as in Pulsed Arc metal transfer modes. The following paragraph is the brief introduction to these developments.

The traditional short circuit transfer mode is described above, where transformers, rectifiers and chokes are used as welding and welding control mechanism. In that the arc is extinguished during the droplets transfer into weld pool. The limitation of the process is realized when the variations in joint design and positions occur, leading to the changing conditions like alteration in arc length and wire feed speed. Such changes result in arc instability, poor joint penetrations, and excessive spatter.

In the new development, this is addressed by inclusion in the welding machine-system the switching frequency over 100 kHz, this is also referred as the frequency of the welding system. The modern systems use digital signal processors with computerized controls, to regulate the current and voltage output. Such controls tend to compensate for the variations in current and arc stability as limitations of short-circuit process discussed earlier.

This principle is utilized by various manufacturers with their own process-focused refinements and as stated above they all have given their own new names. Names like cold metal transfer (CMT), ColdArc®, ColdMIG® and STT® etc., are all trade (Brand) names using their own specific development and patented techniques.

These new developments can be broadly grouped in two.

1. Short circuit arc with constant feed of wire and installation of trigger points in the Voltage and Current path.

   The advantage of this method is that, it allows better control over the energy input pressure on the weld pool which helps reduction of spatters and the resulting joining process becomes very smooth. The trigger points are initiated to control the quick-changing wire feed. This also supports the metal detachment from the electrode wire process. The electrical energy is used for elimination of short arc, as it can be reduced to the minimum. After the detachment of droplet, the wire feed direction is reversed to reignite the arc process. The control mechanism of retraction and forward motion of wire feed allows good electrical contact and good stability of the process. These controls offer a very good welding process for welding with limited heat input and low deposition.

   In the AC versions of the short circuit arc process, the arc varies between negative and positive polarities. During the negative polarity the wire is enveloped under the arc, this leads to the formation of a larger droplet. This can cause and often is the leading cause of low penetration, high deposition rate and low heat input during this phase.
2. Short circuit arc with wire retraction during resignation of the arc process. In this process the computer-controlled power supply has ability to provide switching power.

   These systems have a quick changing wire feed system in the torch that helps the detachment of droplets. In this short circuit arc process the polarity changes to negative polarity which gives a variable negative polarity. This allows for the variations in heat input and the deposition rate to suite the job requirements.

The process combines application of lower current and smaller diameter wire, while as stated in the introduction of short circuit metal transfer process, it is suitable for welding with precise weld metal pool control, for example welding out of position welds and thin section sheet metal. It has one main disadvantage and that is high spatter level due to frequent fluctuation of current and voltage cycle. Which is associated with the squeezing of the droplet detachment stage of the pinch cycle.

To improve this situation, and increase the efficiency of the process better controls of current cycle is included. Different approaches of the newer welding processes as stated above have addressed this issue through electronic regulation installed (either a software or a hardware path is chosen) in their equipment, they approach the issue by developing digitally controlled power sources that gives them distinctive advantage on the control of waveform.

Primarily they all depend on the rapid reduction of the welding current just before the re-ignition stage, this allows a significant reduction in spatter, reduction in heat input, and all this lead to the ability to weld thinner sections, and gap bridging abilities.

4.5 Components of the Welding Arc

Understanding the welding arc is of utmost importance in understanding welding. This statement is universally true and important for all electric arc welding processes, and especially important for a process like GMAW which has number of options or variants that play with one or the other components of the arc to achieve the desired objective.

The area of the welding arc is a region of high complexity that is comprised of physical forces and chemical reactions. The interaction of the components of the arc affects metal transfer and the quality of the finished weld. From the welding perspective, the behavior of the arc is influenced by the following.

i. The welding parameters, — voltage and current.
ii. The type and diameter of the electrode and filler metal.
iii. The interaction of physical forces — gravity, surface tension, jet forces, and electromagnetic force.
iv. The base metal conditions, the cleanliness of the metal surface.
v. The shielding gas - Gas types.

The character of the mode of metal transfer, the penetration profile, and the bead shape are influenced by the forces applied to the metal as it moves from the electrode end to the work-piece. When current flows through a conductor, a magnetic field builds and surrounds the conductor. In GMAW the electro-magnetic forces, which are mathematically proportional to the square of the applied current, affect the mode of metal transfer. The most common term applied to the electromagnetic force is the pinch effect. As the molten drop forms, it is uniformly squeezed from the electrode anode end by the electromagnetic force. The size of the droplet transferred depends upon this force, the applied welding current, and the shielding gas. Surface Tension Forces Surface tension forces are those forces, which are normal to the surface of a molten droplet. They act on both the interior and the exterior surface of the droplet. Together they serve to support the form of a molten droplet. There is always an inward pull of the forces applied to the surface. Jet Forces in the short-circuiting mode of metal transfer, during the shorting portion of the metal transfer cycle, higher currents cause the electrode to heat to the point of melting. The high current drives an increase in the electromagnetic force, which causes the molten metal to detach from the electrode. As the droplet meets the weld pool, the surface tension forces supporting the molten droplet release and the molten droplet then adds itself to the molten weld pool. In the globular transfer mode, a large molten droplet develops. Surface tension forces support the formation of the molten droplet, and jet forces push against the large droplet. The jet forces are responsible for supporting, spinning, and pushing the large droplet in an irregular fashion within the arc. The transfer occurs by the occasional shorting of the large droplet to the weld pool and the force of gravity. Once the droplet contacts the molten pool or work-piece, the surface tension forces in the droplet collapse, and the volume of weld metal is absorbed by the puddle. The shielding gas employed in a welding application has an effect on the surface tension forces. If the energy level within the arc is high, as is the case with a 100% argon gas employed with a carbon steel electrode, then the bead shape will be extremely convex. If the surface tension value is low, because of the addition of carbon dioxide or oxygen, then the bead shape will be less convex, and more acceptable. So, the addition of active gas components will result in improved weld bead and overall arc performance with carbon steel electrodes.

4.5.1 Shielding Gases for GMAW

The selection of the correct shielding gas for a given application is critical to the quality of the finished weld. The criteria used to make the selection may be various, the following list includes some of them but more factors may be used and can be used, in selection of gas type.

- Alloy of wire electrode.
- The mode of GMAW metal transfer
- Desired mechanical properties of the deposited weld metal.
- Material thickness and joint design.
- Material condition – the presence of mill-scale, corrosion, resistant coatings, or oil.
- The welding position.
- Fit-up conditions.

Table 4.5.1 Details the current and the shielding gas type used in spray transfer mode of some of the common materials.

	Wire electrode diameter			Minimum spray arc current
Material	Metric units (mm)	Imperial units (inch)	Shielding gas	Amperes
Carbon steel	0.76	0.03	98% Ar + 2% O_2	150
Carbon steel	0.89	0.035	98% Ar + 2% O_2	165
Carbon steel	1.14	0.045	98% Ar + 2% O_2	220
Carbon steel	1.59	0.062	98% Ar + 2% O_2	275
Stainless steel	0.89	0.035	91% Ar + 1% O_2	170
Stainless steel	1.14	0.045	99% Ar + 1% O_2	225
Stainless steel	1.59	0.062	99% Ar + 1% O_2	285
Aluminum	0.76	0.03	Argon	95
Aluminum	1.14	0.045	Argon	135
Aluminum	1.59	0.062	Argon	180
Copper (deoxidized)	0.89	0.035	Argon	180
Copper (deoxidized)	1.14	0.045	Argon	210
Copper (deoxidized)	1.59	0.062	Argon	310
Silicon bronze	0.89	0.035	Argon	165

- Desired penetration profile.
- Desired final weld bead appearance.
- Cost.

Shielding gases breakdown and react under the heat of the arc. They respond in differently. At the same time the flow of current in the arc, and its magnitude, has a profound effect on the behavior of the molten droplet. In some cases, a given shielding gas will optimally more suited to one transfer mode, but will be incapable of meeting the needs of another. Three basic criteria are useful in understanding the properties of shielding gas:

- Ionization potential of the gas components
- Thermal conductivity of the shielding gas components
- The chemical reactivity of the shielding gas with the molten weld puddle.

The understanding of arc physics of the specific gas or gas blend at the given temperature may help understanding their behavior and optimal selection for shielding.

The two inert gases used for Shielding are Argon and Helium, they are used for protecting the molten weld pool, and they would not chemically react with the molten weld pool. However, in order to become a conductive gas, that is, a plasma, the gas must be ionized. Different gases require different amounts of energy to ionize, and this is measured in terms of the ionization energy. For argon, the ionization energy is 15.7 eV. Helium, on the other hand, has an ionization energy of 24.5 eV.

Note: The electronvolt written as eV, is also written as electron-volt, or electron volt. This is the amount of kinetic energy gained by a single electron accelerating from rest through an electric potential difference of one volt in vacuum. When used as a unit of energy, the numerical value of 1 eV in joules (symbol J) is equivalent to the numerical value of the charge of an electron in coulombs (symbol C).

With lower energy it is easier to ionize argon, than helium. Experienced welders would testify to the fact that argon facilitates better arc starting than helium. The thermal conductivity, or the ability of the gas to transfer thermal energy, is the most important consideration for selecting a shielding gas. High thermal conductivity levels result in more conduction of the thermal energy into the workpiece. The thermal conductivity also affects the shape of the arc and the temperature distribution within the region. Argon has a lower thermal conductivity rate, which is about 10% of the level for both helium and hydrogen. The high thermal conductivity of helium gives it a broader penetration pattern and reduces the depth of penetration. Gas mixtures with high percentages of argon will result in a penetration profile with a finger-like projection into the base material, and this is due to the lower thermal conductivity of argon.

4.5.1.1 Argon Gas

Inert Shielding Gas Argon is the most commonly used inert gas. Compared to helium its thermal conductivity is low. Its energy required to give up an electron, ionization energy, is low, and this results in the finger-like penetration profile associated with its use. Argon supports axial spray transfer.

Nickel, copper, aluminum, titanium, and magnesium alloyed base materials use 100% argon shielding. Because of its lower ionization energy, Argon assists with arc starting. It is the main component gas used in binary (two-part) or ternary (three-part) mixes for GMAW welding. It also increases the molten droplet transfer rate.

4.5.1.2 Helium Gas

Helium is commonly added to the gas mix for stainless and aluminum welding. Its thermal conductivity is very high, resulting in the broad but less deep penetration profile. When in use, arc stability will require additions of arc voltage. Helium additions to argon is useful in reducing the dilution of base material in corrosion resistant applications. Helium/argon blends are commonly used for welding aluminum greater than 25 mm (1") thick. Reactive Shielding Gases Oxygen, hydrogen, nitrogen, and carbon dioxide (CO_2) are reactive gases. Reactive gases combine chemically with the weld pool to produce a desirable effect. Carbon Dioxide (CO_2) is inert at room temperature, but in the presence of the arc plasma, and the molten weld puddle it is reactive. In the high energy of the arc plasma, the CO_2 molecule breaks apart in a process known as dissociation. In this process, free carbon, carbon monoxide, and oxygen release from the CO_2 molecule. This occurs at the DC+ anode region of the arc. At the DC- cathode region, which is invariably the work piece for GMAW, the released elements of the CO_2 molecule undergo the process of recombination. During recombination higher energy level exists, and is responsible for the deep and broad penetration profile that characterizes the use of carbon dioxide.

4.5.2 Dissociation and Recombination

4.5.2.1 Dissociation and Recombination of CO_2 Gas

During the process of dissociation, the free elements of the CO_2 molecule (carbon, carbon monoxide, and oxygen) mix with the molten weld pool or recombine at the colder cathode region of the arc to form, once again, carbon dioxide. The free oxygen combines chemically with the silicon, manganese, and iron to form oxides of silicon, manganese and iron. Formed oxides, commonly referred to as silica islands, float to the surface of the weld pool, then solidify into islands on the surface of the finished weld or collect at the toes of a weld.

Higher levels of carbon dioxide, means higher oxidation potential, and it increases the amount of slag formation on the surface of the weld.

Lower levels of carbon dioxide, provides lower oxidation potential, this increases the amount of alloyed silicon and manganese retained in the weld, and it has the potential increase the ultimate tensile strength, and the yield strength of the weld metal. So, the lower carbon dioxide levels, in a binary or ternary shielding gas blend can be used with awareness that weld-metal strength may have an increased yield and ultimate tensile strength.

4.5.2.2 Oxygen as Shielding Gas

Oxygen (O_2) is an oxidizer, it reacts with components in the molten puddle to form oxides. In small additions (1-5%), with a balance of argon, it provides good arc stability and excellent weld bead appearance. The use of deoxidizers within the chemistry of filler alloys compensates for the oxidizing effect of oxygen. Silicon and manganese combine with oxygen to form oxides. These oxides float to the surface of the weld bead to form small islands, similar to what we discussed when discussing the CO_2 shielding, but the blends of argon and oxygen gas produces less silicon and manganese island than CO_2.

4.5.2.3 Hydrogen Gas

Hydrogen (H_2) in small percentages (1-5%), is added to argon for shielding stainless steel and nickel alloys. Its higher thermal conductivity produces a fluid puddle, which promotes improved toe wetting and permits the use of faster travel speeds.

4.5.3 Binary Shielding Gases

Two-part shielding gas blends are the most common and they are typically made up of either of the following binary gas combinations.

i. Argon + Helium,
ii. Argon + CO_2, or
iii. Argon + Oxygen.

4.5.3.1 Argon + Helium

Argon + Helium binary blends are useful for welding nickel-based alloys and aluminum. The mode of metal transfer used is either axial spray transfer or pulsed spray transfer. The addition of helium provides more puddle fluidity and flatter bead shape. Helium also promotes higher travel speeds.

For welding aluminum using GMAW, helium reduces the finger-like projection found with pure argon. Helium is also linked to reducing the appearance of hydrogen pores in welds that are made using aluminum magnesium fillers with 5XXX series base alloys. The argon component provides excellent arc starting and promotes cleaning action on aluminum.

The binary mix of 75% Argon + 25% Helium is frequently applied to improve the penetration profile for aluminum, copper, and nickel applications. The puddle is more fluid than with 100% argon. With the higher helium content, the thermal conductivity and puddle fluidity increases. The penetration profile is broad, and it exhibits excellent sidewall fusion.

4.5.3.2 Argon + CO_2

This is the most commonly found binary gas blend, and used for welding carbon steel by the GMAW process. All four traditional modes of GMAW metal transfer are used with argon/CO_2 binary blends. They have also enjoyed success in pulsed GMAW applications on stainless steel where the CO_2 does not exceed 4%. Axial spray transfer requires CO_2 contents less than 18%. Argon/CO_2 combinations are preferred where mill-scale is an unavoidable welding condition. As the CO_2 percentage increases, so does the tendency to increase heat input and risk burn-through. Short-circuiting transfer is a low heat input mode of metal transfer that can use argon/CO_2 combinations. Optimally, these modes benefit from CO_2 levels $\geq$ 20%. Caution must be used in using higher levels of argon with short-circuit metal transfer.

4.5.4 Shielding Gases by Transfer Mode

4.5.4.1 Common Short-Circuiting Transfer

a. 75% Argon + 25% CO_2 Shielding Gas mix
This mix or blend of gas very common in use for carbon steel welds, reduces spatter, and improves weld bead appearance.

b. 80% Argon + 20% CO_2 Shielding gas
Used primarily for welding Carbon steel materials. This blend also reduces spatter more than to 75ar+25CO_2 blend described above. This blend also enhances weld bead appearance.

4.5.4.2 Common Axial Spray Transfer

c. 98% Argon + 2% CO_2 Gas mix
For welding with both stainless steel and carbon steel electrodes, the shielding gas blends of 98% Argon + 2% CO_2 with axial or pulsed spray this combination of gas is very common in use.
This blend is very useful for welding HSS (high-speed steel). This gas blend promotes excellent puddle fluidity and fast travel speeds.

d. 95% Argon + 5% CO_2 Gas mix
When welding with the pulsed spray with carbon steel electrodes, the addition of 5% CO_2 provides extra weld metal puddle fluidity, and is useful for heavier fabrication. The 5% CO_2 performs better than blends with 2% CO_2.

e. 92% Argon + 8% CO_2 Gas mix
The blend is used for both axial and pulsed spray applications on carbon steel. Higher energy in axial spray transfer increases puddle fluidity.

f. 90% Argon + 10% CO_2
Used for both the axial spray, or GMAW-P applications on carbon steel. The weld penetration is broader and it reduces the depth of the finger-like penetration exhibited by argon + oxygen mixes.

g. 85% Argon + 15% CO_2
The increased CO_2 level in axial or pulsed spray transfer increases sidewall fusion of the weld material. Generally, produces improved weld-toe wetting on carbon steel with low levels of mill-scale. In GMAW-S, short circuiting transfer, the lower CO_2 level translates to less heat for welding parts with less risk of burn-through.

h. 82% Argon + 18% CO_2
This is the limit of CO_2 gas that can be sued for the axial spray transfer mode of welding. This blend is common I use for a wide range of welding thicknesses. Broad arc enhances penetration profile along the weld interface.

i. Argon + Oxygen
The use of argon + oxygen has historically been associated with high travel speed welding on thin materials. Argon + oxygen blends attain axial spray transfer at lower currents than argon +CO_2 blends. Use of this blend reduces the droplet sizes, and the weld pool is more fluid. Both

stainless steel and carbon steel benefit from the use of argon + oxygen blends.

j. 99% Argon + 1% Oxygen
This high argon blend is used for stainless steel applications. The oxygen in this blend is used as an arc stabilizer, it enhances the fine droplet transfer and maintains the puddle fluidity. Stainless steel welds appear gray in color because of the oxidizing effect on the weld pool.

k. 98% Argon + 2% Oxygen
This blend with 2% oxygen, is used as a shielding gas for either carbon or stainless-steel applications. This blend is possibly one of the earliest blends used for the axial spray transfer on carbon steel welding. This blend applied for axial spray and also for pulsed spray transfer mode allows faster travel speed. Due to the oxidation under 2% oxygen in the blend, the Stainless welds appear gray in color.
The oxidation occurring due to the presence of oxygen in the blend, allows for the silicon and manganese oxides to remain in the weld metal. These oxides enhance the mechanical properties of the low alloy carbon steel welds.

l. 95% Argon + 5% Oxygen
This general-purpose axial spray or pulsed spray transfer shielding gas applied to heavier sections of carbon steel. The base material is usually required to be free of contaminants with a low level of mill-scale.

4.5.5 Ternary Gas Shielding Blends

Three-part shielding gas blends are also used to weld various metals including carbon steel, stainless steel, and, in some specific applications to weld nickel alloys. The ternary gas blend shielding is used for short-circuiting transfer on carbon-steel welding. The blend of 40% helium, to argon and CO_2, as a third component to the shielding gas, provides a broader penetration profile. Helium provides greater thermal conductivity for short-circuiting transfer applications on carbon steel and stainless-steel base materials. The broader penetration profile and increased sidewall fusion reduces the tendency for incomplete fusion, which is a common complaint against the globular transfer type GMAW process. For stainless steel welding, three-part mixes are very common. Helium additions ranging from 55% to 90% added to argon, and 2.5% CO_2 for short-circuiting transfer. They reduce weld spatter, and improve molten metal fluidity, and they produce a flatter weld cross section profile.

4.5.5.1 Common Ternary Gas Shielding Blends

a. Blend with 90% Helium + 7.5% Argon + 2.5% CO_2
This blend is possibly the most popular for the short-circuiting welding of stainless steel. Helium provides the high thermal conductivity, producing a flat bead shape and excellent fusion. This blend is also used in the pulsed spray transfer mode welding. However, this use is limited to the welding of stainless-steel or nickel base materials. The blend allows for greater travel speeds for stainless steel welding.

b. Blend with 55% Helium + 42.5% Argon + 2.5% CO_2
This type of blend features a cooler arc for pulsed spray transfer. It also supports the short-circuiting mode of metal transfer for welding stainless and nickel alloys. The lower helium concentration permits its use with axial spray transfer.
c. Blend with 38% Helium + 65% Argon + 7% CO_2
The short-circuiting transfer mode welding of mild and low alloy steel uses this ternary blend shielding gas. It can also be used on pipe for open root welding. The high thermal conductivity broadens the penetration profile and reduces the tendency to cold lap.
d. Blend with 90% Argon + 8% CO_2 + 2% Oxygen
Another high argon ternary mix suited for the short-circuiting, pulsed spray, and axial spray modes of metal transfer for welding carbon steel. The high inert gas component reduces spatter.

4.6 Effects of Variables on Welding

Welding variables are the key factors that significantly change the results of welds. These changes are not only the aesthetic, it is also about the changes in the intended strength of the weld on the structural members. This is true for all welding processes. However, we are concentrating this discussion to GMAW processes.

Table 4.5.5.1 Gas selection guide.

Base metal	Electrode type AWS class	Mode of transfer	Shielding gas type
Carbon Steel	ER 70S -3	GMAW-S	100% CO_2
	ER 70S - 4	STT™	70 – 90% Ar. + 10 - 25% CO_2
	ER 70S - 6 ER 70 C 6M	Axial Spray or	82 – 98% Ar. + 2% – 18% CO_2
		GMAW-P	95% - 96% Ar. + 2% – 5% O_2
			90% Ar. + 7.5% CO_2 + 2.5% O_2
Low alloy Steel	ER80S-Ni1	GMAW-S	100% CO_2
	ER80S-D2	STT™	75 – 80% Ar. + 20 - 25% CO_2
	ER 100S-G		
	ER 110S-G		
	ER90C-G	Axial Spray or	95% Ar. + 5% CO_2
	ER 110C-G	GMAW-P	95% – 98% Ar. + 2% – 5% O_2

(Continued)

Table 4.5.5.1 Gas selection guide. (*Continued*)

Base metal	Electrode type AWS class	Mode of transfer	Shielding gas type
Aluminum	ER 1100	Axial Spray	100% argon
	ER 4043, ER 4047		75% Helium + 25% Ar.
	ER 5183, ER 5356	GMAW-P	75% Ar. + 25% Helium
	ER 5554, ER 5556		100% Helium
Austenitic Stainless steel	ER 308L Si ER 309L Si ER 316L Si	GMAW-S	98 – 99 Ar. + 1-2% O_2 90% He, + 7.5% Ar. + 2.5 CO_2
		STT	55% He + 42.5 Ar. +2.5 CO_2
		Axial Spray or GMAW-P	98 – 99 Ar. + 1-2% O_2 98% Ar. + 2% CO_2
			97 - 98% Ar. + 1 -3% Hydrogen
			55% He + 42.5 Ar. +2.5 CO_2
Nickel alloys	ER NiCr-3 ER Ni Cr Mo-4 ER Ni Cr Mo-3 ER Ni Cr Mo-10 ER Ni Cr Mo-14 ER Ni Cr Mo-17	GMAW-S or STT	100% argon. 90% He. 7.5Ar. + 2.5 CO_2
			89% Ar. + 10.5% He. + 0.5% CO_2
			66.1% Ar. 33% He + 0.9 CO_2
			75% Ar. + 25% He
			75% He + 25 Ar.
		Axial Spray or GMAW-P	100% argon.
			90% He. 7.5Ar. + 2.5 CO_2
			89% Ar. + 10.5% He. + 0.5% CO_2
			66.1% Ar. 33% He + 0.9 CO_2
			75% Ar. + 25% He
			75% He + 25 Ar.
			97 – 99% Ar. + 1 – 3 % Hydrogen

(*Continued*)

Table 4.5.5.1 Gas selection guide. (*Continued*)

Base metal	Electrode type AWS class	Mode of transfer	Shielding gas type
Duplex Steel	2209 2304	GMAW-S STT™	66.1% Ar. 33% He + 0.9 CO_2
			90% He. + 7.5% Ar. + 2.5% CO_2
			98 – 99% Ar. + 1 – 2% O_2
			98% Ar. + 2 CO_2
		Axial Spray GMAW-P	75% Ar. + 25% He
			75% He. + 25% Ar.
			100% Ar.
			100% He.
			66.1% Ar. 33% He + 0.9 CO_2
Cupro-Nickel (90-10)	ER CuNi Type 70/30	GMAW-P Axial Spray	75% Ar. + 25% He
			75% He. + 25% Ar.
			100% Ar.
			100% He.
Aluminum Bronze	ER Cu Al - A1	GMAW-P Axial Spray	100% Ar.
	ER Cu Al – A2		
	ER Cu Al – A3		
Copper alloys	ER Cu (Deoxidized)	GMAW-P Axial Spray	75% Ar. + 25% He
			75% He. + 25% Ar.
			100% Ar.
Silicon Bronze and Brasses	ER Cu Si	GMAW-P	100% Ar.
		Axial Spray	
		GMAW-S	
		STT™	

Because the variables not only make these changes in the appearances of the weld, their real effect is noted as the changes in the weld's mechanical properties. For example, variables can change weld's hardness and strength, toughness, and ductility. Such unplanned and often undesired changes can sometimes be the cause of catastrophic failures of the constructed structures. On the positive side, as we will see from subsequent paragraphs relating to GMAW process, some changes can be utilized by welding-design engineers to enhance the mechanical properties of the structure being welded.

4.6.1 Current Density

Current density is defined as the current employed with a particular electrode diameter divided by its current carrying cross-sectional area. If the wire feed speed is low, then the current density will be low, and vice versa. From this we can deduce the following,

1. Lower current density applied to a given electrode is associated with the short-circuit mode of metal transfer.
2. Higher current density is associated with the higher energy modes of metal transfer, such as the globular transfer, the axial spray transfer and the pulsed spray metal transfer modes.

During welding at some point the current for a given GMAW solid or metal-cored electrode will reach a maximum density level. Once this level of current density is attained, no additional current can be carried by the electrode. In other words, the electrode has reached its maximum current density. In particular. The current is ascending and relatively linear to a level, but as the current rises further the rise in current becomes exponential. As the current reaches the apex of the current and wire feed speed, the electrode reaches its maximum current density. The electrode at this point becomes saturated with current and the electrode can't carry any additional current. The maximum current density for a given electrode diameter is synonymous with the concept of current saturation.

This phenomenon of current saturation, occurs for all diameters and material types of electrodes used for GMAW process. It is important to note, that once the electrode reaches its maximum current density, the saturation point, any added wire feed speed will provide a higher deposition rate with no increase in current.

4.6.2 Electrode Efficiencies

The electrode efficiency is a numeric value that is assigned to the particular mode of metal transfer. Electrode efficiency is a term that is applied to the percentage of electrode that actually ends up in the weld deposit. Spatter levels, smoke, and slag formers affect the electrode efficiency in GMAW.

1. GMAW-S, short-circuit transfer, shielded with an argon + CO_2 gas blend, will typically operate with an electrode efficiency equal to or greater than 93%.
   - Shielded by 100% CO_2, the electrode efficiency will range from 90 to 93%.
   - Typically, CO_2 increases spatter levels, this reduces the electrode efficiency with CO_2 shielding. Argon with CO_2 blends are typically useful in reducing the wastage through those spatters, but it does not totally eliminate spatter.
2. GMAW-S and its proprietary variation of STT™, have electrode efficiencies of 98%.
3. Globular transfer which is associated with higher spatters, has much lower electrode efficiency. The efficiency of globular transfer varies from 85 to 88%.

if shielded with 100% CO_2. As stated above, the efficiency can be improved to about 90% if argon blends are used.

4. Axial spray has a higher electrode efficiency. This higher energy mode of metal transfer is associated with electrode efficiencies of 98%.
5. The electrode efficiency for GMAW-P varies depending upon the welding application and the sophistication of the power source. The efficiency factor applied for GMAW-P is 98%, like that for axial spray. However, for the GMAW-P the efficiency is not that straight forward, there are variables within the variable. In a situation where higher travel speed is needed that would demand shorter arc lengths, this change will increase higher spatter levels, reducing the efficiency.

4.6.2.1 Calculation of Required Electrode Based on the Electrode Efficiency (EE)

All of this electrode efficiency is related to the amount of electrode that actually ends up in the weld.

If 100 kg. of 0.035" (0.9 mm) diameter electrode is purchased for use on a particular project, and the project calls for the use of GMAW-S, then the effective amount of electrode that will be expected to end up in the welds will be: EE x (lbs. Electrode) = 0.93 x 100 kg. = 93 kg. This calculation is simple and does not include other losses such as wire clipping etc.

4.6.3 Deposition Rate

Unlike the electrode efficiency the melt-off rate for a particular electrode does not include the efficiency of the mode of metal transfer for the process. In simple terms it is how much electrode is being melted. Deposition rate is applied to the amount of electrode, measured in wire feed speed per unit of time, that is fed into the molten puddle. Importantly, its value reflects the use of the factor for electrode efficiency.

Depending upon the mode of metal transfer, as indicated in the Electrode Efficiency section above, the factor for the particular mode of metal transfer employed is applied to the melt-off rate. To determine the deposition rate for a given diameter of solid carbon or low alloy steel wire electrode the following mathematical formula is effectively used.

$$13.1^{*}\ D^{2*}WFS\ ^{*}EE$$

where:

D = electrode diameter
WFS = wire feed speed (inches per minute)
EE = electrode efficiency
13.1 = is a constant that is based upon the density of steel and its cross-sectional area.

- If the melt-off rate is to be determined, the same formula can be used, in that the electrode efficiency (EE) can be removed making it the 13.1*D2*WFS.

- Aluminum is approximately 33% the density of carbon steel, and its constant will be 13.1 x 0.33, or 4.32.
- Stainless steel, typically, is only slightly greater in density than carbon steel, 0.284 lbs/in^3 versus 0.283 lbs/in^3, this difference is significantly low and it can be easily ignored and use the constant 13.1 to determine the melt-off rate.

4.6.4 Electrode Extension and Contact Tip to Work Distance

The extended portion of the electrode wire is called electrode extension. However, it is a very specific distance that is defined as the electrode extension, this is the electrode extended from the end of the contact tip to the arc. The popular non-standard term for the electrode extension is the electrical stick-out (ESO).

In the GMAW process, this is the amount of electrode that is visible to the welder. It is important to point out that the electrode extension is only that length of the electrode which is between the end of the contact tip to the start of the arc. The extension does not include the length of the arc. The use of the term electrode extension is more commonly applied for semiautomatic welding than it is for mechanized or robotic versions of the GMAW process. In that version the term used is different, it is called "Contact tip to work distance" (CTWD). This is the distance between the end of the contact tip to the work piece, that means that CTWD includes the arc length in its measure. In a non-adaptive constant voltage (CV) system the electrode extension or the CTWD acts as a resistor. Varying the length of the electrode affects the current applied to the arc, that is manifest in the following ways, as listed below.

1. Increasing electrode extension increases the resistance to the flow of current in the electrode, and the current in the arc is decreased.
2. Decreasing the electrode extension decreases the resistance to the flow of current in the electrode, and the current in the arc increases.

Because the current can vary with an increase or decrease in extension, the consistency of the extension assumes significance. Its importance relates to the consistency of weld penetration. Welder's steady hand assumes great importance.

1. For short-circuiting metal transfer or GMAW-S, semiautomatic (i.e., manually welding) welding, the electrode extension should be held between 3/8"-1/2" (10 - 12 mm).
2. For either axial spray or GMAW-P, pulsed spray metal transfer, the electrode extension should be held between 3/4" - 1" (19 - 25 mm).

In mechanized or robotic welding also the maintenance of correct CTWD is It is equally as important for quality welding.

Maintaining the correct electrode extension is important to the uniformity of the penetration profile along the length of a weld, and it is considered to be an important variable for any GMAW procedure.

4.7 Advanced Welding Processes for GMAW

Precise controlling the waveform

With the introduction of inverter power source, a new concept was realized wherein it was possible to develop more electronically managed welding power sources for the GMAW process.

This initiated the development of a range of welding processes, mostly proprietary. In the GMAW process an inverter transformer power supply was combined with a computer CPU. This permitted to see the wider spectrum of the electrical wave, and that helped to an amplify the welding power output, in a unique way that was not possible before. The software was developed to drive the output is to enhance and provide superior optimized welding output for a variety of GMAW modes of metal transfer. The most notable of these developments is the proprietary welding system the Surface Tension Transfer™, (STT™), Constant Power™, and a variety of special pulsed spray transfer modes of metal transfer.

These newer power sources have the ability of the power source to interact with the end-users, and permit the programming savvy welder to create their own GMAW-P welding software program. Wave Designer 2000™ software is a commercially available Windows® software program that provides real-time output control of the power source. RS232 connectivity to the power source establishes a communication link with the computer.

In the telecommunication the RS-232, is a standard for the serial communication transmission of data. It formally defines signals connecting between a data terminal equipment (DTE) to, and a data circuit terminal equipment (DCE), such as a modem. The standard defines the electrical characteristics and timing of signals, the meaning of signals, and the physical size and pinout of connectors. The RS-232 standard had been commonly used in computer serial ports and is still widely used in industrial communication devices.

A serial port complying with the RS-232 standard was once a standard feature of many types of computers, including the personal computers, used them for connections not only to modems, but also to printers, mouse, data storage, the uninterruptible power supplies, and other peripheral devices.

RS-232, when compared to later interfaces such as RS 422, RS 485 and Ethernet, has lower transmission speed, shorter maximum cable length, large voltage swing, large standard connectors, no multipoint capability and limited multidrop capability. In modern personal computers, USB, has displaced RS-232 from most of its peripheral interface roles. Few computers come equipped with RS-232 ports today, so one must use either an external USB-to-RS-232 converter or an internal expansion card with one or more serial ports to connect to RS-232 peripherals. Nevertheless, thanks to their simplicity and past ubiquity, RS-232 interfaces are still used—particularly in industrial machines, including welding equipment, networking equipment, and scientific instruments where a short-range, point-to-point, low-speed wired data connection is fully adequate.

For the pulsed spray transfer, short-circuiting transfer and STT™, the output is modulated in response to changes made to the components of the waveform. The use of waveform control software allows further optimization for a given mode of metal transfer. Templates for pulsed spray transfer, short-circuiting transfer and STT™ can be obtained to adjust the system and meet critical weld requirements. The objective of such adjustment could be to improve toe wetting action, reduce dilution levels or to improve high travel speed performance of a pulsed waveform. In any case, the interaction between the arc performance and the adaptable output are central to the success of waveform control approach. Data acquisition tools that are an important part of the software allow the further ability to monitor the waveform during its development. The information collected permits alteration and or final documentation of the suitability of the waveform for the application.

- Synergic control is designed to support all GMAW modes of metal transfer.
- One knob control permits the welder to select the wire feed speed, and then the voltage/trim value automatically follows.
- For all of the synergic modes of metal transfer the concept of synergy eases the use of higher technology on the shop floor.
- The adaptive arc is an arc that quickly adjusts to changes in the electrode extension to maintain the same arc length. The objective for adaptive control is to improve arc performance and maintain finished weld quality.

4.8 The Adaptive Loop

In the robotic and mechanized welding system where a constant current situation is present, as the CTWD is increased, the arc length also increases, and conversely as the CTWD decreases, the arc length also decreases.

As noted earlier the CTWD includes the length of the arc, and to control the length of the arc despite changes in CTWD, an adaptive control becomes a necessity. The adaptive control will add energy to the arc as the CTWD decreases, and it will take energy out of the waveform as the CTWD is increased. This arrangement provides stability to the arc length, and increases the overall usability of the waveform. (1) Frequency, (2) background current, (3) peak time, and (4) peak current are the typical components of the waveform that are used to regulate the arc length. Another term, "Scale factor" is the term attached to arc length regulation, and percentage is the term applied for its relative magnitude. Example, of the above statement,

- If the background current is set to a value of 100 amps and the corresponding scale factor is expressed as 10%, then as the CTWD decreases, and 10% more background current will be added to the present level for background current.
- Conversely, if the CTWD increases, then up to 10% background current will decrease from the original 100 amps.

This is how the arc length regulation operates, and it is coordinated to include the values for the other scalc factor components described in the above examples. The regulation of the arc length occurs automatically, and it is functional within limits of the CTWD. The effective CTWD range for the adaptive loop is 0.50" – 1.25" (12 – 30 mm).

The adjustment of trim relates directly to the scale factors employed in the adaptive loop.

- As the trim decreases from a nominal value of 1.00, then the scale factors apply themselves together to decrease the arc length.
- As the trim is increased to a value greater than 1.00, then the scale factors work together to increase the arc length.

Additionally, the "arc control" feature in the GMAW-P mode is directly tied to the adaptive loop. As the arc control is moved to +1 through +10, then the frequency increases as the background current decreases. This results in the narrowing of the arc column. If the "arc control" feature is moved to negative (–)1 through the negative (–)10, then the result is a wider arc column and a wider finished weld. The absence of the use of scale factors assumes that the arc is stable for a given wire feed speed or for a wide range of wire feed speeds. Arc stability means that the arc will not vary in length with a consistent CTWD. In this scenario, the welding program is non-adaptive, and only by adjusting the length of the CTWD, will there be a variance in arc length. When using a true non-adaptive program, trim and arc control will produce no changes in arc performance or level of arc energy.

4.9 Advanced Waveform Control Technology

4.9.1 Surface Tension Transfer™ (STT™)

The process often used to deposit sound root pass in construction of deep offshore Steel Contrary Risers (SCRs) the Surface Tension Transfer (STT™) welding mode of metal transfer is a low heat input welding mode. It specializes in its ability to provide smooth even rippled weld beads, free of weld spatter, and with consistently good fusion. It is ideal for sheet metal applications requiring excellent weld bead appearance and it is successfully applied for root pass welding of open root pipe joints. The STT™ welding mode is reactive. The power source monitors the arc and responds instantaneously to the changes in the arc dynamics. An attribute most desired in the fixed position pipe welding. A sensing lead attaches to the work piece to provide feedback information to the power source. Uniquely, the STT power source provides current to the electrode independent of the wire feed speed. This feature permits the ability to add or reduce current to meet application requirements.

One unique feature of the system is that the power source that supports STT is neither constant current nor constant voltage. It provides controls for the essential components of the STT waveform. Among these are controls for peak current, background current, and tail-out current. The description of the process is sequentially detailed below.

1. The molten tip of the electrode makes physical contact with the molten pool (note that this is a short circuit system) at the background current level.
2. The background current is reduced to a lower level to prevent the occurrence of a premature molten droplet detachment.

3. The current then ramps up quickly to a point where the pinch force associated with the rise in current (electromagnetic force) starts to neck down the molten column of the electrode. The power source at this point begins to monitor the changes in voltage over time as it relates to the necking of the molten droplet. The molten metal is still in contact with the molten weld pool, through the sensing lead, the power source references the observed voltage, and continuously compares the new voltage value to the previous voltage value.
4. At the point where the molten metal is about to disconnect from the end of the electrode, the power source reduces the current to a lower than background current level. At this point in the waveform, surface tension forces collapse and the molten droplet transfers to the weld pool. This specific step gives the process its name the Surface Tension Transfer, and this precisely controlled detachment of the molten droplet is free of spatter.
5. The power source then rises to the peak current level where a new droplet begins to form. Anode jet forces depress the molten weld puddle to prevent it from reattaching to the electrode. On its descent to the background current, the tail-out current provides the molten droplet with additional energy. The added energy increases puddle fluidity, and the result is improved wetting at weld toes.
6. At this point a plasma boost is applied, which provides the energy to reestablish the arc length, and initiate a new molten droplet, and force the molten puddle away from the molten droplet. The length of time is nominally 1 mS for carbon steel electrodes, and 2 mS for both stainless and nickel alloyed filler metals.
7. The tail-out region is employed in applications where the energy added to the molten droplet provides faster travel speeds and improved finished weld wetting action at the toes. In most pipe root applications, this value is kept to a minimum.

Additionally, the following also happens in the welding cycle.

- The peak current control is responsible for establishing the arc length, and it provides sufficient energy to preheat the work piece to insure good fusion. If it is set too high, the molten droplets will become too large. The molten droplet formed should be equal to about 1.5 times the electrode diameter.
- Background current is the essential component, it is responsible for providing weld penetration into the base material, and it is largely responsible for the overall heat input into the weld. Manipulation of this component controls the level of weld penetration, and it effects the size of the molten droplet.
- Tail-out current is responsible for adding energy to the molten droplet to provide increased droplet fluidity. It applies added energy without effecting droplet size. Increasing the tail-out current permits faster travel speeds and improves weld toc wctting action. The use of tail-out has proven to be a great value in increasing puddle fluidity, and this translates into higher arc travel speeds.

4.10 Equipment for GMAW Process

In the chapter 2 of this book, we have introduced several types of power sources and electrical and electronic control systems relating to arc welding processes. We have also said there that most machines available are of multi-use across number of arc welding processes. For example; welding machines are available that can be used for SMAW, GTAW, GMAW and FCAW.

In this section we will discuss GMAW process machines that may also be useful for other welding process, and in doing that references will be made to the terms and features that have been discussed in chapter 2. The reader may be required to go back and reference those topics if they need to refresh their knowledge of those terms and features.

The GMAW process consists of basic four equipment, as the process gets further evolved more improved versions of these four are introduced and they get more efficient. The basic components of an industrial, GMAW system includes the following principal components:

1. Power source.
2. Wire drive and accessories (drive rolls, guide tubes, reel stand, etc.).
3. GMAW gun and associated cable assembly, suitable for delivering the shielding gas and the electrode to the arc. Cooling system may be added.
4. Shielding gas apparatus and accessories.

There is a very wide range of GMAW components available in the market. They are designed to provide optimum arc performance for the mode of metal transfer they are made for or they may be universal systems where modes could be changed to suit the job requirements. They may be packaged as one unit or may have different modules. The following are some of the descriptions.

- Combination power sources and wire drives, which range in current capacity from 135 – 350 amp. The lower output range power source/wire feeder combinations are intended for limited sheet metal applications.
- Constant voltage fixed power sources with a selection of wire drives and accessories for three-phase input power. They range from 250 – 655 amps of welding output.
- Multi-process power sources capable of the full range of gas metal arc modes of transfer with additional process capabilities. These range from single-phase 300 amps to three-phase 655-amp systems.
- Advanced process power sources are specifically manufactured such as the dedicated sources for the Surface Tension Transfer (STT™) and GMAW-P. They range in output from 225 - 655 amps, and all of these systems require three-phase input power.
- Then there are the engine driven power sources that range from 200 - 600 amps of output.

Further in this section we will discuss some of the welding equipment for the GMAW process.

4.11 GMAW Power Sources

Welding power source for GMAW process need to provide constant voltage, the Transformer–Rectifier type of sources are the most common type. They give DC output, where polarity can be changed for the type of weld being done. Though most of the GMAW welding is done with machines whose output is DC current, primarily the DCEP - direct current where electrode is connected to the positive terminal.

Inverters are also very popular welding source for GMAW process, they are lighter, and compact, the welder has the option to select from constant current to constant voltage mode, and for the GMAW process the selection of constant voltage is required.

4.11.1 The Transformer Rectifiers

The transformer-rectifiers and the generator types are the direct current power source machines. In the transformer rectifier type machine, there are two distinct sections, one is the transformer and the second is the rectifier, together these two make the DC power source.

The transformer section receives the supplied line voltage and current in either 220 volts, 440 volts, and at 60 Hz cycles. The transformer converts the AC line voltage and current to the welding current and voltage of 60 volts to 80 volts, also called open circuit voltage, and the welding current varies to the design of the equipment, and it could be several hundred amperes.

As this low-voltage and high-amperes current exits the transformer it enters in the rectifier section of the machine.

In the rectifier changes the AC to DC. A direct current constant-current output transformer rectifier may be single phase or a three-phase power source.

The rectifiers use devices that are called diodes to convert alternative current to direct current. In diodes the current can flow in one direction only, it does not allow the current to flow in the opposite direction.

Since the flow of current is controlled in one direction, through the diode in the circuit. The alternative current entering into the rectifier is changing direction 120 times per second, the current exiting the rectifier through the diode is in one direction only, and it has changed to direct current. The figure 2.3.1 shows the schematic of single-phase bridge type rectifier – diodes allow the current flow in one direction shown with arrows, this allows the change of AC current to DC current.

The Figure 2.3.2 shows the schematic of a three-phase bridge type rectifier – diodes allow the current flow in one direction shown with arrows, this allows the change of AC current to DC current.

The direct current is produced through a generator. Another option is to use an AC alternator with a rectifier. The construction of a generator is relatively simple as compared to an AC alternator with a rectifier.

The transformer-rectifier machines are designed to control voltage, which is one of the two important variables for welding. The other variable is the welding current that is spoken in terms of the wire feed speed. Both these controls are in the front panels of welding machine, however if the wire feed unit is not in-built in the machine, and an external wire feed unit is used, then the wire-feed control will be located on the wire feeder unit. Welder should familiarize himself with the machine they are going to use.

The wire feed speed is linked to the welding current that is being drawn, a higher feed rate requires higher current to melt the wire, conversely the lower wire feed rate will demand lower current.

4.11.2 Inverters

The modern inverter-based machines offer additional flexibility. This is especially apparent in the use of AC for welding nonferrous materials. Some manufacturers offer power sources with a maximum control over the arc waveform. Some of these newer machines also are considered to be multi-process power sources, these can be used for welding various other electric arc processes. The inverters significantly differ from conventional power sources limited either by constant–current or constant-voltage mode. The inverters however are not limited to either a constant-current or constant-voltage mode.

Other advantage of an inverter type machine is its efficiency in terms of power input versus output. Inverters are energy efficient, and are able to do more work with less energy, thus reducing cost and space. These machines are powerful for their size, but they are also expensive.

Once the application is defined, and power requirements are established, choosing a power source becomes easier. When the output current, duty cycle (refer NEMA side bar), and special requirements, are known then matching them to power source specifications becomes a matter comparison, and selection based on individual preferences.

Arc Welding Power Source classification by NEMA

Welding machines are electrical machines, so they covered under the guidelines of the National Electrical Manufacturers Association (NEMA). NEMA id a trade association of electrical machine manufacturers.

1. NEMA classifies the welding machines primarily on the basis of their rated duty cycle output. The NEMA classifications are given below.
2. NEMA Class I: Machines that deliver 60%, 80%, or 100% duty cycles are classified in this group.
3. NEMA Class II: Machines that deliver output at 30%, 40%, and 50% are classified in this group.
4. NEMA Class III: Machines that deliver output at 20% duty cycle are grouped in this class.

The arc welding machines are described by their following three attributes.

1. Rated current output,

This is the amount of current measured in amperes that a welding machine is rated to supply at a given voltage. NEMA rated output current for different NEMA class rating described above is in the following table.

Rated output current		
Class I	**Class II**	**Class III**
200	150	18-230
250	175	235-295
300	200	
400	225	
500	250	
600	300	
800	350	
1000		
1200		
1500		

2. Duty Cycle,
   Duty cycle is defined as the following.
   The length of time that a welding machine can be used continually at its rated output, in any 10-minute period.
   Most welding machines are not in use 100% of the time, welding is stopped for loading or unloading of the weldment, or for inspection and cleaning of the weld etc. Normally machines used are of 60% duty cycles. However semi-automatic and automatic processes are required to operate at 100% duty cycle. And these machines are developed at 100% duty cycles.
   Most of the hobby welding machines are rated for 20% duty cycles.
3. Power requirement,
   The rated load or the rated voltage or welding voltage are the term often used for power requirement of a welding machine. For the Class I and Class II machines, the rated load voltage is the product of a constant and the rated amperage. The constant for machines rated up to 500 A is 20+0.04 x rated amperes. This formula if applied to a machine that is for 400 A machine will give the rated load voltage of 36 volts.
   The rated load voltage for machines with current rating of 600 A, and higher is determined to be 44 volts.

Manufacturers also offer many different equipment packages, which can be customized based on need. Also consider used equipment may be considered, a well-maintained power source that was not abused, can perform well and for far less money.

The controls on Inverters are similar to the transformer-rectifier units. Inverters often have multi process options, the welder must select the GMAW (or the constant voltage option) and then go to set the required voltage. Other than these few differences, the rest of the control including the wire feed control are same or similar to the transformer rectifier type units. Inverts also may have internal wire feed mechanism in the machine or an external wire feed unit may be used. As always, the welder must familiarize with the machine they are going to use for the welding.

GMAW Power sources incorporate output characteristics designed to optimize the arc performance for a given transfer mode of the welding process. The output characteristics for the GMAW process fall into two main categories.

- constant current
- constant voltage

Each of these two terms references the volt-ampere characteristics of the power source, and in each case, the volt-ampere relationship references the slope of the output.

for comparison output curves of CC (constant current) and CV (constant voltage). Dedicated constant current power sources were more widely used in the early days of GMAW than they are today, but they see continued use in the welding of aluminum. The design uses a drooping output curve, as shown in the Figure 3.7.1.

In constant current, the CTWD (contact tip to work distance) determines the arc length. As the CTWD increases the arc length increases, and as the CTWD decreases the arc length decreases. This presented a problem for semiautomatic welding because it is difficult to maintain the same CTWD. To compensate for this problem an arc-voltage controlled wire feeder is developed to compensate for changes in arc length. In this condition, as the CTWD decreases, the wire feed speed would increase; and as the CTWD increases, the wire feed speed would decrease.

Typically, constant current applications were confined to large diameter, large weld puddle aluminum GMAW applications or large diameter, or large weld puddle carbon steel applications. Constant voltage power sources provide a specific arc voltage for a given pre-selected wire feed speed. The volt-amp curve, or slope, is comparatively flat. As the CTWD increases with these types of power sources, there is a decrease in the welding current. As the CTWD decreases there is an increase in the welding current. The arc in this case becomes a series circuit, and the CTWD provides resistance to current. In either of the situations, the voltage remains constant, and the arc length remains the steady.

Power sources designed for GMAW require a feature to provide inductance. Inductance is an essential component of the short-circuiting transfer mode, and low wire feed speed globular transfer. The inductance however, has very little role to play in the more advanced process equipment used for GMAW-P or STT™, and also for the spray arc transfer. A variable inductance control is important for short-circuiting transfer because it will permit fine-tuning of the arc to minimize spatter and improve weld toe wetting. The traditional GMAW power source has either analog or digital meters used to indicate the voltage and current output of the source. These are essential variables and their accuracy is central to the quality of a finished weld.

Developments in transformer design allows the use of smaller inverter transformers, inverters are discussed in the earlier chapters relating to GTAW and SMAW, these inverters increase the portability of the equipment, and the resulting compactness reduces shop floor occupancy by the welding equipment. the required space needed for the power source. This frees up a lot of real-estate for the other valuable manufacturing activities. Inverter characteristically provide smooth and efficient output to the arc. The Figure 4.11.1 shows the typical lay out of the GMAW machine, the figure shows the external wire feed spool.

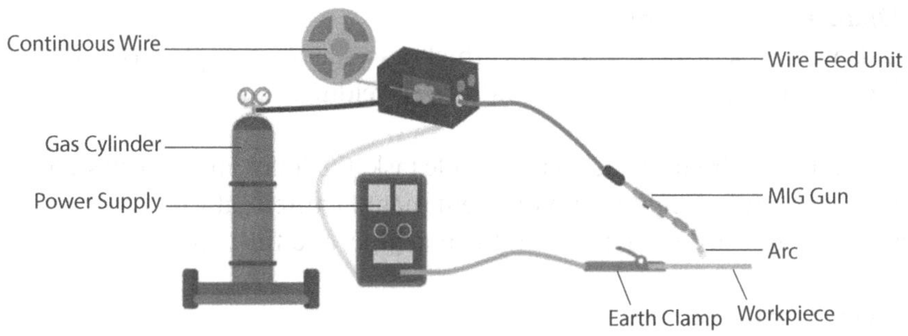

Figure 4.11.1 Typical GMAW (MIG) welding set up with the external wire feed unit.

4.12 Installation of Welding Machines

The welding power sources are often connected to the one phase or three phase power supply points. The power factor that these welding machines develop, these machines tend to disturb the power supply to these machines connected to the same circuit they are connected. This requires that other machines connected to the same supply circuit are provided with some power factor correction devices of their own. This is done by connecting a capacitor to give these machines a boost of power to improve their power factor. This arrangement requires the careful planning and consideration and should be done with the expert analysis and to comply with the local electrical code.

GMAW Set-up

The Figure 4.11.1 shows the schematic illustration of how the GMAW machine set up would look like. In this illustration the external wire feed unit is shown, the connection is not much different if the wire-feed unit is inbuilt in the power source itself.

The shielding gas is in the gas cylinder connected to the wire feeder, so that the gas is delivered to the welding gun and released just before the electrode is exposed out of the seething to strike the arc.

Welding wire, cum-electrode is in the spool and it is installed on the rotator. Wire sizes and types can vary significantly depending on what material, thickness, and what type of weld joint is being used for welding. Some of the common wire sizes used for welding different metals are listed below for reference, other sizes like 0.062" (1.59 mm) are also used. In the following reference use is in terms of mild steel, and carbon steel welding, other metals may have different size for the given thicknesses.

- **Diameter 0.023″ wire**
  This is only for your small welding machines when you're welding thin sheet metal from 24 gauge to around 16 gauge.
- **Diameter 0.03″ wire**
  This wire is a good choice for sheet metal welding, up to around 1/8″ thick.
- **Diameter 0.035″ wire**
  This size of wire is good for welding thicknesses up to 1/4″ (6 mm) and it is preferred size for a beginner to GMAW welding.

- **Diameter 0.045″ wire**
  This wire is best for 1/4″ (6 mm) thicknesses, and multiple pass welding of higher thickness metals, in the industrial welding.

Installing welding wire in the feeder is simple task, the following step by step direction is generic, and it may differ slightly with different manufacturers, always use the manufacturers' instructions while installing the both the machine, and the wire.

i. Open up the cabinet.
ii. Make sure the cap is unscrewed from the spindle and slide the wire spool onto the spindle. Make sure the GMAW welding wire faces in the direction of the drive roll and is on the bottom of the machine.
iii. Flick open the pressure release. If the trigger on the gun is pressed on with this released, nothing will happen. At this point the tightness can also be adjusted, by twisting the pressure roll.
iv. Now slowly feed the welding wire through the drive roll entry making sure the wire is neither twisted, kinked nor bent.
v. Replace back the cap (removed in step 2 above) on the spool of wire. Ensuring that the spring is in place before the cap is attached. The placing of the spring is very important step, if not done correctly, and spring is not placed then the spool will rotate freely and that would damage the welding wire.
vi. Pull out the GMAW nozzle and the contact tip, a pliers may be required for this step.
vii. Press the trigger to start feeding the wire through the gun.
viii. Put the nozzle and contact tip back on.
ix. The GMAW wire is set for welding.

Other accessories that complete the power source to make the complete GMAW welding system are discussed in the following paragraphs.

4.12.1 GMAW Torches

The welding torches are the front end of the welding process. They are also the hardest working visible component of welding machine. Their design and function have to suit the comfort of the welder, while it should also be able to deliver the essential welding components like electrical power, welding wire and electrodes, the shielding gas or gases, and in heavy duty welding the cooling fluid to keep the hot end cool during the welding operation. Figure 4.12.1 shows a typical GMAW welding torch, with trigger type on and off switch on the handle, for the welder to have full control over the start and stop of the welding operations.

The selection of the proper GMAW torch, which often called aa a MIG gun, depends upon the following key factors.

1. Type of welding: semiautomatic, hard automation or robotic automation.
2. Level of current (amps) required by the welding application and capacity of the torch.

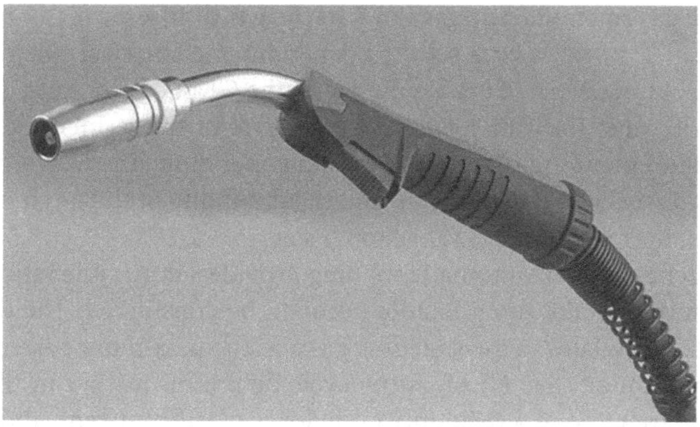

Figure 4.12.1 A typical GMAW torch with trigger type on-off switch on the handle.

3. Shielding gas selected.
4. Duty cycle of the torch.
5. Preference of an air-cooled or water-cooled torch.

As introduced earlier the GMAW torch provides a conduit for the welding current, the shielding gas, and the electrode. The torch is connected to a specifically designed cable connector that facilitates the flow of electrical current, the gas and wire for welding. This is shown in the Figure 4.12.3.

The welding current is picked up at the torch power block located on the wire drive. Current transfers from the welding cable to the electrode through the contact tip. Contact tips are available in a range of sizes designed to accommodate the electrode diameter in use, and they usually attach to the gas diffuser via a threaded connection.

Figure 4.12.2 shows the blow out of the GMAW torch that shows some of the components that make up a welding torch.

Portable Engine Driven GMAW System The electrode is fed through an internal liner usually located internal to the power cable see Figure 4.12.2. The shielding gas connections are located at the welding gun mounting block on the wire drive. They connect to the

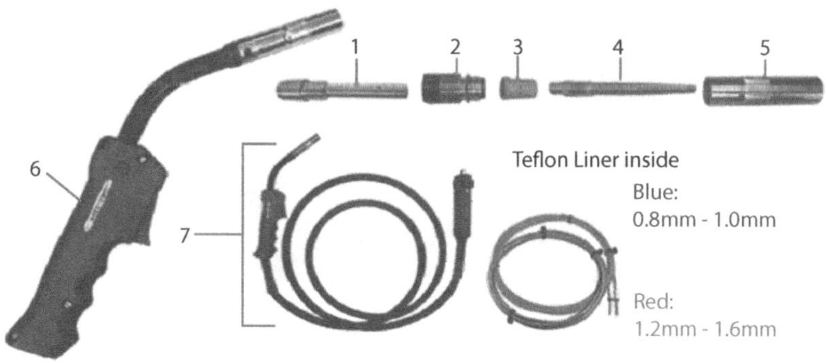

Figure 4.12.2 Blow out of the GMAW torch that shows some of the components that make up a welding torch.

output side of the gas solenoid. The gas flows to the gas diffuser, which uniformly delivers the gas to the arc. The nozzle size is selected depending on the electrode diameter and the shielding gas rate of flow. Most of the welding with the GMAW process requires a selection of a torch that will meet the anticipated comfort level of the welder and simultaneously meet the wear requirement imposed by the welding operation. The welding current used in the application is primary to the selection, and the durability of the torch under conditions of the arc dictates the appropriate GMAW torch size.

All GMAW torches for semiautomatic welding provide a duty cycle rating. The heat generated and transferred to the torch handle needs to be considered. The duty cycle of the GMAW torch selected relates to the shielding gas and the maximum current that is specific to the welding application. See NEMA duty cycle definition further in the chapter. Most air-cooled torches are rated at a 60% duty cycle for a specific current, and their operation is based upon the use of 100% CO_2 shielding. If argon-based blends are indicated, then the torch duty cycle is reduced by 50%. Most GMAW torches come with cable connectors Figure 4.12.3, in the lengths of 10 to 25 ft. (3 to 8 meters) and the length selected should provide no compromise for delivery of the shielding gas and the electrode to the arc.

The selection of a water-cooled torch for GMAW has several advantages. They are rated 100% duty cycle for their given capacity. They increase the life of the consumable components of the torch by approximately 50%. Water–cooled torches have operator appeal because they reduce the heat transferred to the GMAW torch handle. The downside of a water-cooled torch is that they tend to require more maintenance. Additionally, the use of a water-cooled torch requires the purchase of a water cooler. Implementation of a water-cooled GMAW torch depends largely upon the work load and the size of the electrode used, the amount of time a welder spends at the arc, and the projected cost of welding torch consumables.

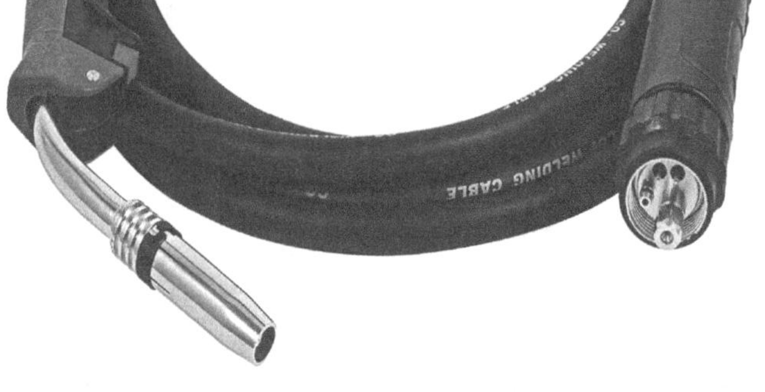

Figure 4.12.3 The GMAW torch and the cable connector.

4.12.1.1 *Welding Torches for Automation and Robotic GMAW*

The criteria for selecting GMAW torches for robotic welding operations are the same criteria as for both semiautomatic welding.

The torch must be of a physical size to move between tooling, holding clamps, and it must also be flexible enough to access hard to reach locations. To meet the demands of the heavy-duty production associated with the robotic welding, a number of torch configurations are available that incorporate long torch barrels, small diameter nozzles, and torch exchange systems. Whether or not to proceed with a water-cooled or air-cooled torch depends, again, on the demands of productivity, and the preventive maintenance program employed.

A unique feature typical for the robotic welding is the collision sensor and a breakaway mounting to the end of the robot arm. These are standard attachments for all robotic welding systems, they are fitted to limit damage to a system in the event of a crash.

4.12.1.2 *The Wire Drive and Accessories*

GMAW wire drive designs provide for the use of a wide range of solid or metal-cored electrodes, ranging from 0.025" to 0.0625" (0.6 to 1.6 mm). The wire feed speed can be pre-set via a digital readout or a calibrated marking system on the wire feed speed control. The ability to provide a precise wire speed is important to good a welding procedure control. Most standard wire drives for GMAW provide a permanent magnet motor, which in turn provides for fast starting and stopping of the feed system.

The effective range of wire feed speed is important, and most wire-drivers can give a range of 70 to 800 inches per minute, (or 2 to 20 M/min.) of wire feed speed. Higher wire feed speed wire-drivers are also available for applications requiring the delivery of wire feed speed up to 1200 ipm (or 30 M/min.). Optional considerations for wire-driver controls include timers for setting pre-flow and post-flow for the shielding gas. A burn-back control is an important addition and should be considered, to prevent the electrode from sticking to the crater at the end of the weld. Some wire-drivers may provide a cold inch control for safely inching wire electrode through a GMAW torch to the work.

An optional purge control for the shielding gas system, if used provides gas flow in advance of the arc, and displaces the air that may have entered the system. The wire drive system provides a gas solenoid, which activates when the GMAW torch trigger is depressed. Shielding gas pre-flow and post-flow conditions control the solenoid circuit and add gas before and after the arc is established.

Optional water connections are available for use with water-cooled GMAW guns. Many newer machines are on electronic circuitry, and not much moving mechanical parts yet, they generate lot of heat and especially if they are in heavy duty use for log time. Even the welding machines get hot in their operation, not due to the welding heat, but due to the operation within the welding machines itself. Most welding machines are naturally air cooled, smaller machines get the gravity feed air to flow through the machine to prevent them getting from overheated.

The lager machines need some help for the air to flow through all parts of the machine to keep them cool, they use forced air circulation. An electrical motor is used run a fan that provides the air to cool the machine. Same cooling system can also support the nozzle assembly too.

For effective cooling, the air passageway should be designed for free flow of the air from entry to exit ends. Periodic maintenance of the duct and openings must be inspected, and cleaned of dust and any other dirt from the system. In extreme situations a cooling fluid often water is circulated with the help of pump.

Two- or four-roll drive systems deliver the electrode to the welding torch. Two-roll systems are standard with smaller non-industrial systems, but the four-roll system is popular for industrial applications.

A mounting block for the power cable permanently fixes to the GMAW torch receiver of the wire drive. The use of a wire straightening device incorporates the ability of the wire drive to provide three important features for the arc:

1. The straightening device reduces the cast of the spooled, coiled, or bulk electrode used for welding. This is especially important for nickel alloys.
2. Reducing the cast improves electrode placement of the arc in the weld joint.
3. Reducing the cast helps reduce drag in the liner (conduit) of the GMAW torch, and, therefore, will reduce premature wear.

Typical platform mount installations, where the wire drive is located on the top of the power source, will require the selection of a wire reel stand. There are a variety of enclosures designed to provide protection for the electrode in use. The spool mounting spindle design provides variable tension for the electrode package in use, and it incorporates a braking action to prevent unspooling of the electrode at the reel stand. A lift bale provides a mounting feature for extending the reach in a work cell using boom type arms.

4.12.1.3 Special Wire Feeding Considerations

For smaller jobs, smaller wire-spools designed and to match them Spool-guns are available especially for welding lighter metals like aluminum, these spools weigh about 1 and 2 lb. (0.45 and 0.90 kg). Like their bigger cousins, the spool-gun have all the same system for operation, including the wire drive motor, a wire feed speed control, and an electrode enclosure in a comfortable lightweight design.

Aluminum fillers are characterized as softer than steel electrodes, and they have lower stiffness, called column strength. The smaller the diameter of aluminum electrode, 0.030 to 0.047" (0.8 to 1.2 mm), the more difficult it is to feed. As a result of the softer and less stiff characteristics, they generally benefit from either a push-pull or a spool gun feeding system. A spool gun only has to push the electrode for 8" to 10" (200 to 250 mm) to the arc, and a push-pull system is designed with the same principle in mind. In either case, these systems more reliably feed aluminum filler metals than a standard hand held GMAW gun. The robotic welding automation system uses the push-pull systems. Reliable feeding is best accomplished with an assist type of system that reduces the distance that the electrode has to travel from the wire drive to the arc. A standard GMAW gun will require the use of Teflon™ or nylon™ gun liners to permit the delivery of the electrode to the work see Figure 4.12.2 for liners. The GMAW gun cable is kept short, usually they are about 10 ft. (3 meters), this helps trouble free electrode delivery. Pushing aluminum through a GMAW gun is usually restricted to ER 5XXX type aluminum filler alloys 0.0469" (1.2 mm) diameter or greater.

4.12.1.4 Shielding Gas Regulation

Shielding gas or a blend of gases is an essential part of the GMAW process. The gas protects the weld metal, weld pool, and adjoining parent metal from atmospheric contamination, the gases sometimes also add some specific properties to the weld metal. Hence the delivery of a shielding gas to the arc is important to the quality of the finished weld.

To deliver shielding gas at precise amount and at right flow rate is very important. In GMAW process as in any other welding process a regulator is fitted to measure the flow rate of gas. Gas may be supplied through commercially supplied individual cylinders or from a bank of cylinders as a manifold system. IN major consumption situation bulk inert gas is also distributed through a piping system, or even a small plant is installed within the manufacturing facility itself.

At the welding point, a Flow-meter regulator provides two readings to the welder, (1) from a meter that measures internal cylinder pressure, and it allows the welder to recognize how much gas remains in the cylinder. (2) The second measures the rate of flow of the shielding gas as it exits the cylinder. The rate of gas flow is measured in either cubic feet per hour (cfh) or liters per minute (L/min).

A hose connects the regulating device to the gas solenoid contained in the wire drive. A connecting hose extends from the front of the wire drive to a brass nipple located at the GMAW torch. Bulk gas systems or manifold systems connected to piped-in mixes of shielding gas usually include a pressure regulator, which controls line pressure. An adjustable flow meter is then added as a separate item.

The rate of flow for short-circuiting transfer with either CO_2 or a mixed shielding gas is usually 25 - 30 cfh, (12-17 L/min).

For globular transfer or axial spray transfer, the flow rate is nominally set at 35 - 50 cfh (17 - 21 L/min).

Special procedures designed to meet the requirements for electrode diameters greater than 0.0625" (1.6 mm) requires a higher rate of flow.

Helium has lower density. Hence it requires a higher flow rate than those indicated above.

4.12.1.5 Welding Cables and Other Accessories

Welding cables used to carry current from the welding machine to the work and back are also called welding leads, these leads are super flexible large diameter electrical cables.

The lead that is connected to the GMAW torch is described above.

The cable that is connected to the work place often by a clamp, is called **ground cable** or **workpiece lead.**

These work leads need to be very flexible to meet the demands of welding activities, they also need to be very well protected from damage, since they are carriers of heavy current. The flexibility and insulation are provided by a thick rubber covering which is often supported by a layer of reinforcement, by woven fabric layer to provide some rigidity and protection from damage.

Copper leads are more suitable for carrying higher currents, this attribute of copper cable also reduces the diameter cables. But copper cables are heavier than aluminum cables. Aluminum cables are lager in diameter as compared to the copper cable for carrying same current capacity, this is because aluminum can carry only up to 61% current. But the advantage of aluminum is that these cables are lighter in weight.

The electrical capacity is the most key difference between application of copper and aluminum cable. The alloy mix is also determined by the intended use of the welding cables. Copper is considered a better conductor with a higher capacity per volume. However, aluminum has higher capability per weight. The weight difference also is determined by the specified material used.

The leads are produced in various sizes, and identified by universal numbering, the number indicates the diameter of the lead, which in turn indicates the current carrying capacity of the cable. Larger the number, thicker the diameter and lower the current carrying capacity. A set of various types of cable heads are shown in Figure 4.12.1.4 below. The current carrying capacity by diameter and length is clearly brought out in the Table 4.12.1.4 below. This table also brings out another factor, which is the drop in current as the length of the cable is increased. The length shown in the table includes the length of electrode lead and the workpiece lead. So, the point to note here is that, the reduction in lead diameter, and the increase in the lead length reduces the current capacity of the welding lead.

Figure 4.12.1.4 Copper and aluminum welding leads: note the number of fine wires that compose a cable, and the rubber sheathing that covers them.

Table 4.12.1 4 Welding lead current carrying capacity.

Significance of cable (welding lead) diameter and length and current carrying capacity					
	Lead diameter		**Cable length**	**Cable length**	**Cable length**
Lead no.	**Inch**	**mm**	**Amperes**	**Amperes**	**Amperes**
4/0	0.959	24.4	600	600	400
3/0	0.827	21.0	500	400	300
2/0	0.754	19.2	400	350	300
1/0	0.720	18.3	300	300	200
1	0.644	16.4	250	200	175
2	0.604	15.3	200	195	150
3	0.568	14.4	150	150	100
4	0.531	13.5	125	100	75

Note the drop in current as the length of the lead increases.

The corresponding drop in voltage is very low, if all connections are tight and secured, the drop in a copper cable is about 4 volts.

4.12.1.6 Welding Personal Protecting Equipment

Welding Helmets

The most important PPE for a GMAW welder and operators can be easily said to ne the welding Helmet. The arc welding helmet is used to protect the face and eyes from sparks, spatters and the heat and ultraviolet rays emanating from the electric arc of the welding. While the purpose of both shield and helmet is same, both these terms are often interchangeably used. The difference in shield and helmet is that shields are often hand held, while the welding helmet is worn on the head, leaving both hands of welder free to carry out their work. Obviously, the shields are suitable for smaller work, where the welders have to inspect their work frequently, and work that requires frequent viewing of the work for example, when fitting and aligning components prior to welding, or tacking as preparation for welding etc.

The helmet consists of a head mount that is supported by a head cover either full or partial, and a head band that wraps around the head on the forehead. The side knobs on the head band allow adjustment to fit different sizes of head. The head and side coving protects the face in general, while the front of the helmet/shield has a window where a dark filter glass of suitable rating is fitted between two plain glass to filter UV rays, and heat to enter in the eye and protects the eye and face from any damage to the welders' eyes.

While welding activity is very safe if proper precaution is taken, the welding arc contains some very damaging rays, they can burn the skin, and damage the eye. The exposure to UV rays in the welding arc can cause eye pain, eye watering, and swelling with irritation as if sand is in the eye, and pain and discomfort can last for about 10 to 20 hours after arc exposure. However, exposure to infrared rays, can injure eyesight. Hence protection is extremely necessary.

The filter glass or lens is an important a part of the shield or helmet, it is required to protect eye of the welder from damaging UV rays, and heat of the welding, while the welder is able to see the arc and progress of his work. Due to its very critical role much study has been done and determined that various shades of lens are required to protect differing intensities of welding arcs.

The density of filter shades is such that the welder cannot see through it until arc is struck.

Helmets have been developed with some "automatic" features, battery is used to introduce photoelectric cells inbuilt in the helmet, and this allows the lens to be clear until the arc is struck. This is a great advantage to the welders precisely locate their electrode tip on the weld location and the trike the welding arc. This helps produce cleaner welds, reduce damage to the parent metal near welds, less of arc strike outside the weld zone.

For further eye protection from back flashes some welders also wear a pair of ordinary welding glasses that has # 1 or # 2 lenses on them, this allows them to inspect the weld before or after welding, chip slag etc., from the weld.

Clarity of the vison especially that of the arc and weld pool is of utmost importance. More developed technologies and concepts have been adopted by manufacturers. In that a is the concept of 1/1/1/2 optical clarity rating, this allows for a lighter state while not welding, and keeping the helmet down thus maximizing safety and productivity. The helmet has four arc sensors and four modes: weld, cut, grind, and X-Mode. X-Mode electromagnetically senses the weld to eliminate sunlight interference and continuously detects the arc even if sensors are blocked. Gen 3.5 headgear with comfort cushion has an ergonomic design that provides extensive adjustability, settings, and enhanced support. Digital controls easily allow welder to adjust shade, delay and sensitivity. AutoSense™ eliminates issues related to setting helmet sensitivity by allowing the welder to push and hold the AutoSense™ button to automatically set the helmet sensitivity for their environment. Auto-on/off power control triggers lens at the strike of an arc.

More complex jobs require more complex safety equipment and PPE. Helmets have been developed and are used where air quality around the welding work is improved by introducing fresh air into the helmet, and some are fitted with air filters that provides clean air for the welder to breath.

The following table gives the recommended safe shade numbers for various welding operations. The user must experiment with the most suitable and safe shade number that suits, the Table 2.4.1 gives an indicator and may work as the safe trials tart point in selecting most suitable shade for individual safety in welding.

4.12.1.7 Other Essential Clothing for Welders

In welding operation, the molten weld metal often in the form of spatter fly all over and can easily land on the person of the welder. This can cause sever burn and injuries. Welders need protection from such injuries. Most of these accessories are made from leathers, hence they are often collectively referred as *leathers.*

- Welding gloves,
- Welding Gauntlet sleeves,
- Aprons,
- Leggings,
- A Jacket, especially if welding in overhead position
- A cape, especially if welding in overhead position
- Heat protected gloves, these are insulated gloves in addition to the leather welding mentioned above, and used when welding on hot surfaces and for longer time.

4.13 Welding Various Metals

The GMAW process is sued for welding most of the commercially bused engineering metals. That includes carbon steel, low alloy steels, stainless steels of various grades, various grades of aluminum, copper and copper alloys, and nickel alloys.

In the subsequent paragraphs a basic discussion on welding various metals is given. The tables are specifically tailored to initiate training effort, but the information in those tables is much more than just for training, it is an introduction to the GMAW welding of various by number of transfer modes.

4.13.1 Carbon Steel

For the steel welding, the welding machine is usually setup for DC positive polarity. The shielding gas, which is usually carbon dioxide or mixtures of carbon dioxide and argon, see the reference tables for more accuracy, these shielding gases protects the molten metal from reacting with the atmosphere. Shielding gas flows through the gun and cable assembly and out the gun nozzle with the welding wire to shield and protect the molten weld pool. If the molten metal and the immediate surrounding metal is exposed to the atmosphere consisting of the reactive gases like oxygen, nitrogen, and hydrogen in the atmosphere, the weld will not be a metallurgically compatible joint, with the desired strength. The inert gas usually continues to flow for some time after welding is completed, this is to keep protecting the metal as it cools. A slight breeze can blow the shielding away and cause porosity, therefore welding outdoors is usually avoided unless special windscreens are erected.

When correctly welded, the weld appearance and the quality is excellent, and or this reason alone the GMAW process is welders' favorite process to weld. Use of good welding technique brings out excellent results. The properly made finished weld has no slag and virtually no spatter. A "push" gun angle of about 10° is normally used, this enhances the gas coverage and gets the best result. Cleanliness of the material being welded is of utmost importance, If the material is greasy, dirty, rusty, or painted it must be cleaned by grinding until bare metal shines up. GMAW process welding is used to join most of the carbon and alloy steels.

Table 4.4.1.1 also functions as the basic WPS for welding carbon steel and low alloy steel with short-circuit mode, and similarly the Table 4.4.1.4.1 as WPS for welding carbon steel and low alloy steels using spray transfer mode and other charts that give guidance on selection of type of shielding gas or gas mix are presented earlier. They should be referred to select suitable Current range, the size and type of welding wire suited to the grade of steel being welded, and the shielding gases to complement the training efforts. The welder may be required to adjust the initial setup to exactly suit their welding conditions and abilities.

4.13.2 Aluminum and Aluminum Welding

Metallurgically aluminum is a very different material as compared to any type of steel, it is not different just in its physical strength and appearance, it is also metallurgically different. That difference makes it to exhibit some very different behavior when worked especially when heated.

It is appropriate to get some understanding of aluminum metallurgy, along with the welding behavior if subjected to heat as is done in welding. Once we have some understanding of these basics working with and welding aluminum would be much easier.

4.13.2.1 Understanding Aluminum

Steel is often considered the 'default' metal to for structural construction and therefore welding. During welding the steel gives stage wise indication of what is happening to it by the application of heat, by change of color, sweating, and then melting and slowing building a weld pool. There is some however small, time laps in these stages, but aluminum has none of these stages. The human brain is used to expect somewhat similar temperature indications from the aluminum when heated during welding aluminum. But aluminum does not show such changes with applied heat. Therein lies the fundamental challenge of welding aluminum.

Welding aluminum is different than welding steel because of one very basic property, when heated aluminum does not change color as does steel, this lack of color change, visible weld pool fails the brain, it does not give heat perspective to the welder especially to the new welder, welder does not know what is happening with the material as it is heated, and it simply collapses under the heat.

Aluminum is much greater conductor of heat, that changes the behavior of heat dwell on the parts to be welded.

Aluminum is an active metal and it forms oxides while being heated for welding, it is harder to create a welding pool, suitable for welding aluminum. When combined with the metal's high heat conductivity and low melting point, it is very easy for a new welder to completely melt the aluminum pieces involved in the process. As a result, two very important steps need to be taken.

1. First step to arc welding aluminum is to clean the base metal of any oxides or solvent oils, and prevent oxide formation during welding. And
2. The second step is to be mindful of aluminum's behavior under heat,
    i. Aluminum does not change color,
    ii. Aluminum does not show molten pool,
    iii. Aluminum has high conductivity of heat,
    iv. Aluminum has very low melting point as compared to the steel.

On the positive side, if the welding technique is mastered, the welding of aluminum is less energy intensive, and therefore easier than welding steel.

Another important point to note is that most of the welding machines are tailored to weld steel, they are mostly calibrated and set with those parameters. So, it is important to know the features of the welding machine, and if required reprogram the machine to suite aluminum welding.

Alloy 4043 is one of the oldest and most widely used welding and brazing alloys. ER 4043 wire-electrode can be classed as the general-purpose welding electrode that can be used for welding a variety of aluminum grades, like, 3003, 3004, 5052, 6061-T4, 6061-T6, 6063-T6 and 2014-T6.

The alloy has about 4% to 6 % silicon in it, silicon is a wetting agent, and it allows the weld metal fluidity and this property makes the electrode use very welder-friendly. The silicon additions result in improved fluidity (wetting action) to make the alloy a preferred choice by welders. The weld metal is less sensitive to weld cracking, and produces brighter, almost smut free welds.

4.13.2.2 Designation of Aluminum Welding Wires

For aluminum the electrode and welding wire designations system differs from the steel welding wire system, for example, a commonly used welding wire ER 4043 has somewhat different description. The designation simply replicates the grade of aluminum the specific welding wire is used for welding, Table 3.20.1 lists the aluminum welding alloy designation system. As we will note, the metallurgically alloys of aluminum are as different in properties as we learned and know the steel and its various alloys are. A near matching welding wire may be used for welding an aluminum alloy, however a welding wire may be used to weld a number of different aluminum alloys. Not all aluminum alloys would have exactly matching welding wires.

ER 4043 welding wire contains approximately 5% silicon. The presence of silicon gives good wetting, properties, and better fluidity to the weld metal, resulting in better weld penetration, and arc control. The ER 4043 wire is frequently used to weld a wide spectrum of aluminum grades such as, alloys 3003, 3004, 5052, 6061, 6063 and Aluminum casting alloys 43, 355, 356, and 214.

4.13.3 Aluminum Metallurgy and Grades

Aluminum as a metal is graded in different number system, below is a brief description aluminum and its alloys. The numbering system predominantly indicates the primary alloying element in the alloy. For the alloys with the suffixes like 'T' for example alloy 5052T, indicates that the alloy is in heat treated condition.

Aluminum Alloys and Their Characteristics

There are seven series of wrought aluminum alloys, and as a welder, or a welding engineer, or manager of aluminum fabrication, it is imperative to know these alloys and their differences and understand their applications and characteristics.

4.13.3.1 1xxx Series Alloys

These are non-heat treatable alloys. In unheat-treated condition the ultimate tensile strength of these alloy have range from 10ksi, to 27 ksi. The 1xxx series is often referred to as the pure aluminum series, this is because they are required to have a minimum of 99.0% aluminum.

These grades are weldable. However, because of their purity level of aluminum, they have very narrow melting range. This demands both welders' ability, engineering considerations to develop an acceptable welding procedure. When considered for fabrication, these alloys are selected primarily for their superior corrosion resistance such as in specialized chemical tanks and piping, or for their excellent electrical conductivity as in electrical bus-bar applications. These alloys have relatively poor mechanical properties, and would seldom be considered for general structural applications. These base alloys are often welded with matching filler material or with 4xxx filler alloys dependent on application and performance requirements.

4.13.3.2 2xxx Series Alloys

2xxx series alloys are heat treatable alloys. Their ultimate tensile strength ranges from 27 ksi, to 62 ksi. This increase of strength is attributed to the alloying element, like copper. These are aluminum and copper alloys where copper additions vary from 0.7% to 6.8%. The high strength, and better temperature resistance, these high-performance alloys are often used for aerospace and aircraft structural applications. They have excellent strength over a wide range of temperature.

Some alloy grades with in this series are considered non-weldable by the arc welding processes because of their susceptibility to hot cracking and stress corrosion cracking, both due to the presence of copper in them. However, there are others that can be successfully welded with the correct welding procedures. These base materials are often welded with high strength 2xxx series filler alloys, designed to match their performance, but can sometimes be welded with the 4xxx series fillers containing silicon or silicon and copper, dependent on the application and service requirements.

4.13.3.3 3xxx Series Alloys

The 3xxx Series alloys are an alloy of aluminum and manganese. 3xxx series alloys are non-heat treatable. These alloys have relatively low tensile strength, where their ultimate tensile strength ranges between 16 ksi and 41 ksi. These are the aluminum and manganese alloys contain from 0.05% to 1.8% manganese. With their moderate strength, they have good corrosion resistance, good formability and are suited for use at elevated temperatures. Among their various uses are the industrial component like heat exchangers in vehicles and power plants, and of course their suitability for manufacturing cooking utensils cannot be ignored. Their moderate strength, however, often precludes their consideration for structural applications. These base alloys are welded with 1xxx, 4xxx and 5xxx series filler alloys, dependent on their specific chemistry and particular application and service requirements.

4.13.3.4 4xxx Series Alloys

4xxx series alloys are alloys of aluminum and silicon, silicon content varies from 0.6% to 21.5% in various alloys in this series. This series contains both heat treatable as well as non-heat treatable alloys. Their ultimate tensile strength varies from 25 ksi, to 55 ksi.

The effect of silicon added to aluminum reduces its melting point of the alloy, and improves the fluidity of the molten metal, making is very suitable for welding. These characteristics are desirable for filler materials used for both fusion welding and brazing. Consequently, this series of alloys is predominantly found as filler material. Silicon, independently in aluminum, is non-heat treatable; however, when paired with magnesium or copper the alloys can be heat treated for improved strength. Number of these silicon alloys have been designed to have additions of magnesium or copper, which provides them with the ability to respond favorably to solution heat treatment. Typically, these heat treatable filler alloys are used only when a welded component is to be subjected to post weld thermal treatments.

4.13.3.5 5xxx Series Alloys

5xxx series alloys are non-heat treatable alloys. These are alloy of aluminum and magnesium, where magnesium is from 0.2% to 6.2%. The alloys have very high strength in all non-heat treatable aluminum alloys. They have ultimate tensile strength from 18 ksi to 51 ksi.

Alloys in this series are readily weldable, combined with good strength, and good weldability they are used for variety of fabrication and construction. They are used for a wide variety of applications such as Tankers, Boats, Shipbuilding, Transportation, pressure vessels, bridges and buildings. The magnesium base alloys are often welded with filler alloys, which are selected after consideration of the magnesium content of the base material, and the application and service conditions of the welded component. Alloys in this series with more than 3.0% magnesium are not recommended for elevated temperature service above 65°C (150° F) because of their potential for sensitization and subsequent susceptibility to stress corrosion cracking. Base alloys with less than approximately 2.5% magnesium are often welded successfully with the 5xxx or 4xxx series filler alloys. The base alloy 5052 is generally recognized as the maximum magnesium content base alloy that can be welded with a 4xxx series filler alloy. Because of problems associated with eutectic melting and associated poor as-welded mechanical properties, it is not recommended to weld material in this alloy series, which contain higher amounts of magnesium with the 4xxx series fillers. The higher magnesium base materials are only welded with 5xxx filler alloys, which generally match the base alloy composition.

4.13.3.6 6XXX Series Alloys

Alloys in 6xxx series are heat treatable, and have the ultimate tensile strength from 18 ksi to 58 ksi. These are the aluminum, magnesium and silicon alloys, where magnesium and silicon additions for each element is around 1.0%, generally less than 1%. These alloys are the most welding and fabrication friendly alloys, they are predominantly used in the form of aluminum extrusions, for use in strength bearing structural components.

The addition of magnesium and silicon to aluminum produces a compound of magnesium-silicide, which provides this material its ability to become solution heat treated for improved strength. These alloys are naturally solidification crack sensitive, and for this reason, autogenous welding of these is not recommended, the addition of adequate amounts of filler material during the arc welding process is essential in order to provide dilution of the base material, to preventing the hot cracking problem. They are welded with both 4xxx and 5xxx filler materials, dependent on the application and service requirements.

4.13.3.7 7XXX Series Alloys

7XXX series alloys are heat treatable. These are the aluminum and zinc alloys, where zinc additions range from 0.8% to 12.0%. This gives the alloy group the highest strength aluminum alloys with ultimate tensile strength ranging from 32 ksi, to 88 ksi. These alloys are often used in high strength applications such as aircraft, aerospace, and competitive sporting equipment. Like the 2xxx series of alloys, this series incorporates alloys which are considered unsuitable candidates for arc welding, and others, which are often arc welded successfully. The commonly welded alloys in this series, such as 7005, are predominantly welded with the 5xxx series filler alloys.

4.13.4 The Aluminum Alloy Temper and Designation System

Since we are discussing about welding of aluminum and its various alloys, it is important to learn about the commercial availability of the material, in that the first step would be to know how the material is designated and identified. The Aluminum Association Inc. https://www.aluminum.org registers, maintains, and allocates designations to aluminum alloys, and it updates the register with any new developments. Currently, it is estimated that there are over 400 wrought aluminum, and wrought aluminum alloys, and over 200 aluminum alloys in the form of castings and ingots registered with the Aluminum Association.

The chemical composition limits for all alloys are that are registered are included in the Aluminum Association's two books. As described below, these books are named by the colors.

1. **Teal Book** entitled "International Alloy Designations and Chemical Composition Limits for Wrought Aluminum and Wrought Aluminum Alloys" and
2. The **Pink Book** entitled "Designations and Chemical Composition Limits for Aluminum Alloys in the Form of Castings and Ingot".

These two publications can be extremely useful to the welding engineer, when developing aluminum welding procedures, and when the consideration of chemistry and its association with crack sensitivity is of importance. They can be very useful in selection of material grade for specific service conditions.

Aluminum alloys can be categorized into a number of groups based on the particular material's specific characteristics, theses could be its ability to respond to thermal and mechanical treatment, or the primary alloying element added to the aluminum to make a specific alloy. When we consider the numbering and identification system used for aluminum alloys, the characteristics are identified through these numbers. The wrought and cast aluminums have different identification systems. The wrought alloys having a 4-digit system, and the castings have a combination of a 3-digit and 1-decimal place system.

4.13.5 Wrought Alloy Designation System

The 4-digit (XXXX) wrought aluminum alloy identification system is explained as the following.

The first digit (Xxxx) indicates the principal alloying element, which has been added to the aluminum alloy and is often used to describe the aluminum alloy series, i.e., 1000 series, 2000 series, 3000 series, up to 8000 series as shown in Table 3.20.1.

The second single digit (xXxx), if different from 0, indicates a modification of the specific alloy, and the third and fourth digits (xxXX) are arbitrary numbers given to identify a specific alloy in the series.

For the example of the above explanation let us consider alloy 5183, the number 5 indicates that it is of the magnesium alloy series, the 1 indicates that it is the 1st modification to the original alloy 5083, and the 83 identifies it in the 5xxx series.

The only exception to this alloy numbering system is with the 1xxx series aluminum alloys, known as pure aluminums grades, in which case, the last 2 digits provide the minimum aluminum percentage beyond the basic 99% purity of aluminum. So, in the alloy 1350 would mean that the alloy contains 0.50% more purity above the 99% aluminum that means that alloy 1350 has aluminum purity of 99.50%.

4.13.6 Cast Alloy Designation

The cast alloy designation system is based on a 3 digit-plus decimal designation xxx.x, for example, 356.0. The first digit (Xxx.x) indicates the principal alloying element, which has been added to the aluminum alloy as shown in Table 3.21.2.

The second and third digits (xXX.x) are arbitrary numbers given to identify a specific alloy in the series. The number following the decimal point is a system of identification where .0 is casting and 0.1 or 0.2 are ingots. If a capital letter prefix is used it indicates a modification to a specific alloy. The example of the above description, Alloy - A356.0 the capital A (Axxx.x) indicates that the original alloy 356.0 has had one modification. The number 3 (A3xx.x) indicates that it is of the silicon plus copper and, or magnesium series. The 56 (Ax56.0) identifies the alloy within the 3xx.x series, and the .0 (Axxx.0) indicates that it is a final shape casting and not an ingot.

4.13.7 The Aluminum Temper Designation System

From the description and table presented above, one can see that there are different series of aluminum alloys, what is important to note is the considerable differences in their characteristics. The first point to recognize, after understanding the identification system, is that there are two distinctly different types of aluminum within the series described above. These are the Heat Treatable Aluminum alloys that can be imparted strength through the treatment of heat and cooling cycle. The other group consist of the non-heat Treatable Aluminum alloys. This distinction is particularly important when considering the effects of arc welding on these two types of materials.

- The 1xxx, 3xxx, and 5xxx series wrought aluminum alloys are non-heat treatable, but they are strain hardenable only.
- The 2xxx, 6xxx, and 7xxx series wrought aluminum alloys are heat treatable and,
- The 4xxx series consist of both heat treatable and non-heat treatable alloys.
- The Cast alloys 2xx.x, 3xx.x, 4xx.x and 7xx.x series are heat treatable. Strain hardening is not generally applied to castings.

The heat treatable alloys acquire their optimum mechanical properties through a process of heat treatment, the most common heat treatments are the Solution Heat Treatment, and the Artificial Aging.

Solution Heat Treatment is the process of heating the alloy to an elevated temperature to around 482°C (about 990°F) in order to put the alloying elements in the metal into solution. This is followed by quenching, usually in water, to produce a supersaturated solution at room temperature. Solution heat treatment is usually followed by a process called aging. Aging is the precipitation of a portion of the elements or compounds obtained from a supersaturated solution in order to yield desirable properties. The aging process is divided into two types: aging at room temperature, which is termed natural aging, and aging at elevated temperatures termed artificial aging. Artificial aging temperatures are typically about 160°C (about 320°F). Many heat treatable aluminum alloys are used for welding fabrication in their solution heat treated and artificially aged condition. So, unless stated otherwise, it is assumed that the post weld properties may be required to match that of the base metal welded.

The non-heat treatable alloys acquire their optimum mechanical properties through Strain Hardening. Strain hardening is the method of increasing strength through the application of cold working. The Temper Designation System addresses the material conditions called tempers. The Temper Designation System is an extension of the alloy numbering system and consists of a series of letters and numbers which follow the alloy designation number and are connected by a hyphen. Examples: 6061-T6, 6063-T4, 5052-H32, 5083-H112. Table 3.21.3 details the letters used for temper designations, and meaning of those letters.

Further to the basic temper designation, shown in the Table 3.21.3 above, there are two subdivision categories.

i. addressing the "H" Temper – Strain Hardening, and
ii. The other addressing the "T" Temper – Thermally Treated designation.
iii. H Temper – Strain Hardened

- The first digit after the H indicates a basic operation:

  H1 – Strain Hardened Only.
  H2 – Strain Hardened and Partially Annealed.
  H3 – Strain Hardened and Stabilized.
  H4 – Strain Hardened and Lacquered or Painted.

- The second digit after the H indicates the degree of strain hardening:

  HX2 – Quarter Hard,
  HX4 – Half Hard
  HX6 – Three-Quarters Hard
  HX8 – Full Hard
  HX9 – Extra Hard

- Subdivisions of T Temper – Thermally Treated

  T1 - Naturally aged after cooling from an elevated temperature shaping process, such as extruding.
  T2 - Cold worked after cooling from an elevated temperature shaping process and then naturally aged.
  T3 - Solution heat treated, cold worked and naturally aged.
  T4 - Solution heat treated and naturally aged.
  T5 - Artificially aged after cooling from an elevated temperature shaping process.
  T6 - Solution heat treated and artificially aged.
  T7 - Solution heat treated and stabilized (overaged).
  T8 - Solution heat treated, cold worked and artificially aged.
  T9 - Solution heat treated, artificially aged and cold worked.
  T10 - Cold worked after cooling from an elevated temperature shaping process and then artificially aged.

Additional digits indicate stress relief.
Examples:

TX51 or TXX51 – Stress relieved by stretching.
TX52 or TXX52 – Stress relieved by compressing.

4.13.8 Welding Aluminum

As said in the introduction of the aluminum welding. Welding aluminum requires much more than just changing to aluminum wire.

Getting used to welding steel is one step that can make the welder used to welding in general, however getting used welding steel could come as a limitation too, when moving to welding aluminum they need to realize with the fact that aluminum behaves much different than steel under heat.

Cleanliness of the wire and base metal are critical. Wiping the material with acetone on a clean shop rag is the best way to clean the metal for welding. Stainless steel wire brushes that have only been used on aluminum, are the tools essential for welding aluminum.

Drive roll tension and gun length should be reduced to the optimum. A Teflon, nylon or similar gun liner is used to reduce friction in feeding the wire. Aluminum is a very soft metal, damage to the wire should be avoided and for that end, it requires aluminum drive rolls that have a U-groove to guide the welding wire through.

Shielding gas used to weld most of the aluminum alloys is the 100% pure Argon, also see the Tables 4.5.1 and 4.5.5.1 more information on shielding gases. Technique of manipulating the welding gun movement during the welding is required to be learned.

4.13.8.1 Electrode Selection

The filler metal must produce a weld deposit that either closely matches the mechanical properties of the base metal or provide enhancement to the base metal property. The deposit must also be free of discontinuities.

There can be several job-related advantages, and also restrictions that may push to select one welding wire from the other. These may include if the weld can be heat treated after welding to improve the properties, or if the weld needs color matching to the parent metal. All these considerations are to be evaluated for selecting suitable welding electrode for specific job. The Tables 4.5.1 and 4.5.5.1 and the discussions above, give some guidance for the election of weld metal. AWS Specification A5.10 includes filler wires for aluminum and aluminum alloy welding. Referencing the specification is always the better way to select the best weld metal for the job at hand.

4.13.9 Welding Stainless Steel with the Gas Metal Arc Process

Stainless steels may be welded by the gas metal arc process, using either spray arc, short - circuiting or pulsed arc transfer. Copper backup strips are often used for welding thin sections of stainless steel. And some Backup is also needed when welding 0.25-inch (6 mm) or thicker plate if welded from only one side. Use of backing strips are common when welding on thinner sections, square butt welds, this practice prevents the weld-metal drop-through. The backing strip use is recommended for thinner sections and when the joint fit-up is poor.

Protecting the weld and heat affected zone from atmospheric contamination is of utmost importance when welding stainless steel. This step protects the weld from losing the material's basic properties especially the corrosion resistance properties. The shielding gas protection must be maintained till the weld-metal has solidified.

Oxygen picked up by the molten metal may reduce the corrosion resistance and ductility of the stainless steel as it cools. To prevent this, the underside of the weld should be shielded by an inert gas such as argon. The supply of the shielding gas is made through a fixture, that is inserted in the pipe and gas is connected to it.

Electrode diameters as large as 0.9 inch, but usually less than 0.062 inch, are used with relatively high currents to create the spray arc transfer. A current of approximately 300-350 amperes is required for a 0.062-inch electrode, depending on the shielding gas and type of stainless wire being used. See Table 4.5.5.1 for more details.

The degree of spatter is dependent upon the composition and flow rate of the shielding gas, wire feed speed and the characteristics of the welding power supply. Refer the discussion on the shielding gases earlier in the chapter.

DCEP current is used for most stainless steel GMAW process, shielding gas mix of argon with 1 or 2%-oxygen gas is most common practice however, refer Table 4.5.5.1 for more information.

Suggested procedures for GMAW 200 and 300 series stainless steels in the spray transfer mode are given in those referenced tables.

When welding with the semiautomatic gun, the pushing techniques is more helpful. In this process the operator's hand is exposed to more radiated heat, but it provides better visibility of the weld. Other techniques may include moving the gun back and forth in the direction of the joint and at the same time moved slightly from side to side. This technique is more useful for thicker metal section for example 0.25 inch (6 mm) or more.

For the overhead, and horizontal position short circuit weld metal transfer mode is highly recommended. The short circuit mode may be changed after the first passe like root and one or two more passe are successfully done.

Power supply units with slope, voltage and inductance controls are recommended for the welding of stainless steel with short-circuiting transfer. Inductance, in particular, plays an important part in obtaining proper puddle fluidity.

Welding with short circuit mode

The shielding gas often recommended for short-circuiting welding of stainless steel contains 90% helium, 7.5% argon and 2.5% carbon dioxide. This gas mix gives the most desirable bead contour while keeping the CO_2 level low enough to prevent influence the corrosion resistance properties of the metal. The Figure 4.13.8.1 is the indicative representation of bead contour in different weld positions.

High inductance in the power supply output is beneficial when using this gas mixture. Also refer Table 4.5.1 and Table 4.5.5.1 and discussions on selection of gas mixes for more detail on gas mix selection.

Single pass welds may also be made using argon-oxygen and argon-CO_2 gas mixes. However, arc voltage for steady short-circuiting transfer may be about 6 volts lower than for

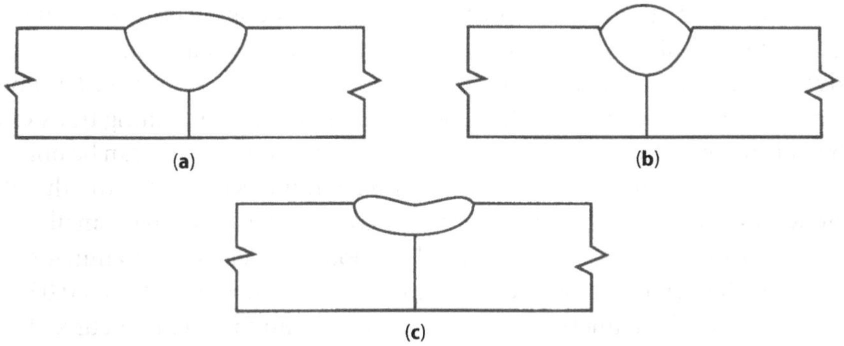

Figure 4.13.8.1 (a) Contour of a weld bead in the flat position with the work horizontal; (b) welding slightly uphill; (c) welding slightly downhill.

the helium-based gas. The colder arc may lead to lack of fusion defects, adjust the voltage as required.

Note that in short circuit transfer process, chances of carbon pick-up increase due to the presence of CO_2 gas in the shielding gas, this can severely affect the corrosion resistance properties of stain less steel in multi-pass welds.

For welding stainless steel by the short circuit mode, the wire extension or stick-out should be kept as short as possible. Other recommended techniques are given below, they can be modified to match the actual welding conditions.

- Backhand progression welding is usually easier on fillet welds and it produces much cleaner weld.
- Forehand progression welding is recommended for the butt welds.
- Outside corner welds may be made with a straight motion.
- A slight backward and forward motion along the axis of the joint can be practiced to get better weld penetration and bead profile.
- Short circuiting transfer welds on stainless steel made with a shielding gas of 90% He, 7.5% Ar., 2.5% CO_2 mix, gives better corrosion resistance and coalescence.

Welding with the pulsed arc mode

Normally Spray -pulsed arc is used for welding stainless steels. In this mode one small drop of molten metal is transferred across the arc for each high current pulse of weld current. The high current pulse must be of sufficient magnitude and duration to cause at least one small drop of molten metal is detached from the welding wire by the pinch effect caused by the magnetic forces of the arc, the droplet is then propelled through the arc to the weld puddle.

During the low current portion of the weld cycle, the arc is maintained and the wire is heated, but the heat developed is not adequate to transfer any metal. For this reason, the time duration at the low current value must be limited otherwise metal would be transferred in the globular mode.

Wire diameters of 0.045 and 0.035 inch are most commonly used with his process. Gases for pulsed arc welding, such as argon plus 1% oxygen are popular, the same as used for spray arc welding. These and other wire sizes can be welded in the spray transfer mode at a lower average current with pulsed current than with continuous weld current.

The advantage of this is that thin material can be welded in the spray transfer mode which produces a smooth weld with less spatter compared to the short-circuiting transfer mode.

Another advantage is that for a given average current, spray transfer can be obtained with a larger diameter wire than could be obtained with continuous currents. And this affects the cost of the welding, as the larger diameter wires are less expensive than smaller sizes, and the lower ratio of surface to volume reduces the amount of deposit contamination.

The electrode diameters for gas metal arc welding are generally between 0.030 and 0.09 inch. For each electrode diameter, there is a certain minimum welding current that must be exceeded to achieve spray transfer, see Table 5.5.1. For example, when welding stainless steel in an argon-oxygen atmosphere with 0.045-inch diameter stainless steel electrode, spray transfer will be obtained at a welding current of about 220-amp DCRP.

It must be kept in mind that, along with the minimum current, a minimum arc voltage must also be obtained. This is generally between 22 and 30 volts.

Welding wire or electrodes come on spools of varying in weight ranging between 2 and 60 lb. The electrodes for welding the straight chromium stainless steels and austenitic electrodes that contain more than the usual amount of silicon, are better wetting properties, especially when used with the short-circuiting transfer process.

Some stainless-steel weld metals during welding have a tendency toward hot shortness or tearing when they contain little or no ferrite Type 347, is one such alloy. When welding these types, Hot cracking is possibility. To avoid Hot Cracking, stringer beads are strongly recommended, this may increase the number of weld passes required to complete the weld. Stringer beads use lower current, low heat, and that reduces the contraction stresses, hence cooling is more rapid through the hot short temperature range. By the use of stringer bead will change the weld profile, the weld would be more convex than the normal, this also helps in reducing the cooling rate. Additional care should be taken to fill craters.

If weld metal hot cracking is a serious concern, then welding with the short-circuiting transfer mode must be selected. This mode due to its lower current welding, lowers the dilution from the base metal. Excessive dilution may produce a highly austenitic weld metal, low in ferrite, which can promote weld metal cracking.

When welding ferritic and martensitic types stainless steels, to the austenitic types it is desirable to following recommendations.

a. Use a single bevel joint to obtain minimum joint reinforcement.
b. Use low heat input short-circuiting transfer to minimize the arc deflection encountered when welding magnetic to nonmagnetic steels.
c. For uniform fusion, ensure that the wire is kept centered over the nonbeveled edge of the joint.

4.13.10 Introduction to and Understanding Stainless Steel

Stainless steels contain least 10% chromium. The term stainless comes from the metal's clean and bright surface appearance, that does not corrode from outside. This resistance to corrosion is possible due to the uniformly spread formation of a thin layer, but dense chromium oxide film. This oxide film provides corrosion resistance and prevents further oxidation.

Based on the alloying elements, and degree of corrosion resistance in various corrosive environments, there are number of varieties of stainless steels. These varieties are often referred as "Type". Depending on the metallurgical structure of the type, they are grouped as the austenitic, martensitic, ferritic or even a combination of metallurgical structures called Duplex steels, and by virtue of their corrosion resistance properties attained by specific type of heat treatment.

1. Austenitic stainless steels include the 200 and 300 series of which type 304 is the most common. The primary alloying elements in this group are the chromium and nickel.
2. Ferritic stainless steels are non-hardenable Fe-Cr alloys. Types 405, 409, 430, 422 and 446 are the prime examples of this group.
3. Martensitic stainless steels are similar in composition to the ferritic group but contain higher carbon, and lower chromium to permit hardening by heat treatment. Types 403, 410, 416 and 420 represents this group.
4. Duplex stainless steels are supplied with a microstructure of approximately equal amounts of ferrite and austenite. They contain roughly 24% chromium and 5% nickel, but this varies as various newer grades have been developed. This special type of metallurgical structure is very difficult to maintain on heating and welding. They are not part of any of the previously described stainless steel types of numbering system.
5. Precipitation hardening (stainless) steels contain alloying elements that include aluminum. Aluminum permits hardening of these alloys by heat treatment, a process in that the steel is heated to bring the steels structure in a solution stage, and then capturing that solution structure often by quick cooling. The properties so attained by the heat treatment may be further enhanced by the aging heat treatment. These precipitation hardening alloys have sub group of type where the structure may fall in to either the martensitic, semi-austenitic, and austenitic, structures. Steels in this group are identified as the 600-series of stainless steels, types 630, 631, 660 represent this group.

The weldability and more details relating to welding of these groups are described furrier in the chapter.

4.13.11 Alloying Elements and Their Impact on Stainless Steel

In stainless steel, among others, the primary metallurgical structure are austenite and ferrite. This single aspect has very important impact on the weldability, selection of suitable welding wire or electrode. And estimation of corrosion resistance, and cracking properties of the weld and heated materials.

The balance of alloying elements which promote formation of two structures decide the specific type of alloy they are. In this respect we concentrate on following aspects directly related to welding. One aspect that has most important is the content of ferrite structure in the weld metal.

4.13.11.1 *The Elements that Promote Ferrite are*

- Chromium, is the basic element that is responsible for steels' corrosion resistance properties.
- Molybdenum, promotes structures that imparts the high temperature strength, and increases corrosion resistance.
- Niobium (Columbium), and Titanium are strong carbide formers.

4.13.11.2 *The Elements that Promote Austenite are*

- Nickel imparts high temperature strength, and promotes toughness, and ductility.
- Carbon is a carbide former, and carbides increases strength.
- Nitrogen increases strength, reduces toughness.

4.13.11.3 *Neutral Effect Regarding Austenite & Ferrite*

- Manganese is a sulfide former.
- Silicon is a wetting agent. From the welding point of view that property is good in promoting fusion, and penetration of the weld.
- Sulfur and Selenium improve machinability, but may cause hot cracking in welds.

4.13.12 Weldability of Stainless Steels

Most stainless steels are considered to have good weldability and may be welded by several welding processes including the GMAW processes. And other like other arc welding processes, and Electric Resistance Welding (ERW), Electron Beam Welding (EBW), Laser Beam Welding (LBW), Friction Stir Welding (FSW) and joining by brazing.

The coefficient of thermal expansion for the austenitic types of stainless steel is about 50% greater than that of carbon steel, and this property in particular distinguishes stainless steel welding from carbon and low alloy welding. When welding stainless steels, this property must be kept in mind, and must be considered while developing a welding procedure. Not acknowledging this may result in serious damage to steel by excessive localized heating, that also causes serious distortion. If a welding procedure is properly developed considering the low thermal and electrical conductivity of austenitic stainless steel, the restriction may be converted in a helpful property for welding. As this restriction may help welding at much lower heat. This is possible because the heat is not conducted away from a joint, it is helpful in melting is at much lower heat.

4.13.12.1 *Welding Austenitic Steels*

The austenitic stainless steel's structure contains a big portion as austenitic structure. This is possible due to the alloying elements. Austenitic stainless steels contain, Chromium about 16 to 26%, Nickel about 8 to 24%. Manganese about up to 0.40% and carbon as primary elements, like, Molybdenum(Mo), Titanium (Ti), Niobium (Nb)or Columbium (Cb), and tantalum (Ta) that impart the specific grades their signature properties.

The balance between the Cr and Ni + Mn is normally adjusted to provide a microstructure of 90 - 100% austenite. These alloys are characterized by good strength and high toughness over a wide temperature range and oxidation resistance to over 1000°F (538°C).

This group includes Types 302, 304, 310, 316, 321 and 347. Welding filler metals for these alloys should generally matched to the base metal, but for most alloys, they should also be able to introduce in the microstructure some ferrite. Ferrite helps control the possibility of hot cracking. This topic is discussed further in the austenitic steel welding.

To achieve certain level of ferrite structure in the weld-metal, with near 100% austenite structure of the parent metal, the ER 308 welding electrode is used to weld Type 302 and 304 parent metals, and stabilized grades parent metals like Type 347 and Type 321 are welded using ER 347 electrode.

The other austenitic grades of steel are welded with matching filler wire.

Type 347 can also be welded with ER 308H filler. The suffix H indicates higher percentage of carbon content in the welding consumable i.e., electrode and wires. These filler materials are available as coated electrodes, solid bare wire and cored wire. ER or Type 321 electrodes are also available as solid and cored wires.

4.13.12.2 Challenges of Welding Austenitic Steels

Heating austenitic steels in the temperature range of 427°C and 870°C causes some serious damaging effect on stainless steel. This temperature range is called sensitization range. The welding temperature as is obvious, exceed this sensitization range and passes through it both while welding, and gain while cooling.

The resulting problems associated with welding the austenitic stainless steels can be summed up in two terms.

1. Sensitization of the weld heat affected zone, and
2. Hot cracking of weld metal.

4.13.12.3 Sensitization

Sensitization is caused by chromium carbide formation and precipitation at grain boundaries in the heat affected zone when heated to sensitization range of temperature. This week in strength and hard, phase of chromium carbide becomes the cause of cracking on very nominal stress.

From welding point of view, the Sensitization leads to intergranular corrosion in the heat affected zone. This happens as the carbon found near grain boundaries, combines with chromium to form chromium carbide, leaving less chromium in the austenitic solution near the grain boundaries to resist corrosion, at the grain boundaries. This problem can be remedied by using low carbon base material and filler material to reduce the amount of carbon available to combine with chromium. Or to alloy other elements that have more affinity to carbon to form less harmful compounds, and leaving chromium to remain in austenitic structure to resist corrosion.

Welding should be done without preheat, and with minimum heat input to shorten the time in the sensitization temperature range. Stringer beads, with no oscillation or weaving is strongly advised.

The degree of carbide precipitation increases if alloy type contains higher amount of carbon and if they are heated and held for longer duration in the sensitizing range. For welding in such situation, the welding wire should be such that it contains higher chromium, has elements like Niobium or columbium and strict control is practiced on heat control by using low heat input, low arc dwell time, especially in the critical sensitizing temperature zone.

4.13.12.4 Intergranular Corrosion in the Heat Affected Zone Control of Carbide Precipitation

The amount of carbide precipitation is reduced by promoting rapid cooling. Several approaches are practiced their application and limitations are determined by the job specific conditions. Some of the commonly used weld techniques applied to reduce the possibility of sensitization are listed below.

- Copper chill bars,
- Skip welding,
- Stringer beads,
- Low heat input.

Metallurgical solutions are also practiced both in selecting suitable welding wire electrode with low carbon, and other elements in the welding consumable that have greater affinity to carbon than chromium has to combine with carbon. And post weld heat treatment (PWHT) like solution annealing etc.

The PWHT has limited application and sometimes just not possible if the fabricated structure is too large for any heat treatment, or for cost of heat treatment.

The solution annealing if possible, should be done at 1900°F (1038°C) or higher to bring the weldments' structure to a solution where formed chromium carbide is broken in to carbon and chromium and that point, quenching the weldment in water to arrest the solution structure.

The other metallurgical solution is to select such grades of stainless steels that have very low carbon content, or they are stabilized grades like Type 321 and Type 347, and further more they are also weld them with low carbon content electrode or welding wires to prevent chromium carbide formation.

Molybdenum Grades of austenitic stainless steels

Molybdenum containing Grades that include Type 316, 316L, 317, 317L, etc. are also similarly welded with low carbon content where sensitization may be an issue, either during welding or in service. Primarily the molybdenum in stainless steel increases the localized corrosion resistance to water related contaminates, especially marine conditions, causing corrosion many chemicals. These steels are particularly effective in combatting pitting corrosion. Their most frequent use is in industrial processing equipment. Types 316 and 316L are welded with AWS ER316L-XX type welding wires or electrodes. The type 316L and 317L are low carbon grades that must be welded with low carbon electrodes to maintain resistance to carbide precipitation. Both 317 and 317L are generally welded with E317 or E317L electrodes respectively. They can be welded with AWS ER316-XX wire or electrode, but the welds are slightly lower in molybdenum content than the base metal with a corresponding lower corrosion resistance. Use of higher chromium content wire and electrodes and best among them are the ER309 or ER309Cb electrodes, the latter is also stabilized with columbium, may be better option to choose from. If practical to heat threat the weldment then type 316 and 317 type may be PWHT to restore corrosion resistance properties, the following is the nominal temperature range.

- Type 316 or 317: Full anneal at 1950°F to 2050°F (1066°C to 1121°C).
- Type 316L and 317L: Stress relieve at 1600°F (870°C).

High Temperature Grades

Some stainless-steel grades are specifically developed for service in higher temperature. These grades include Type 302B, 304H, 309, 309S, 310, 310S constitute high temperature service alloys in the austenitic steel group. These high alloy grades maintain strength at high temperatures and have good scaling resistance properties. They are primarily used in industrial equipment at high service temperatures, sometimes over 2000°F (1093°C).

- Type 302B and Type 309 grades are generally welded with ER309-XX wire electrodes.
- Type 304H is generally welded with ER308H-XX wire electrodes.
- AWS ER310-XX wire and electrodes are needed to match the high temperature properties and scaling resistance of grades Type 310 and Type 310S.
- ER310-XX electrodes can be used on light plate. ER310-XX welds on heavy plate tend to be more crack sensitive than E309-XX weld metals.

Free Machining Grades (303, 303Se)

Type 303 and type 303Se are free machining austenitic steels.

Production welding of these grades is not recommended, because the sulfur or selenium and phosphorus cause severe porosity and hot short cracking. If welding is necessary, special E312-XX or E309-XX electrodes are recommended because they develop higher ferrite content, and ferrite reduces cracking tendencies.

The welding techniques that reduce admixture of base metal into the weld metal and produce convex bead shapes, are advised.

4.13.12.5 Hot Cracking

Hot cracking is caused by low melting materials such as metallic compounds of sulfur, and phosphorous. These low melting compounds tend to gather around the grain boundaries. And as weld cools the cooling stress develop, and they cause cracks along these week grain boundaries primarily located in the weld and HAZ.

Hot cracking can be prevented by adjusting the composition of the base material and filler material to obtain a microstructure with a small amount of ferrite in the austenite matrix. The ferrite provides ferrite-austenite grain boundaries which are able to control the sulfur and phosphorous compounds so they do not permit hot cracking. This problem could be avoided by reducing the sulfur and phosphorus to very low amounts, but this would increase significantly the cost of making the steel. Normally, a ferrite level of 4 FN minimum is recommended to avoid hot cracking.

4.13.12.6 Design for Welding Stainless Steels

Since the coefficient of thermal expansion for austenitic stainless steels is relatively high, the control of distortion must be considered in designing weldments of these alloys. The volume of weld metal in joints must be limited to the smallest size which will provide the necessary properties. For welding thick sections, plates and where possible a "U" groove design is better than a "V groove. The "U" groove reduces the weld volume compared to the "V" groove.

Another way to reduce the volume, and heat input and possibility of distortion. Where it is possible, the design should consider to weld from both side of the joint, a double "U" or "V" groove joint preparation should be designed. Accurate joint fit-up, careful joint preparation, and weld sequencing, are essential for high quality welds, these preparations also help minimize heat input and distortion.

Strong tooling and fixturing that hold job in place, also resist tendencies for components to move during welding. The tooling should also provide an inert gas backup to the root of the weld to prevent oxidation when the root pass is being made.

If the production application involves long joints in relatively thick material or a large number of parts, the GMAW process with solid or metal cored electrodes may be the best choice. Solid or metal cored electrodes will provide the fastest deposition rates with the GMAW process but wire feeding equipment, power supplies and the requirement for inert gas shielding add to the cost of using these fillers. However, there is little need to remove slag between passes. Solid and metal cored electrodes can be used in short-circuiting, globular and spray modes of arc operation which gives a wide range of deposition rates and heat input levels. Solid and metal cored electrodes can therefore be used for welding a wide range of thicknesses. Gas metal arc welding with spray transfer is used to join sections thicker than about 0.25" (6.4 mm) because deposition rates are higher than with other transfer modes. Welding procedures are similar for conventional austenitic and PH stainless steels. The shielding gas is generally argon with 1 to 2% oxygen added for arc stability. Mixtures of argon and helium are employed if a hotter arc is desired. A small oxygen addition can be added to provide a stable arc, but some aluminum or titanium can be lost from certain PH filler metals during transfer across the arc as a result of oxidation. Response of the weld metal to heat treatment might be less because of this action. Stainless steels may be welded by the gas metal arc process, using either spray arc, short-circuiting or pulsed arc transfer. For flat position welding, spray transfer is usually preferred. For other welding positions, short-circuiting transfer is often used with helium-rich gas such as 90% He, 7.5% Ar, 2.5% CO_2. Pulsed spray transfer can be employed using argon or an argon/helium mixture with a small addition of oxygen or carbon dioxide. Copper backup strips are necessary for welding stainless steel sections up to 0.0625" (1.6 mm) thick. Backup is also needed when welding 0.25" (6.4 mm) and thicker plate from one side only. No air must be permitted to reach the underside of the weld while the weld puddle is solidifying. Oxygen picked up by the molten metal may reduce the corrosion resistance and ductility of the stainless steel as it cools. To prevent this, the underside of the weld should be shielded by an inert gas such as argon.

Selecting suitable Solid or Metal-Cored Stainless-Steel wire-electrodes for GMAW to weld any grade of stainless steel is recommended. Use of AWS specifications for Filler Metal A5.4, A5.9, A5.22 are advised.

4.13.12.7 Determining and Measuring the Ferrite in Welds

Ferrite is best determined by measurement with a magnetic instrument calibrated to AWS A4.2 or ISO 8249. It can also be estimated from the composition of the base material and filler material with the use of any of several constitution diagrams. The oldest of these is the 1948 Schaeffler Diagram. The Schaeffler and DeLong diagrams are in the chapter 2 and 3 of this book.

Despite long use, the Schaeffler Diagram has now become outdated, primarily because it does not consider effects of nitrogen in the steel or weld metal, as a result the values obtained by this method are not consistent and often not accurately reproducible.

An improvement on the Schaeffler Diagram was made by the WRC-DeLong Diagram, the WRC diagram is an improvement on the original DeLong diagram. The WRC diagram is most relevant graph to use to estimate ferrite level, Figure 4.13.12.1 is the recent WRC diagram.

The main differences are that the DeLong Diagram includes nitrogen (N) in determining the Ni equivalent, and shows Ferrite Numbers in addition to "percent ferrite." Ferrite Numbers at low levels may approximate "percent ferrite." The 1992 version of the WRC diagram is an improvement over the DeLong diagram it has modified the weightage on the components that make up both the Creq and Ni eq calculations.

The WRC diagram is easy to use once the Cr.eq and Ni.eq are calculated based on the chemical analysis of the metal. Once these two numbers are available, then it is simply the matter of finding the point where the two lines meet on the diagram. The Ferrite Number is

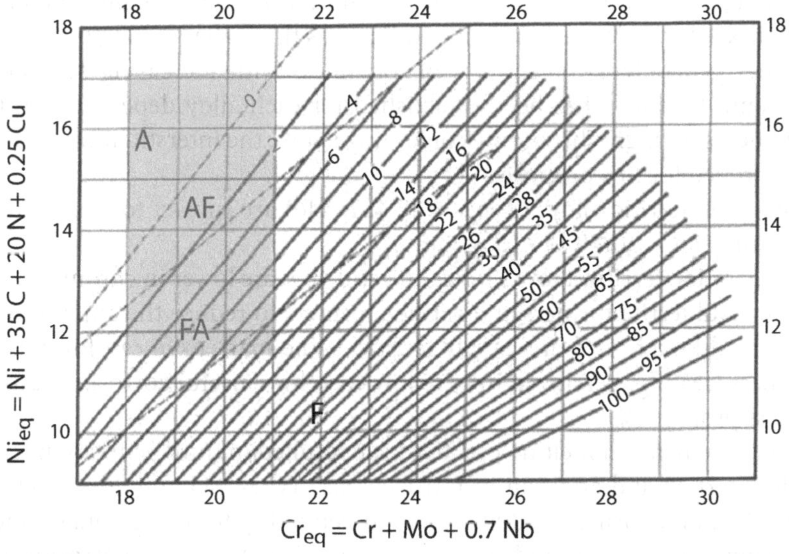

Figure 4.13.12.1 WRC diagram.

indicated by the diagonal line which passes through the intersection of the horizontal and vertical lines.

Predictions by the WRC-1992 and WRC-DeLong diagrams for common grades like 308 are similar, but the WRC-1992 diagram generally is more accurate for higher alloy and less common grades like high manganese austenitic or duplex ferritic-austenitic stainless steels.

In situ measurement of Ferrite Number on the weld is possible by the use of specially developed instruments that work primarily on the principles of magnetic permeability of the weld metal. These instruments read the ferrite number directly on the gauge. The amount of ferrite normally should not be greater than necessary to prevent hot cracking with some margin of safety. The presence of more than necessary ferrite can reduce corrosion resistance, and it can impair ductility and toughness of the weldment.

4.13.12.8 Welding Ferritic Stainless Steels

Ferritic stainless steels are another metallurgical variation the stainless-steel grades. As the name suggest these steels are corrosion resistant and retain the ferrite as primary grain structure.

4.13.12.9 Properties and Application

Ferritic stainless steels are Fe-Cr-C alloys with ferrite stabilizers such as aluminum (Al), columbium (Cb), molybdenum (Mo), and titanium (Ti) to inhibit the formation of austenite on heating. Therefore, they are non-hardenable. In annealed conditions, Lower-alloy ferritic stainless steels have mechanical properties somewhat similar to the low-alloy austenitic stainless steels like type 304. The typical yield strength is in the rage of 30 to 50 ksi (205 to 345 MPa). Alloys with increased chromium, molybdenum, and nickel content have higher strengths.

In the high-chromium-containing alloys such as UNS S44626, the welding procedure typically developed to minimize interstitial pickup during welding and to retain material toughness. These alloys are predominantly utilized as thinwalled tubing products.

These alloys generally exhibit a loss of toughness with increasing section thickness, and a maximum thickness has been established for each alloy depending on the toughness requirements. In high-chromium-containing alloys, the interstitial contents have been carefully controlled for this purpose.

First generation ferritic alloys (Types 430, 422, 446) are subject to intergranular corrosion after welding and exhibit low toughness.

Second generation ferritic alloys (Types 405 and 409) are lower in chromium and carbon and have powerful ferrite formers and carbide formers to reduce the amount of carbon in solid solution. Although they are largely ferritic, some martensite can form as a result of welding or heat treating. Ferritic alloys are low cost, have useful corrosion resistance with low toughness properties.

Recent improvements in melting practice have resulted in third generation ferritic alloys with very low carbon and addition of nitrogen e.g., Types 444 and 26-1 steel. Stabilizing with powerful carbide formers reduces their susceptibility to intergranular cracking after welding, improves toughness, and reduces susceptibility to pitting corrosion in chloride environments and to stress corrosion cracking.

The most important metallurgical characteristic of the ferritic alloy is the presence of enough chromium and other stabilizers to effectively prevent the formation of austenite at elevated temperature. Most grades do form some small amount of austenite since interstitials are present.

Since austenite does not form and the ferrite is stable at all temperatures up to melting, these steels cannot be hardened by quenching. The small amounts of austenite which may be present and transform to martensite are easily accommodated by the soft ferrite.

Annealing treatment at 760°C to 815°C (1 400°F to 1 500°F) is required to restore optimum corrosion resistance after welding. Ferritic stainless steels cannot be strengthened appreciably by heat treatment. These steels are generally used in the annealed condition. The cooling rate from the annealing temperature chosen depends on the particular alloy. The importance of proper heat treatment is emphasized by the fact that the higher-chromium-containing alloys are subject to embrittlement by sigma or alpha prime phase if not properly heat-treated.

All ferrites if heated above 927°C (1 700°F) are susceptible to severe grain growth, due to this the material toughness is reduced and it can only be restored by cold working and annealing.

4.13.12.10 Welding Ferritic Steel

Types 430, 434, 442, and 446 are susceptible to cold cracking when welds are made under heavy restraint. A 150°C (300°F) preheat can minimize residual stresses that contribute to cracking. These steel grades are also susceptible to intergranular corrosion.

Filler material selection would include any of the three available options.

1. Matching compositions,
2. Use of austenitic stainless steels consumables, and
3. Use of nickel alloy consumables.

Matching fillers are normally used only for Types 409 and 430. Austenitic stainless steels electrode or filler wire matching ER 309 or ER 312 grade or nickel alloys are often selected for dissimilar welds.

The need for preheating is determined by the chemical composition, desired mechanical properties, thickness, and conditions of restraint. High temperatures can cause excessive grain growth and heat affected zone cracking can occur in some grades. Low 150°C (300°F) and interpass temperatures are usually recommended.

If post weld heat treatment is deemed necessary, it is done in the 700°C (1 300°F) to 843°C (1 550°F) range to prevent excessive grain growth. Rapid cooling through the 538°C (1 000°F) to 371°C (700°F) range is necessary to prevent embrittlement.

4.13.12.11 Precipitation Hardening Stainless Steels

Precipitation hardening (PH) stainless steels can develop high strength with simple heat treatments. They have good corrosion and oxidation resistance without the loss of toughness and ductility that is normally associated with high strength materials. Precipitation hardening is promoted by alloying elements such as copper (Cu), titanium (Ti), columbium (Cb), and aluminum (Al). Submicroscopic precipitates formed during the ageing treatment increase hardness and strength.

Martensitic PH steels provide a martensitic structure which is then aged for additional strength. Semi austenitic precipitation hardened steels are re-heated to form martensite and also aged. Austenitic precipitation hardened steels remain austenitic after cooling and strength is obtained by the ageing treatment.

As a group, the precipitation hardened steels have corrosion resistance comparable to the more common austenitic stainless steels. Corrosion resistance is dependent on the heat treatment and the resulting microstructure. Welding can reduce corrosion resistance by overaging and sensitization.

Precipitation hardened steels tend to become embrittled after exposure to temperatures above 300°C (580°F), particularly if heated for long periods of time in the range of 370°C to 427°C (700°F to 800°F) temperature. After welding, the maximum mechanical and corrosion resistance properties can be obtained by solution heat treatment followed by ageing. For some applications, only ageing treatment is sufficient.

Martensite precipitation hardened steels are often fabricated in the annealed or overaged condition to minimize restraint cracking. Solution heat treatment and ageing is performed after fabrication.

4.13.12.12 Welding Precipitation Hardened (PH) Steels

The Semi-austenitic precipitation hardened steels are welded in all conditions. Austenitic conditioning and ageing are performed after welding for maximum mechanical properties. Austenitic precipitation hardened steels are difficult to weld because of cracking problems. Matching, nickel alloy, or austenitic filler materials are used. The selection of suitable filler metal is dependent on the post-weld heat treatment and final property requirements.

The following are the key points that must be kept in mind for selection of material as well as welding of all stainless steels discussed thus far.

- Thermal expansion, thermal conductivity, and electrical resistivity have significant effects on the weldability of stainless steels.
- The relatively high coefficient of thermal expansion and low thermal conductivity of austenitic stainless steel require better control of distortion during welding.
- Low thermal conductivity for all stainless steels indicate that less heat input is required.
- The weldability of the martensitic stainless steels is affected mainly by hardenability that can lead to cold cracking.
- Welded joints in ferritic stainless steels have low ductility as a result of grain coarsening related to the absence of an allotropic transformation.
- The weldability of the austenitic stainless steels is governed by their susceptibility to hot cracking.
- The precipitation hardening stainless steels have welding difficulties associated with transformation (hardening) reactions.
- Stainless steels which contain aluminum or titanium can only be welded with gas-shielded processes.

- Joint properties of stainless-steel weldments will vary considerably as a result of their dependence on welding process and technique variables.
- Suitability for service conditions such as elevated temperature, pressure, creep, impact, and corrosion resistance must be carefully evaluated. The complex metallurgy of stainless steels must be accounted for.

4.13.12.13 Martensitic Stainless Steels

Martensitic stainless steels are metallurgically different from austenitic stainless steels discussed above. The weldability for this grade differs slightly and different grades of martensitic steels may present different challenges depending on the alloying elements.

4.13.12.14 Properties and Application

Martensitic stainless steels are Fe-Cr-C alloys that are capable of the austenite-martensite transformation under all cooling conditions. Compositions for most of martensitic steel alloys are covered by number of specifications, such as ASTM A 420 or API 13 Cr L80 and 420 M with additional small amounts of Ni and/or Mo. Although 9Cr1Mo is not strictly a martensitic stainless steel; it is often included in this alloy group, especially because of challenges associated with welding of 9-Cr-Mo steel is in many ways similar to this group of materials.

The martensitic stainless steels are generally used in the quenched and tempered, or normalized and tempered condition. For services where hydrogen evolution or presence of Sulphur is expected as in sour gas services in oil and gas industry a maximum hardness of 22 HRC is specified by most of the specifications, and for most of the alloys. Some of the alloys like type 410 and 420 develop quench crack if quenched in water, so they are quenched only in oil, or polymer, or air-cooled before tempering.

Some alloys like type 410, type 415, and J91540 (CA6NM) receive a second temper treatment called "Double tempering" at a temperature lower than the first tempering temperature, to reduce the untampered martensite in type 410, type 415, and J91540 (CA6NM). Double tempering has not been shown to improve resistance to stress corrosion cracking in type 420 tubular products and for 9Cr1Mo tubular or forgings.

The mechanical properties of typical base metal strength (SMYS) are grouped as 414 MPa (60 ksi), 517 MPa (75 ksi), 552 MPa (80 ksi), and 586 MPa (85 ksi), with hardness controlled to the maximum 22 or 23 HRC, and often have specified to maximum yield strength of 95 to 100 ksi (660 to 690 MPa). For sour service applications the tubular products are generally used according to the API Specification 5CT, or L80, strength level; forgings and castings are generally specified with hardness not exceeding 22 on Rockwell C scale. Higher strengths are used in sweet service; however, corrosion resistance and ductility are adversely affected as the strength of steel is increased.

4.13.12.15 Welding Martensitic Stainless Steels

Weld design strength levels range from 414 MPa (60 ksi) upward, but they can be different than the parent metal; for example, a 552 MPa (80 ksi) mandrel could be welded with a duplex or austenitic stainless filler metal that results in a lower weld joint strength, provided this has been considered to meet the design and operation demands.

The martensitic stainless steels are easy to work with, including welding, the welding processes used include GMAW, FCAW, SMAW,GTAW (TIG), SAW, EBW, and laser beam welding (LBW) Typical welding consumables include 410 Ni Mo (matching weld metal), or 2209, 309LSi (overmatching consumables), while some limited application of autogenous welding is also practiced. These alloys are not used in the as-welded condition in more demanding environments like, sour service.

Extreme care is typically required when these alloys are welded, because they are susceptible to high hardness. Tubing and casing are generally not welded.

When welding type 410, high pre-heat temperatures are used. The alloys classified as type 410, type 415 (F6NM), and J91540 (CA6NM) are tempered again as a post weld heat treatment after welding to ensure that they have maximum specified strength and hardness. These alloys have been welded using nominally matching filler metals. The use of non-matching austenitic types consumables can increase the risk of fusion boundary cracking in sour service, this increase in fusion boundary cracking is irrespective of the hardness limits in the weld area.

These alloys are known for moderate corrosion resistance, heat resistance up to 535°C (1 000°F), relatively low cost, and the ability to develop a wide range of properties by heat treatment.

If left in as-welded condition, the intergranular (sensitization) cracking is common occurrence in both sweet (CO_2 containing) and sour conditions. These problems also arise as a result of poor PWHT cycles, where the treatment has been ineffective in refining structure and reducing HAZ hardness.

They are capable of air hardening from temperatures above 815°C (1 500°F) for nearly all section thicknesses. Maximum hardness is achieved by quenching from above 950°C (1 750°F). They lack toughness in the as-hardened condition and are usually tempered.

Martensitic alloys can be welded in any heat treat condition. Hardened materials will lose strength in the portion of the heat affected zone. With a carbon content of 0.08% and 12% Cr (Type 410), the heat affected zone will have a fully martensitic structure after welding. The steep thermal gradients and low thermal conductivity combined with volumetric changes during phase transformation can cause cold cracking.

The hardness of the heat-affected zone depends primarily on the carbon content and can be controlled to some degree by developing an effective welding procedure. As the hardness of the heat affected zone increases, its susceptibility to cold cracking become greater and its toughness decreases.

Weldability is improved when austenitic stainless-steel filler is used because it will have low yield strength and good ductility. This also minimizes the strain imposed on the heat-affected zone.

Martensitic steels are subject to hydrogen-induced cracking like low alloy steels. Covered electrodes used for welding must be low-hydrogen and maintained in dry condition.

Preheating and good interpass temperature control is the best means to avoid cracking. Preheating is normally done in the 200°C to 315°C (400°F to 600°F) range. Carbon content, joint thickness, filler metal, welding process, and degree of restraint are all factors in determining the pre-heat, heat input, and post weld heat treatment requirements.

Post weld heat treatment is performed to temper or anneal the weld metal and heat affected zone with aim to decrease hardness and improve toughness, and to decrease the residual stresses associated with welding. Subcritical annealing and annealing are performed. When matching filler metal is used, the weldments can be quench hardened and tempered to produce uniform mechanical properties.

Types 416 and 416Se are free machining grades that must be welded with caution to minimize the hydrogen pickup. ER312 austenitic filler metal is recommended for welding type 416 and 416Se alloys, since it can tolerate the sulfur and selenium additions.

Type 431 stainless can have high enough carbon to cause heat affected zone cracking if proper preheat, preheat maintenance, and slow cooling procedures are not followed.

4.13.12.16 Duplex Stainless Steels

Duplex stainless steels solidify as 100% ferrite, but about half of the ferrite transforms to austenite during cooling through temperatures above approximately 1900°F (1040°C). This behavior is accomplished by increasing chromium and decreasing nickel as compared to austenitic grades. Nitrogen is deliberately added to speed up the rate of austenite formation during cooling. Duplex stainless steels are ferromagnetic. They combine both the higher strength and fabrication properties of austenite with the resistance to chloride stress corrosion cracking of ferritic stainless steels.

This alloy group was developed over past 30 years; the development progress has resulted in a range of compositions including lean 22% chromium (Cr) and 25% chromium, listed in the table 2.8.1. These alloys have high strength, good toughness, good corrosion resistance, good weldability, and formability, all of which ease manufacturing. These alloys combine the strength characteristics of ferritic stainless steels and the corrosion resistance of austenitic stainless steels. They have higher resistance to environmental corrosion than austenitic stainless steels. Dual phase alloying requires relatively lower Ni and Mo contents than single-phase austenitic alloys.

The alloys with higher F_{PREN} values, this is possible as a result of adding nitrogen in the alloy.

The duplex stainless steel contains up to 22% chromium. The key property that is of value to industry is the material's pitting resistance F_{PREN}, which is typically in the rage of 35 to 40.

The chromium content of super duplex steel is up to 25% and its pitting resistance F_{PREN} is typically in the rage of 40 and 45.

4.13.12.17 Mechanical Properties

The mechanical properties of the different types of duplex stainless steel are shown in Table 2.8.2. The mechanical properties of the cast versions of these alloys (e.g., UNS J93380, J92205, etc.) are lower than their wrought counterparts. ASTM A 995 "Standard Specification for Castings, Austenitic-Ferritic (Duplex) Stainless Steel, for Pressure-Containing Parts" details the compositions and mechanical properties of cast duplex alloys that are used for pressure-containing parts.

The duplex stainless steels used by the oil and gas industry have a roughly 50/50 austenite/ferrite, in general the duplex steel of various types would present a phase balance within the range 35% to 65% ferrite. They have adequate toughness at low temperatures, the alloy is commonly used to temperatures as low as minus 60°C (−76 °F).

Super duplex stainless steel (UNS S32760) has been successfully used up to minus 120 °C (−184 °F), but this requires a well-developed welding procedure and closely monitored welding parameters during the production process.

On long exposure to temperatures above 320°C (608°F) and up to about 550°C (1 022°F), the ferrite decomposes to precipitate alpha prime. This phase causes a significant loss of ductility; hence, duplex stainless steels are not normally used above 300°C (572°F).

In oil and gas service applications these alloys have fared very well in both sour and sweet environmental conditions.

4.13.12.18 Heat Treatment

Generally, these alloys are used in the annealed or annealed and cold worked condition. Prolonged heating at temperatures between 260 and 925 °C (500 and 1 700 °F) can cause the precipitation of number of phases, including sigma, which reduces toughness and can reduce SCC resistance. Any prolonged heating below the minimum solution-heating temperature is to be normally avoided. Low-temperature toughness generally decreases with decreasing cooling rates in annealing.

Cold-worked alloys are usually not welded because the mechanical strength of the weld would be lower than the base metal. Annealed alloys are easily welded. The weld filler metal is chosen to produce a desired volume fraction of ferrite and austenite. Hence, fabrication using autogenous (without filler) metal can result in welds that are poorer in mechanical and corrosion-resistant properties. The welding procedure is typically developed to control and balance the ferrite/austenite phase, this is essential to prevent deleterious phases or intermetallics.

These alloys are readily weldable, by GMAW, GTAW, SMAW and SAW processes. Other processes are also successfully used. Where the weld is to be heat treated after completion it is usual practice to weld with matching composition filler metal.

In as welded application, it is normal to use an over alloyed filler metal with an extra 2 to 2.5% nickel (Ni). This helps in getting the austenite/ferrite phase balance of about 50/50, if the weld is cooled rapidly. The lean duplex grades are welded with the filler metal used for 22% Cr duplex stainless steels. Except for thin sheets of up to 2 mm thickness autogenous welding is normally not recommended for duplex stainless steels.

Normally, argon gas is used for both shielding and backing gases, and welding does not begin until the oxygen level is dropped below 0.1%. As with high alloy stainless steels, care is to be taken in welding duplex alloys, and adherence to good stainless-steel welding practice discussed earlier in this chapter a good practice. Good joint design, control of interpass temperatures and keeping low heat inputs are other essential variable for good welding. Preheat and post weld heat treatment are not required for duplex stainless steels.

Maximum permissible heat input and interpass temperature increase with section thickness. The values for these parameters generally decrease as the alloy content increases. If heat inputs or interpass temperatures are too high, there is a risk of precipitating sigma (Σ) or chi (X) phases in the heat affected zone (HAZ) or weld metal. These are intermetallic phases, rich in chromium and molybdenum that leave a denuded area around them, which reduces the localized corrosion resistance. Sigma and chi phases also reduce impact toughness properties. In many applications especially in some oil and gas applications, the low temperature toughness is compromised for the corrosion resistance properties.

After welding, there is usually a heat tint in the weld and heat affected zone (HAZ), and it is normal to remove this, by manual brushing, by mechanical abrasives, or by a suitable pickling solutions or gels.

While developing welding procedures it is common and advised to include a corrosion test for example testing according to ASTM G 48, (http://www.astm.org/) as part of the weld qualification procedure.

The corrosion test sets important weld parameters; hence it is essential that the qualified parameters of welding are followed very closely during production welding. The experience tells us that sometimes "less experienced" welders have difficulty passing the corrosion test. Although the weld made by these less experienced welders would meet the mechanical requirements, it may not meet the corrosion tests as specified above. This increases the importance of welders/operators' qualification test and production weld parameters monitored by inspectors. In a very limited way this problem is resolved with use of 2% nitrogen gas along with argon as shielding gas. The reasons for the corrosion test failure can be due to the development of third phases, which are the result of poor supervision and control over the heat input and the inter-pass temperature.

Duplex Stainless-steel welds usually have lower impact toughness than their parent metals. The welding process used often affects the level of toughness, the GTAW and GMAW welds bring out the toughest weld-metal toughness as compared to SMAW and the SAW. The weld-metal toughness is a function of both the heat input and the type of flux used, since GMAW and GTAW processes do not use fluxes, their ability to weld at low heat input allows them to produce better toughness properties. The GMAW variants like short circuit, and spray and pulsed arc modes use significantly lower heat input for welding. These modes are successfully used for welding Duplex steels.

The experience suggests that a minimum of 70-Joule Charpy impact toughness in the parent metal ensures adequate toughness in a duplex weld, and it is easily achieved when correctly welded. To improve the low temperatures toughness requirements, especially for very low temperature services, it is worth considering the use of a nickel alloy filler metals, taking into account that other properties are not compromised for example, the nickel alloy weld must have the same strength as the parent duplex stainless steel. A practical example of above would be the selection of C-276 (UNS N10276) filler metal to improve the impact toughness of cast super-duplex (UNS J93380) at minus120 °C (-184 °F) service.

Some specifications for duplex material that are used in subsea environments with cathodic protection (CP) require maximum austenite spacing. In a weld, this cannot be controlled and the result cannot be changed by any heat treatment. However, duplex welds usually have a fine microstructure and meeting a maximum austenite spacing of 30 μm, is usually not difficult. Although the welding in itself does not necessarily degrade the resistance of duplex stainless steel against HISC, the presence of higher stress and stress raisers like weld toe, poses a significant problem when uncoated duplex stainless steels or steels with defective coating are exposed to CP under mechanical stress. Failures have occurred as a result of this effect, and guidance to avoid them can be sought from industrial specifications.

4.14 Welding Nickel Alloys

Nickel is a very versatile material with excellent weldability. Nickel and nickel alloys are major corrosion resistant materials in use in chemical and petrochemical industries. They are also used in other industries like, marine engineering, aeronautical and automobile making.

Nickel is an element, with the symbol Ni and atomic number 28, as shown in the periodic table below.

In appearance, nickel is a silvery-white, lustrous metal with a slight golden tinge. Nickel belongs to the transition metals and it is hard, and ductile. Nickel is slow to react with air because of passivation, an oxide layer forms on the surface and prevents further corrosion. See Figure 2.9.2 for nickel's place on periodic table and its atomic number.

Nickel is slow to oxidize in air and at room temperature and is considered corrosion resistant. It is used for plating iron and copper alloys like brass, and coating chemical equipment, and manufacturing alloys that retain a high silvery polish, one such alloy is called German silver, which contains about 60% copper, 20% nickel and 20% zinc.

Nickel like iron, cobalt, and gadolinium is ferromagnetic at room temperature. That property allows its use to make permanent magnets. The metal is valuable in modern times primarily to make various industrial alloys, primary among them are various grades of stainless steels, and cupro-nickel alloys.

Nickel 200 is the purest of nickel commercially available in wrought condition. The metal is weldable. Among the various alloys of nickel most are commercially named with number identifiers, some of them are discussed below for introduction purpose.

Nickel and Nickel alloys are welded by number of welding processes including GMAW process. The wrought nickel alloys can be welded under conditions similar to those used to weld austenitic stainless steels. Cast nickel alloys, particularly those with a high silicon content, present difficulties in welding.

The most widely employed processes for welding non-age-hardenable (solid-solution-strengthened) wrought nickel alloys include gas-metal arc welding (GMAW), shielded metal arc welding (SMAW), and gas-tungsten arc welding (GTAW).

Nickel alloys are usually welded in the solution-treated condition. Precipitation-hardenable (PH) alloys should be annealed before welding if they have undergone any operations that introduce high residual stresses.

Most of the time these materials do not require post weld chemical or heat treatment, however in some cases a full solution anneal is desired to improve corrosion resistance. Heat treatment may be necessary to meet specification requirements, such as stress relief of a fabricated structure to avoid age hardening or stress-corrosion cracking (SCC) of the weldment in hydrofluoric acid vapor or caustic soda. If welding induces moderate-to-high residual stresses, then the PH alloys would require a stress-relief after welding and before aging.

Nickel and nickel alloys are susceptible to embrittlement by low-melting-point elements like, lead, sulfur, phosphorus, and other. These materials can exist in contaminants like grease, oil, paint, marking crayons or inks, forming lubricants, cutting fluids, shop dirt, and other processing chemicals. Hence thorough cleaning of parts to be welded is very essential.

Work-pieces should be thoroughly cleaned of all foreign material before they are heated or welded. Shop dirt, oil and grease can be removed by either vapor degreasing or swabbing with acetone or another nontoxic solvent. Paint and other materials that are not soluble in degreasing solvents may require the use of methylene chloride, alkaline cleaners, or special proprietary compounds. If alkaline cleaners that contain sodium carbonate are used, then the cleaners themselves must be removed clean of the material, prior to welding. Spraying or scrubbing with hot water is recommended. Marking ink can usually be removed with alcohol.

Processing material that has become embedded in the work metal can be removed by grinding, abrasive blasting, and swabbing with 10% HCl solution, followed by a thorough water wash. Oxides must also be removed from the area involved in the welding operation, primarily because oxides get imbedded in the weld as inclusions, because the oxide have much higher meting point than the base metal melting points. Oxides are normally removed by grinding, machining, abrasive blasting or pickling.

Nickel alloys, both cast and wrought and either solid-solution-strengthened or precipitation-hardenable, can be welded by the GMAW process.

4.14.1 Welding of Precipitation Hardenable Nickel Alloy

The PH alloys, alloy 718 is the primary of them are very good weldability, but they also have higher susceptibility of cracking, both in the base-metal, and heat affected zone, hence they require special welding procedures for successful welding.

Cracks also occur post weld, and during the operation if the service temperature is greater than the aging temperature, and residual stresses developed during welding or stresses induced during the precipitation. Before welding these alloys, a full-solution anneal is usually performed. After welding, the appropriate aging heat treatment is performed. To further improve alloy properties, a full anneal after welding, followed by a post-weld heat treatment, can be incorporated in the welding procedure.

Any part that has been subjected to severe bending, drawing or other forming operations should be annealed before welding. If possible, heating should be done in a controlled atmosphere furnace to limit oxidation and minimize subsequent surface cleaning.

Heat input during the welding operations should be held to a moderately low level in order to obtain the highest possible joint efficiency and minimize the extent of the HAZ. For multiple-bead or multiple-layer welds, many narrow stringer beads should be used, rather than a few large, heavy beads. Any oxides that form during welding should be removed by abrasive blasting or grinding. If such films are not removed as they accumulate on multiple-pass welds, they can become thick enough to inhibit weld fusion and produce unacceptable laminar type oxide stringers along the weld axis.

4.14.2 Welding of Cast Nickel Alloy

Cast nickel alloys can be joined by the GMAW processes. For optimum results, casting should be solution annealed before welding to relieve some of the casting stresses and provide some homogenization of the cast structure. Light peening of solidified metal after the first pass will relieve stresses and, thus, reduce cracking at the fusion line, or the interface of the weld metal and the cast metal. The peening of the subsequent passes is of little, if any, benefit. Stress relieving after welding is also recommended.

4.14.3 Nickel – Chromium Alloys

- Alloy 600,
  This alloy has nominally about 72% nickel and 16% Chromium, among other alloying elements. It is resistant to oxidation at high temperatures.
  Weldability: Has good weldability.
  Recommended GMAW welding wire-electrode: ER NiCrFe-3

- Alloy 601,
  This alloy has nominally about 60% nickel and 25% Chromium, 1% Aluminum, Iron among others. It has higher strength and is excellent resistant to oxidation at high temperatures.
  Weldability: Has good weldability.
  Recommended GMAW welding wire-electrode: ER NiCrFe-3
- Alloy 617,
  This nickel, chromium, molybdenum, and cobalt alloy is metallurgically stable alloy, that serves well as corrosion resistant in wide range of corrosive environment, and in high temperature environment, while maintain it strength.
  Weldability: Has good weldability.
  Recommended GMAW welding wire-electrode: ER NiCrCoMo-1
- Alloy 625,
  This Nickel, chromium, and molybdenum alloy is designed to be an excellent corrosion resistant material by adding niobium, which stabilizes the structure matrix and the strength of the material at high temperature applications. One of the most significant properties is the high resistance to pitting in vary corrosive environment.
  Weldability: Excellent weldability.
  Recommended GMAW welding wire-electrode: ER NiCrMo-3
- Alloy 718,
  This precipitation hardenable, nickel and chromium alloy with iron and niobium, and molybdenum presents excellent creep resistance properties, and resists cracking after welding.
  Weldability: Excellent weldability.
  Recommended GMAW wire-electrode: ER NiCrFe-3

4.14.4 Nickel - Copper (Cupro-Nickle Alloys)

- Monel alloy 400,
  This alloy with nominal, about 60% nickel, and 30% copper, is excellent material for service in sea water, and other acids like sulfuric acid, and hydrofluoric acid environment.
  Weldability: Good weldability.
  Recommended GMAW wire-electrode: ER NiCu-7
- Monel alloy 401,
  This copper and nickel alloy with nominal, about 45% nickel, and reminder copper, is excellent material for electrical application service.
  Weldability: Not considered.
- Monel alloy 450,
  This cupro-nickel alloy with nominally 70% copper, and 30 % nickel is resistant to biofouling and sea water corrosion.
  Weldability: Superior weldability.
  Recommended GMAW welding wire-electrode: AWS Class, ER CuNi

- Monel alloy K-500,
  This precipitation hardenable alloy of nominally 60% nickel, and 30% copper with other alloying elements is a version of Monel 400 discussed above, but it has greater hardness, and strength.
  Weldability: Not considered.
  Recommended GMAW welding wire-electrode: Not recommended

4.14.5 Nickel - Iron - Chromium Alloys

- Alloy 800,
  This Nickel-Iron-Chromium, with good creep properties is excellent alloy for service in oxidizing and carburizing environment in high temperature atmosphere service.
  Weldability: Good
  Recommended GMAW welding wire-electrode: ER NiCrFe-2
- Alloy 825,
  The nickel-iron chromium with added molybdenum and copper has excellent resistance to both oxidizing and reducing acids. Resists SCC and has good pitting and crevice resistance.
  Weldability: Very Good,
  Recommended GMAW welding wire-electrode: ER NiCrMo-3
- Alloy 902
  The nickel-iron chromium is designed for precipitation hardening by adding aluminum and titanium.
  Weldability: Not considered.
- Alloy 330
  The nickel-iron chromium with added silicon for increased resistance to oxidation, is good alloy in high temperature service in both oxidizing and carburizing environment.
  Weldability: Good.
  Recommended GMAW welding wire-electrode: ER NiCrFe-1
- Alloy 020
  The nickel-iron chromium with added copper and molybdenum and stabilizer niobium. The alloy has good resistance to general corrosion, and localized corrosion forms like pitting and crevice corrosion occurring in chlorides and sulfuric, nitric, and phosphoric acids.
  Weldability: Good.
  Recommended GMAW welding wire-electrode: ER NiCrFe-1

4.15 Minimizing Discontinuities in Nickel and Alloys Welds

The discontinuities including the metallurgical issues encountered in the arc welding of nickel include can be listed as the following.

1. Porosity

2. Susceptibility to high-temperature embrittlement by sulfur and other contaminants
3. Cracking in the weld bead, caused by high heat input and excessive welding speeds
4. Stress-corrosion cracking in service.

4.15.1 Porosity

From welders' point of view, cleaning of the parent metal, and surrounding area can reduce the possibility of porosity. However, from the metallurgical angle more reactions are possible that can cause porosity, gases that are either intentionally present in the weld area, or present due to the service environment of the material being welded can react in welding heat and cause discontinuities. Oxygen causes oxidation, carbon dioxide is a reducing agent, nitrogen allows formation of nitrides, or hydrogen reacts with atmospheric oxygen to form water vapor that cause porosity, all of these gases can cause porosity in welds. In the SMAW processes porosity can be minimized by using electrodes that contain deoxidizing or nitride forming elements, such as aluminum and titanium. These elements have a strong affinity for oxygen and nitrogen and form stable compounds with them. Presence of deoxidizers in either type of electrode serves to reduce porosity. In addition, porosity is much less likely to occur in chromium-bearing nickel alloys than in non-chromium-bearing alloys.

4.15.2 Weld Cracking

Hot shortness of welds can result from contamination by sulfur, lead, phosphorus, cadmium, zinc, tin, silver, boron, bismuth, or any other low-melting-point elements, which form intergranular films and cause severe liquid-metal embrittlement at elevated temperatures. Many of these elements are found in soldering and brazing filler metals.

Hot cracking of the weld metal usually results from such contamination. Cracking in heat-affected zone is often caused by intergranular penetration of contaminants from the base-metal surface. Sulfur, which is present in most cutting oils used for machining, is a common cause of cracking in nickel alloys.

Weld metal cracking also can be caused by heat input that is too high, as a result of higher current and slower travel speed. Welding speeds have a large effect on the solidification pattern of the weld. High welding speeds create a tear-drop molten weld pool, which leads to uncompetitive grain solidification at the center of the weld. At the weld centerline, residual elements will collect and cause centerline hot cracking or lower transverse tensile properties.

In addition, cracking may result from undue restraint. When conditions of the high restraint are present, as in circumferential welds that are self-restraining, all bead surfaces should be slightly convex. Although convex beads are virtually immune to centerline splitting, concave beads are particularly susceptible to centerline cracking. In addition, excessive width-to-depth or depth-to-width ratios can result in cracking may be internal (that is subsurface cracking).

4.15.3 Stress Corrosion Cracking

Nickel and nickel alloys do not experience any metallurgical changes, either in the weld metal or in the HAZ that affects normal corrosion resistance. When the alloys are intended to contact substances such as concentrated caustic soda, fluorosilicates, and some mercury salts, however, the welds may need to be stress relieved to avoid stress corrosion cracking. Nickel alloys have good resistance to dilute alkali and chloride solutions. Because resistance to stress-corrosion cracking increases with nickel content, the stress relieving of welds in the high-nickel-content alloys is not usually needed.

4.15.4 Effect of Slag on Weld Metal

Because fabricated nickel alloys are ordinarily used in high-temperature service and in aqueous corrosive environments, all slag should be removed from finished weldments. If slag is not removed in these types of application, then crevices and accelerated corrosion can result. Slag inclusions between weld beads reduce the strength of the weld. Fluorides in the slag can react with moisture or elements in the environment to create highly corrosive compounds.

4.16 Calculating Heat Input in Pulsed Arc GMAW

The importance of heat input is highlighted several times while describing the welding of various alloys. Heat input is an important aspect of all welding, where the control on the resulting microstructure of steel is important. Given the variations in different weld-metal transfer modes the complexity of calculating heat input is no longer a simple method used in SMAW or GTAW welding processes.

The importance of heat input is especially of significance to steels that are sensitive to heat input, alloy steels that are required to have good toughness properties, especially with low temperature ductility, and quenched and tempered alloy steel are the examples where controlled heat input is desired and welding engineers specify the limits.

Heat input is related to the welding voltage, welding current arc, and welding arc travel time or dwell time of the welding arc which is related to the travel speed in unit time the arc is at a spot during welding. The calculated value is often reported in units of energy input.

$$\text{Heat input (in kj/mm)} = \{V \times I \times 60 \times \text{Process efficiency (Table 2-3-2)}\} \div \{\text{Travel speed (in mm per minute)} \times 1000\}$$

The above calculation is easy for the processes that have constant current output and arc voltage, where measurement taken from the meters is the data input for the calculation. This is due to the non-pulsed current and conditions, simple arithmetic mean of power values can be most effective way to calculate the heat input.

However due to the non-constant as is the output in pulsed arc process this simplistic approach is not effective.

As we have seen above description of pulsed arc welding systems, the current fluctuates between high peak amperage for the pulse mode and a low background current for regular cycles. This mode gives an apparent square wave. I have called it apparent squire wave because it is not exactly squire wave, due to the inductance and resistivity effect; the wave adopts a slightly roundish appearance. Several research works have been carried out and papers published (By M.R.Bosworth; WRC Supplement to 111-s and AWS Welding Journal, May 1991) that suggest that to obtain correct heat input the power delivered to the arc should be measured as an arithmetic mean of instantaneous power values. The arc voltage is measured as close as possible to the arc between the work-piece and the welding gun.

The arithmetic mean is calculated using following relationship between current and voltages of various passes.

$$P_{av} = \Sigma_{i=1}^{n} (V_j I_i) \div n$$

The root mean square (RMS) calculation of same data could be carried out using following relations.

$$RMS = \{\Sigma_{I=1}^{n} (X_i)^2 \div n\}^{0.5}$$

Where X_i is the first value, X_n is the n values.

Power values can be further derived from using the above obtained data and substituting in following relations.

$$P = V_{av} I_{av}$$

Where the P is the average power and V_{av} is the arithmetic mean voltage.

$$RMS = V_{RMS} I_{RMS}$$

Where P_{RMS} is the root mean square power and I_{RMS} is the root mean square current.

Thus, the determination of more accurate thermal efficiency of pulsed GMAW process the RMS values are more effective over the arithmetic mean values.

There are however several variables consideration of those will make this approach more effective in calculation of realistic values these variables could be the weld design, the shielding gases and position of welding.

4.17 Review Your Knowledge

To review the knowledge, take the test on the question set on the content of the book. After wring the answers check the chapters again and where necessary read those chapters again, and correct your answers. After a couple of hours take the test again, till you answered all questions correctly.

1. What are different types of DC welding machines, suitable for GMAW process?
2. What are the various modes of metal transfer in GMAW process?

3. What type of metal transfer mode is the STT®™ process?
4. What is the normal no-load voltage range in DC welding arc?
5. How does the drooping characteristics of a welding machine maintain constant voltage?
6. Explain the DCEN and DCEP, in each circuit what direction the electrons flow?
7. What is the current threshold for spray transfer mode?
8. What are the dangers of arc flash?
9. What how is welding wire for GMAW classified, give AWS classification example of a carbon steel or a low alloy steel wire?
10. What are the key characteristics of welding austenitic steel?
11. Describe pulsed arc transfer mode, how does pulse arc created in a welding machine?
12. Describe what you understand by the electrode designation ER 309?
13. State the importance of wire/electrode extension.
14. Give the effect of argon +CO_2, Vs Argon + O_2 blend of gases on welding carbon and low alloy material.
15. When welding stainless steel with wire size 0.045-inch with spray transfer mode what amperes and voltage setting is the optimal?
16. What is the choice of shielding gas for the above weld conditions?
17. Differentiate what will be the effect on weld if 75% Argon + 25% Helium or 75% Helium + 25% Argon gas mix is sued for shielding to weld aluminum?
18. What is porosity, and how to control it? Give all that may cause the porosity.
19. What is the difference between pre-heat and Interpass temperatures?
20. What do you understand by Post weld heat treatment, and why is it necessary for PH alloys?

5

Flux Cored Arc Welding (FCAW) Process

5.1 Synopsis

This chapter about flux core arc welding (FCAW) process, discusses the application and limitations of the process. The type of electrode used for self-shielding, FCAW-S and gas shielding FCAW-G variants.

The chapter discusses the importance and limitations of gas shielded and also the advantaged of un-shielding processes. It also details the effects of gas used for shielding on the mechanical properties of steel being welded. How the gas decomposition in the heat of the arc column affects the alloying composition in the weld pool.

Various types of welding electrode wire are discussed and listed for specific material type, a table is included for carbon and alloy steel wires with recommended shielding and current polarity.

5.2 Keywords

Constant current, constant voltage, polarity, shielding gas, wire identification and classification, mechanical strength, CTWD, ESO.

5.3 Introduction to Flux Cored Arc Welding (FCAW) Process

The Flux-Cored Arc Welding (FCAW) process is very similar to GMAW process except that it uses a cored electrode wire that is filled with flux. The continuously fed electrode is both a filler metal and an electrode to initiate and maintain the welding arc. The tubular electrode is hollow, and is filled with flux as a core inside the hollow of the tube.

In the flux core arc welding (FCAW) process, the arc is maintained between a continuous fed filler metal electrode and the parent metal being welded. Shielding is obtained from a flux contained within the tubular electrode in which case it is called self-shielding FCAW (FCAW-S), the Figure 5.3.1 below details of the welds area details. The other version of the process is the FCAW-G in which the shielding provided by the gas is supplied externally. The welding wire and the shielding gas both are supplied through the specifically designed welding torch unit as shown in the Figure 5.3.2 below, the shielding gas is delivered at the point of weld, to cover the weld area comprising of weld pool, and the HAZ.

Ramesh Singh. Arc Welding Processes Handbook (299–328) © 2021 Scrivener Publishing LLC

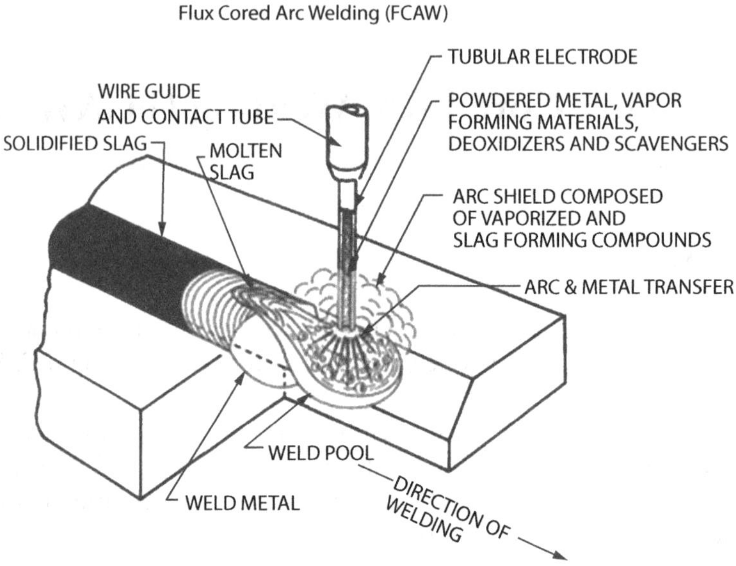

Figure 5.3.1 FCAW-S self-shielding tubular wire process.

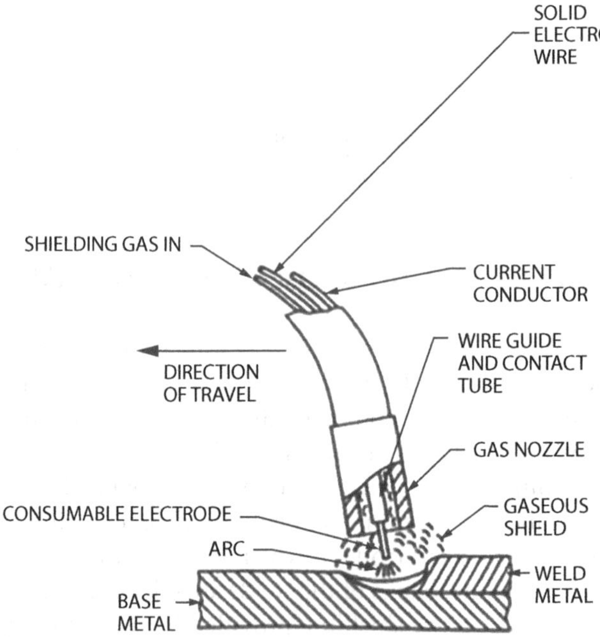

Figure 5.3.2 FCAW-G, gas shielding solid wire process.

The flux-cored electrode is a composite tubular filler metal electrode with a metal sheath and a core of various powdered materials. An extensive slag cover is produced during welding. Self-shielded FCAW-S protects the molten metal through the decomposition and vaporization of the flux core by the heat of the arc. Gas shielded FCAW-G uses a protective gas flow in addition to the flux core action, and for the same objective as the FCAW-S version.

The FCAW process combines the productivity of continuous feed wire welding and the metallurgical benefits that are derived from the use of flux and the support of slag in shaping and protecting the weld bead from contamination and providing controlled cooling of weld metal.

The FCAW process has higher productivity comparable to SMAW process. This can be the chief benefit of FCAW for most of the applications. Equipment costs are higher, setup and the operation are more complex, and there is a limit on the operating distance from the electrode wire feeder. The application of this process may be limited by the availability of suitable filler wire and flux combinations for various metals and alloys to be welded.

FCAW process generates large volumes of fumes, which is unhealthy and must be removed for the safety of people in the vicinity of the work, similarly the slag must be removed between passes to keep the weld free from slag inclusions.

Generally speaking, the self-shielded (FCAW-S) version of the process can be used for nearly all applications that would normally be done with the SMAW process, and the gas-shielded (FCAW-G) process would cover most of the applications that would be done with GMAW process. The limitation in both comparisons is the availability of suitable filler wire and the flux combination for the specific metal to be welded.

In the complex welding, such as projects involving serious metallurgical or mechanical issues it may be limited to fill and cap passes, as a more reliable process may be more desired for root and hot passes. This is especially true if the weld is accessible only from one side, and the structure is sensitive to fatigue from dynamic loading, and resulting stresses or from environmental conditions.

5.4 Process Description

Flux-Cored Welding (FCAW) uses a spool of filler wire that is either housed inside the power source or fed from an external wire feeder. This wire or filler material is fed through a welding gun. The power source is used to start and maintain the arc between the wire and the base metal.

FCAW uses a hollow wire filled with a flux powder that may or may not need an external shielding gas, because the gas produced from the flux may itself work as the shielding gas. This gas is of course produced from the flux contained in the wire and as it burns in the arc the gas is produced. The flux in the wire serves many of the same purposes as the electrode coating does in the SMAW process, such as adding alloying elements to the weld metal.

FCAW processes does not require the degree of operator skill like GTAW, or SMAW welders. Since other processes typically require very specific electrode positioning and manipulation. In the FCAW process the wire-feeder linked to the machine feeds the wire, the operator needs to do is to maintain the wire stick-out and maintain the arc length. The welding operator holds the gun in one hand, squeezes the trigger, and welds. The shielding gas helps make a very smooth, and stable arc maintenance.

Welding speeds is also higher, because of the continuously fed electrode at relatively higher speed, this results in higher filler metal deposition rates. Its operating factor is typically 30-50 percent, the operating factor is measured of the time spent creating and maintaining the arc.

Currently the FCAW is primarily used to weld ferrous metals like steels and stainless steels, over a wide range of material thickness and operate in all positions. For these reasons, FCAW is usually the welding processes of choice for most fabrication and production shops.

On the downside, the equipment for FCAW is more complex, more costly, and traditionally less portable than SMAW processes. For general maintenance welding, purposes some new portable models are available in the market but with limited capacities. Welding is typically done within a 10 to 12 feet radius of the wire feeder, for this reason or limitations, the work is usually brought to the weld station.

The FCAW process is further divided into two fundamentally different sub-processes.

1. The Self-Shielded Flux-Cored Arc Welding (FCAW-S) process, See Figure 5.3.1, and
2. Gas-Shielded Flux-Cored Arc Welding (FCAW-G) process. See Figure 5.3.2.

While the electrodes for both these sub processes produce a slag covering over the weld, which prevents atmospheric contaminants and also helps reduce the cooling rate. However, their main method of preventing atmospheric contamination is very different.

In the self-shielding FCAW-S process, reactionary agents necessary to shield the arc and cleanse the molten weld pool are placed inside the tubular wire as the flux. And these agents when burned in the heat of the arc, produce gas that provides the shielding gas, and no additional shielding is required.

In the gas shielding FCAW-G process, the core of the electrode has different chemistry, and it does not develop gas that can depend provide the required shielding, for shielding this process depends completely on an external shielding gas supply.

5.4.1 Self Shielding Flux Cored Arc Welding (FCAW-S) Process

As described in the introduction of this chapter, the FCAW process is different in that it uses a hollow-wire which contains materials in its core. As the arc is truck, the material in the core burns in the intense heat of the arc that produces shielding gases and fluxing agents to help produce a sound weld. No further need for any additional shielding gas is required. In that respect this is self-contained shielding process.

The shielding is very positive, and can endure a strong breeze. The arc is forceful, but does produce spatter. When finished, the weld is covered with a slag that usually needs to be removed, by light chipping.

For welding technique, a "drag" angle for the gun is specified, this angle is normally between 10° to 15°. which improves operator visibility. The settings on the wire feeder on the welding machine assumes more importance for this process. Improper technique can cause bad welds. FCAW-S process is primarily preferred to weld mild-steel and suited for welding outdoors.

The FCAW-S welding is primarily different from FCAW-G process on the basis of the electrode, and the content of the core in the electrode.

The process relies on the arc being exposed to the atmosphere, and the resulting reactions of the atmosphere in the heat of the arc, with the elements within the core, which then cleanse the weld pool and help create the protective slag. This reaction occurs because the FCAW-S process predominantly uses an aluminum-magnesium deoxidizing and de-nitriding cleansing system. The weld-metal is typically composed of an average of 1% aluminum in compounds, a volume that is much more than is present in the weld-metal from the other welding processes. However, the aluminum is not in a pure state, the aluminum exists in the form of beneficial compounds. They work as oxygen and nitrogen removers. Aluminum, and magnesium atoms enter the weld pool where they attract oxygen and nitrogen atoms and react chemically to form aluminum oxide, aluminum nitride, and magnesium oxide. These compounds, particularly the magnesium oxide, have high melting temperatures. They are fast freezing, what that means is that as the molten weld pool starts to cool, they solidify more rapidly than other elements in the pool. These lightweight, fast freezing compounds quickly float as slag to the weld surface and protect the weld-metal from further atmospheric contamination. This slag formation system effectively transforms oxygen and nitrogen, two potential contaminants, into chemical compounds that protect the weld.

FCAW-S is an important welding process for steel fabrication in many industries, particularly when the work is done outside in fabrication yards, and other outdoors situations such as offshore structures, pipeline welding, and construction of buildings, bridges, and other infrastructure members. The self-shielding process holds very well against the wind.

FCAW-S welds often produces good mechanical properties, and can be used in all position welding. Generally, these electrodes are designed with good operating characteristics, like they have necessary stiffness with high column strength. They have higher deposition rates where the electrode efficiency is up to 83%. After each weld, the flux-cored wires naturally form a ball of slag at the end of the electrode, this ball acts as an insulator and protects the hot end of the wire from contamination, and this slag ball should be removed for proper arc striking of the next weld.

Additional deoxidizing and scavenging elements are added to the core ingredients, in addition to what is present in the steel sheath. The FCAW process has harsher arc, more spatter, higher fume generation levels are very high as compared to the Gas-Shielded welding processes.

The Self-Shielded electrodes are optimal for outdoor procedures since the flux is built into the wire for positive shielding even in windy conditions. An external shielding gas and additional equipment are not needed, so setting up is simpler, faster and easier.

5.4.2 Flux Core Arc Welding (FCAW-G) Gas Shielding Process

The FCAW-G process uses an inner core that primarily has alloying and arc facilitating elements and chemical compounds. The Gas-shielded flux-cored electrodes produce a light slag, within that are two types of cored wires, wires that have metal core and wires that have some flux components in the core. While the solid and metal- core electrodes leave no slag, requiring no post weld cleaning. The weld is faster and much cleaner, due to these advantages the electrode efficiency increases to 86% to 97%, which becomes the ultimate advantage of the FCAW-G process.

The primary elements in the core include silicon-manganese system used by the other main arc welding processes. The silicon improves the wetting ability, and manganese imparts the strength (Figure 5.4.1).

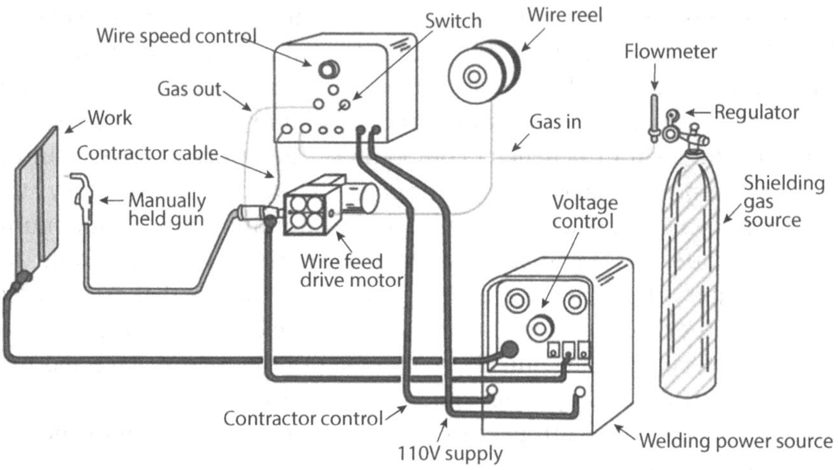

Figure 5.4.1 Typical FCAW setup.

5.5 Welding Wires/Electrodes

Since a wide variety of ingredients can be enclosed in the tubular electrode, FCAW presents a good versatility for welding alloy steels. Alloying ingredients are often included in the core of the wire. AWS Specification A5.20 includes the mild steel electrodes. E70T-1 is an electrode with 70 ksi tensile strength, of tubular construction, suitable for flat and horizontal position welding, with a specific chemical composition. AWS Specification A5.29 includes the low alloy steel electrodes. FCAW electrodes are also specified for surfacing, for stainless steel (AWS A5.22), and for nickel alloys (AWS A5.34). The figure below describes the electrode classification guide for carbon steel, and alloy steel. The alfa numeric system has specific meaning for each number and the letters used. The figure uses examples of E70T-5C and E 71T-9M-JH4 to explain the classification system.

In the Table 5.5.1 typical carbon steel FCAW electrode are listed by their AWS classification, the table includes the electrodes use in welding positions, use of shielding gases, or if they are self-shielding electrodes, and if they are suitable for single pass or for multiple weld passes. The G and GS designated electrode are those electrodes that are not covered under any other designations, for example the multiple-pass electrode classified as EXXT-G they are open for agreement between manufacturers and buyer. Similarly, the classification listed as EXXT-GS is a single-pass type, electrodes not covered under any of the presently defined classification. Since the electrodes covered under these generalized classifications, they are subject to change and most recent AWS classifications system should always be referred for accuracy.

Similar classifications and details can be found in AWS electrode specifications A5.29 for alloy steels, A5.22 for stainless steels, and A5.24 for nickel alloys. The latest versions should be referred for accurate information.

The electrode indicated for use as the single-pass welding contain more highly deoxidizing compounds to address metals that have some rust or mill scale on them. These electrodes help clean these contaminants, while welding the first pass. If these electrodes are used further than the single pass, these deoxidizers tend to build up as alloying elements

Table 5.5.1 Carbon steel electrodes their use descriptions.

Electrode classification (AWS)	Suitable for positions	Shielding gas	Polarity	Application[1]
E 70T-1	H & F	CO_2	DCEP	M
E 70T-1M	H & F	75-80% Ar. + CO_2 Balance	DCEP	M
E 71T-1	H, F, VU, OH	CO_2	DCEP	M
E 71T-1M	H, F, VU, OH	75-80% Ar.+CO_2	DCEP	M
E70T-2	H & F	CO_2	DCEP	S
E70T-2M	H & F	70-80% Ar. + CO_2 Balance	DCEP	S
E71T-2	H, F, VU, OH	CO_2	DCEP	S
E71T-2M	H, F, VU, OH	70-80% Ar. + CO_2 Balance	DCEP	S
E70T-3	H & F	Self-shielding electrode	DCEP	S
E70T-4	H & F	Self-shielding electrode	DCEP	M
E70T-5	H & F	CO_2	DCEP	M
E70T-5M	H & F	75-80% Ar. + CO_2 balance	DCEP	M
E71T-5	H, F, VU, OH	CO_2	DCEP or DCEN	M
E71T-5M	H, F, VU, OH	75-80% Ar. + CO_2 balance	DCEP or DCEN	M
E70T-6	H & F	Self-Shielding electrode	DCEP	M
E70T-7	H & F	Self-Shielding electrode	DCEN	M
E71T-7	H, F, VU, OH	Self-Shielding electrode	DCEN	M
E 70T-8	H & F	Self-Shielding electrode	DCEN	M
E71T-8	H, F, VU, OH	Self-Shielding electrode	DCEN	M
E70T-9	H & F	CO_2	DCEP	M
E70T-9M	H & F	75-80% + CO_2 Balance	DCEP	M
E70T-10	H & F	Self-Shielding electrode	DCEN	S
E70T-11	H & F	Self-Shielding electrode	DCEN	M
E71T-11	H, F, VU, OH	Self-Shielding electrode	DCEN	M
E70T-12	H & F	CO_2	DCEP	M
E70T-12M	H & F	75-80% Ar. + CO_2 Balance	DCEP	M

(Continued)

Table 5.5.1 Carbon steel electrodes their use descriptions. (*Continued*)

Electrode classification (AWS)	Suitable for positions	Shielding gas	Polarity	Application[1]
E71T-12	H, F, VU, OH	CO_2	DCEP	M
E71T-12M	H, F, VU, OH	75-80% Ar. + CO_2 Balance	DCEP	M
E61T-13	H, F, VD, OH	Self-Shielding electrode	DCEN	S
E71T-13	H, F, VD, OH	Self-Shielding electrode	DCEN	S
E71T-14	H, F, VD, OH	Self-Shielding electrode	DCEN	S
EX0T-G	H & F	Not specified		M
EX1T-G	H, F, VD or VU, OH			M
EX0T-GS	H & F			S
EX1T-GS	H, F, VD or VU, OH			S

Legend: H = Horizontal, F = Flat, OH = Overhead, VD = Vertical downward progression, VU = Vertical upward progression. M = Multiple pass, S = Single pass.
[1] Read description in the text.

and act to increase the strength and hardness of the weld metal. These same effects will also be observed when an electrode classified to be used with CO_2 shielding gas is used with a less reactive gas for example, argon or combinations containing argon.

5.5.1 Construction of FCAW Electrodes

Flux-cored electrodes are considered "fabricated" wires. This is because they are not manufactured by the typical drawing process as solid wires, they are produced by forming process from a strip of sheet that is rolled into a tube. The basic manufacturing steps include:

1. Flux-cored electrodes start off in one of two forms of raw steel, round "green rod" or flat "strip".
2. The steel is drawn down and rolled into a "U" shape.
3. Flux ingredients are then precisely and uniformly poured into the U-shape. Monitoring equipment ensures that 100% of the electrode has the proper fill rate.
4. The electrode is then rolled together to form the tube, with a tight seam. The seam could be either a tightly closed butt seam or edge over lapping on each other, to form a lap seam. The outer steel tube is called the "sheath" or "jacket" and the inner portion is the "flux core".
5. The assembled electrode is then drawn down to its final diameter, and a light density lubricant is applied to the surface. This light but uniformly applied

lubricant is chemically so designed that it is not considered as a contaminant in welding. This lubricant helps in the smooth wire feeding. The lubricant on electrodes is also a rust inhibitor, for this reason these cored-electrodes are not copper coated as are the solid electrodes, used in GMAW and GTAW process.

More recently the seamless flux-cored electrodes have come in the market, these seamless cored wires are often copper coated.

5.5.2 Sheath Thickness Variations

The flux ingredients in the core are often referred as the "fill". The ratio of the weight of the fill, as compared to the total weight of the electrode is called the percent fill, which varies slightly between cored wires, depending on their density and volume of alloying elements in the core. As the outside diameter of all cored wires of a particular size designation is fixed as 1.1 mm or 1.6 mm etc., the wall thickness of the sheath varies to accommodate the varying fill percentage. Carbon steel wires generally have lower percent filler and therefore a thicker sheath. While higher alloy wires generally have a higher percent fill and therefore a thinner sheath, this is due to the fact that the alloying elements are tried to infuse in the weld through the wire, and only supplemental or the elements that cannot be infused in weld through the wire metal are included in the core fill.

Hard-facing wires particularly have thinner sheaths. As sheath thickness can also vary slightly from cored wire manufacturer to manufacturer, it is not uncommon for different brands of cored wires of equivalent types and sizes to weld at slightly different procedures. A very important point to note when developing a critical welding procedure.

5.5.3 Important FCAW Variables

The following aspects relating to FCAW are important variables for welding. Their knowledge and understanding will help master the art of FCAW process.

5.5.4 Contact Tip to Work Distance (CTWD)

The contact tip to work distance (CTWD) is a very important part of the FCAW welding. Though CTWD is often considered as a non-essential variable by welding codes, it is very important variable for effectively making and maintain the arc and the very welding itself. The normal recommended CTWD for flux-cored electrodes is 0.75 to 1 in. (20 - 25 mm). This is much longer compared to short circuit GMAW process which is normally about 10 mm (0.375 -in). Welding problems can occur if CTWD is too short or too long. More about CTWD and related terms are discussed further in the chapter.

5.5.5 Travel Angle

As a part of good welding technique, the proper flow of shielding gas either added as in FCAW-G or from the weld flux itself as in FCAW-S, it is always an important part of the welding process, similarly important is the visibility of the weld pool to the welder. Both these aspects of good welding are addressed by keeping a good angle of the torch during the welding.

Always use a drag travel angle with flux-cored electrodes, trailing the slag behind the puddle. Do not use a push travel angle, as this greatly increases the chance of trapping slag and/or decreasing penetration.

5.5.6 Single Pass Limitations

Certain FCAW-S electrodes are limited to single pass welding only. They rely on admixture or dilution with the base metal to produce the weld deposit. If used for multiple pass welding, then after the first few passes you would begin to have an all filler-metal or all weld metal deposit. The resulting alloy content of the weld bead could be undesirable and weld cracking could potentially result.

5.5.7 Thickness Restrictions

Certain FCAW-S electrodes are limited to a maximum steel plate thickness in which they can be used. If used with plate thicknesses beyond these recommended limits, then the cooling rate of the weld metal could be faster than desired, due to the thermal conductivity or weld quenching effect of the thicker plate. This in turn could potentially create an undesirable microstructure in the weld metal, from which weld cracking issues could potentially result. If joining plates of different thickness, ALL the plates should be within the maximum thickness restriction. The plate thickness restrictions also apply regardless if making single pass or multiple pass welds.

5.5.8 Charpy V-Notch Toughness Properties

Well evaluated, and based also on past experiences, the selection of filler wire should be made, in case of any doubt a test specimen may be welded at the given parameters, and this weld should be tested for the required properties. For example, some of the FCAW-S flux-cored electrode classifications are not specified for good toughness values as indicated by the Charpy V-Notch (CVN) impact energy values. If good CVN values are demanded of the weld, the selection of compatible electrode should be made.

Most FCAW welding is done semiautomatically, requiring a power source, wire feeder and welding gun. Some others may be welded automatically, utilizing automatic welding equipment with special welding nozzles, wire straighteners, and other accessories. These are discussed further in the chapter.

5.5.9 Electrode Care and Packaging

Electrodes come in several package sizes, from small spools and coils to bulk packaging in reels and drums. Some packages are hermetically sealed in vacuum sealed foil bags or pails. Small spools are easily loaded or installed in the wire feeder following the machine manufacturer's instructions. They are generally in the following order.

1. A coil adapter is required that fits in the wire feeders with a 2 in. (51 mm) OD spindle.
2. Unpack the coil of wire. Straighten any metal tabs that may have been bent.

3. Lay coil of wire into adapter. Remove the start end of the wire. While always maintaining tension on the wire with one hand, straighten the first six inches. Cut off the first inch. Be sure the cut end is round and burr free. Otherwise, wire may hang up while initially feeding through gun, causing a "bird nest" of tangled wire at the drive rolls. Place the other side of adapter over the coil of wire and adjust until the two sides fit tightly together. There should only be a small gap between the two halves.
4. While still maintaining tension(1) on the start end of wire, place the loaded coil adapter over the wire feeder spindle, lining it up with the spindle pin. To prevent the wire from de-reeling, do not allow the coil and coil adapter to spin. Install the spindle locking collar onto the spindle to hold the coil adapter firmly in place.
5. Feed the start end of wire into the wire feeder's inlet guide and push tight against the drive rolls. Hit the cold feed button or gun trigger and feed just a few inches of wire into the drive rolls. For some other wire feeders, it may be necessary to thread the wire through the feeding liner until about four inches of wire is exposed.
6. Feed the wire the rest of the way through the drive rolls and gun. Use the wire feeder's "cold feed" option, if so equipped. For best initial wire feeding through the gun, keep the gun as straight as possible and remove the contact tip before the wire feeds past the gun tube. Then inspect and reinstall the contact tip. Also check wire braking tension at the spindle and adjust if necessary.

If tension is lost on the wire, that can happen when the welder loses the control of the electrode wire, and let go of start end, this will quickly unravel several loops of wire from the coil. If this unraveling of wire occurs, carefully pull off all loose loops until a new straight starting point is found. While maintaining tension on wire, cut loose loops, and proceed further.

Heavier coils of 50 lb. (22.7 kg) are loaded as per the instructions provide with the machine and wire manufacturers. The general steps for loading the wire are given below.

1. A specific type of coil adapter is required for wire feeders with a 2 in. (51 mm) OD spindle. Wire feeders with the old style 1 in. (25.4 mm) OD shaft require a different type of wire reel assembly coil adapter. Once the adopter type is resolved and suitably fitted, following steps are taken.
2. Unscrew the front side of adapter and set aside. Lay the back side of adapter on a flat surface. Place new coil of wire onto adapter, being careful that the adapter's four spring loaded arms are not lined up on a tie wire or hiding the wire label. The wire should feed from the top or bottom of coil, depending on the type of the wire feeder's configuration.
3. Place the front side of adapter over coil, lining up with four spring loaded arms. Screw top adapter onto bottom adapter. Tighten by hand as much as possible. Force tightening is not advised.
4. Place the loaded coil adapter over the wire feeder spindle, lining it up with the spindle pin, and put on the spindle locking collar. Cut and remove the tie

wires on the coil. Maintain tension on the start end of wire before the last tie wire is cut. Straighten end of wire and feed through drive rolls and gun, as described above.

Most feeding problems are caused by wrong drive roll tension or improper handling of the gun cable or electrode. To ensure these are not the case, refer to the user manual of the supplier or take following steps as minimum.

1. Loosen drive roll tension. While squeezing the wire with your gloved hand between your thumb and forefinger, cold inch the wire and push against it at the contact tip, making the drive rolls slip. Adjust the drive roll tension such that there is just enough feeding force that you can no longer stop the wire from feeding.
2. Do not kink, coil up or pull the gun cable around sharp corners. Keep the gun cable as straight as possible when welding.
3. Do not allow two-wheeled hand trucks, fork lift trucks, etc. to run over the gun cable or other damage to occur to the gun cable.
4. Keep the gun cable clean per instructions in the wire feeder operating manual.
5. Most sheathed and cored electrodes have proper surface lubrication on them.
6. Replace the gun's contact tip when it becomes worn or the end appears fused or deformed.

5.6 Power Sources

FCAW Power sources are similar to the GMAW process power sources discussed in the earlier chapter. The manufacturers now provide machines that give the operators a selection of process on the machine console itself, just by switching the knob. Most electrode diameter of 1.6mm or larger and operate at fairly high current levels. Therefore, an industrial, three-phase power source of at least 350 amps capacity is generally required. Common power sources include transformer/rectifier-based machines, inverter-based machines, or engine driven machines.

The smaller diameter electrode 1 mm (0.045 inch) and smaller diameters and can be used with smaller, single phase, compact wire feeder/welder units. Note all power sources for use with electrodes should have constant voltage (CV) output.

Flux-cored electrodes, can also be used on power sources with pulse welding capability like Power Wave® etc. However, the use of pulse waveforms with flux-cored electrodes has not generally proven to show any measurable benefits, as they have with solid and metal-cored GMAW electrodes. Therefore, wires are generally used with the power source's constant voltage (CV) modes and not with pulsing modes.

5.6.1 Arc Voltage (Constant Voltage)

FCAW machine output is only the constant voltage (CV). The CV ensures a stable arc length, which is very critical for the FCAW process. Power sources with CV output also work the wire feeders, and produce a very consistent arc length. This produces a stable, well balanced arc. If in place of constant voltage (CV), a constant current (CC) is used, then that means that the voltage is variable, this CC will result in the variations in arc length. An erratic arc would result, and potentially cause porosity in the weld, and other defects, including the reduced Charpy V-notch toughness properties. Details about the constant voltage is discussed in chapters 3 and 4 of this book.

As stated above, the voltage affects the length of the arc. As voltage decreases, arc length gets shorter and the resulting arc cone gets narrower and smaller, this results in an excessively convex or ropey weld-bead. If the voltage is increased, the arc length gets longer, resulting wider and larger arc cone. The excessive arc voltage leads to increase in the arc cone surface area. Exposing the arc to the atmospheric air. The limited quantity and limited number of core elements inside the tubular electrode to protect the weld pool from contamination, but in this exposer to air these elements are no longer capable of protecting the weld pool and general heat exposed weld area. At this point nitrogen from air is absorbed into the weld metal. This increases hardness of the weld metal and it leads to the decreased toughness, this can be assessed by both measurement of weld hardness, as well as test for CVN values at the given temperature.

The choice of the constant Voltage (CV) power source for FCAW process to ensure arc voltage for welding there are other situations that should be properly maintained, and periodically inspected to prevent voltage drop across the arc, these are;

- use of very long cable lengths,
- poor work cable connections,
- undersized or damaged cables, and
- poor cable clamps

All the above can cause a significant drop between the set voltage at the power source and the actual voltage at the arc.

5.6.2 CTWD, ESO and WFS

Related to the topic of arc voltage is the subject of contact tip to work distance (CTWD), we have used this term earlier in the paragraph 5.5.4 while discussing the welding variables. This is the distance from the end of the contact tip to the work piece.

Another term often used is the electrical stick-out (ESO). It is not the same as the CTWD and we need to understand the difference.

The ESO is the distance from the end of the contact tip to the top of the arc. In general, ESO is set to about 0.25 in. or 6.4 mm, shorter than CTWD.

It is very important to hold a consistent CTWD with the gun while welding for good arc stability. Maintain this length within ±0.125 in. (3.2 mm) for CTWD of ≤1 in. (25 mm) during welding. The normal recommended CTWDs for flux-cored wires are longer than the CTWD used for the short circuit GMAW process, which is normally about 0.375 in. or about 10 mm.

The wire becomes electrically hot as soon as it touches the inside of the contact tip. This longer CTWD for cored wires results in a split second of extra time of resistance heating in the wire, which allows the core elements to fully react or activate and provide proper protection of the arc.

If the CTWD is too short, incomplete activation of the core elements may occur, potentially resulting in gas marks or porosity on the surface and in the weld. Conversely, too long of CTWD, with no change to wire feed speed, can cause an unstable arc, increased spatter, and decreased penetration.

When inching, the wire is always electrically "hot" to ground, except when using the "cold inch" feature on wire feeders equipped with this option.

Some wires and procedures utilize extended stick-out distances of 1.5 in. to 3.75 in. about 38mm to 95 mm, for higher productivity. The longer CTWDs and resulting increase in resistance heating increases the melt off rate of the wire. Therefore, much faster wire feed speeds should be used, which greatly increases deposition rates. To consistently maintain these long CTWDs at ± 0.25 in. or 6.4 mm, nozzle "insulated guides" are used, theses insulated guides of various lengths are available from the suppliers. While still using extended CTWDs, the visible portion of wire extended beyond the end of the insulated guide, called the visible stick-out (VSO), is much shorter and easier to maintain at a consistent distance. These insulated guides threaded onto the end of the gun tube.

When using the insulated guide, the contact tip to work distance, (CTWD) must be maintained to obtain this distance the insulated guide is removed from the end of the gun tube. Inch the wire out beyond the end of the contact tip until the desired CTWD is obtained. Then the insulated guide is replaced. When using normal CTWDs, "thread protector" insulators are used. This allows the end of the contact tip to extend beyond them, while protecting the gun tube's threads from spatter buildup.

Wire Feed Speed Wire feed speed (WFS) is the rate at which the electrode is fed into the weld pool. This feed rate is measured in inches per minute written as in/min or ipm, or meters per minute or m/min. As WFS increases, so does the deposition rate which is measured in lbs/hr or kg/hr. Welding current and resulting penetration levels are directly related to wire feed speed rates. Higher WFS results in more amperage and more penetration, while lower WFS results in less amperage and less penetration. With CV output, WFS is a more precise weld parameter or setting to use than current, as current varies on CV. Consumable literature often lists approximate currents for specific wire diameters at various WFS levels.

The wire speed is adjusted with the using the WFS control on the wire feeder. If the wire feeder does not have a WFS meter or a scaled WFS knob, the wire speed may be measured manually.

For the manual measure of WFS, following step wise action is required.

1. Disconnect the work clamp from the weld circuit.
2. Pull the gun trigger and feed the wire for six seconds, then measure the length fed out and multiply by ten to get the WFS in inches per minute (in/min) or meters per minute (m/min)
3. Set WFS to the suggested rate as recommended in the wire suppliers manual.
4. The approximate amperage corresponding to each WFS at the specified CTWD is also listed in manual.

Amperage depends on wire feed speed and CTWD. If CTWD is shortened, amperage will increase.

5.7 Other Accessories to Power Source

5.7.1 Welding Cable

Welding cable is needed for both the electrode and work sides of the welding circuit. Choose the appropriate AWG cable size per the chart.

5.7.2 Semiautomatic Wire Feeders

Since most electrodes are fairly large in diameter, this means that heavy duty, industrial, constant speed, semiautomatic wire feeders are generally required. Cored electrodes also require the use of "knurled" drive rolls. A cored electrode cannot withstand as much drive roll tension or squeezing force as a solid electrode can with smooth drive rolls. The electrode would be crushed or deformed. The drive roll's knurls (i.e., teeth) help grip the cored electrode, providing equivalent pushing force, but with less drive roll tension. Note: Knurled drive rolls are not recommended for solid electrodes. The knurls can potentially chew or flake off some of the electrode's external copper coating, causing liner clogging issues. However, cored electrodes with a seam do not have an external copper coating and thus do not have this potential wire feeding issue.

5.7.3 Welding Guns

Welding guns designed specifically for FCAW electrodes, the GMAW wire feeders are not suited for use with FCAW wires. Some feeders do have interchangeable guides to accommodate the FCAW wires, but the welder must ensure to change the suitable guides before using the common feeders from one process to another. Also, the guns have liners that is specific to FCAW welding.

Included with gun. Many guns have additional gun tube options with different angles. (2) Gun liner and contact tip can be changed for use with 0.035 and 0.045 in. (0.9 and 1.1 mm) wire.

Welding Gun tubes A variety of gun tubes (aka nozzles or goosenecks) of various angles, lengths and jacket types (classic braided or stainless steel) are available. The back ends all guns have the same 1/2 in. (12.7 mm) outside diameter.

5.7.4 Reverse Bend Gun Tubes

The majority of the electrodes are very stiff with great column strength and feed-ability. They also feed out of the curved gun tube with a stiff bend, potentially prematurely wearing one side of the contact tip. Therefore, to help straighten the wire and achieve less and uniform tip wear, the gun tube has a second, "reverse bend" in it.

Non-Reverse Bend Gun Tubes Some self-shielded flux-cored electrodes may have a thinner outer steel sheath and/or may be softer. This reduces their stiffness and resulting feed-ability. This is particularly the case with hard-facing flux-cored electrodes. Sometimes feeding them through standard reverse bend gun tubes can be troublesome. Therefore, "non-reverse bend" gun tubes are also available for use if feeding problems are encountered. These gun tubes do not have the second bend in them.

5.7.5 Gun Angles

The angles at which the welder holds the electrode and gun is an important to weld variable. These includes both the travel angle, and the work angle.

The travel angle is the angle between the electrode and a line perpendicular to the surface of work, as measured from the weld side view. For FCAW-S process the drag travel angle is used. The general rule is "drag slag and push gas". Therefore, always use a drag travel angle of typically 20° to 30° with flux cored electrodes. Push travel angle is not recommended, as this greatly increases the chance of rolling slag ahead of the puddle and trapping it underneath. Pushing can also cause the arc to ride on top of puddle instead of in front of it, resulting in less penetration.

Work angle is the angle between the electrode and surface of work, as measured from the weld end view. For a butt joint, typically it is 90° angle, and for a corner, tee or lap joint, typically an angle of 40° to 45° is used. Note that this is not a fixed angle as it varies with the position of the weld, at different passes in a multiple pass weld.

Generally, the travel and work angles used with FCAW electrodes are generally the same as used with SMAW electrodes.

5.7.6 Polarity

The specific core elements or arc stabilizers used in a particular electrode determines the welding polarity in which the arc is the most stable, refer Table 5.5.1. Most of the self-shielding FCAW electrodes operate both on DC electrode negative (DCEN) polarity. And the gas shielded electrodes operate on DC electrode positive (DCEP) polarity. But that is not a fixed rule, the choice of polarity is dependent on the several factors, that includes the chemical constituents of the wire, the position of use, and desired properties of the weld. Refer Table 5.5.1 above and note that electrode in the calcification EX0T-5.EX0T-7, EX1-7, EX0T-8 EX1T-8, EX0T-10, EX0T-11, EX1T-11, EX0T-13. EX1T-13 and EX0T-14 are all on DCEN polarity.

In the same table there are others that are usable on DCEP polarity, a few others that can be used on both polarities, especially to increase deposition rate if the position supports their use.

5.8 Shielding Gases

The self-shielding FCAW-S process does not require additional gas shielding, however the FCAW-G version does.

Carbon dioxide (CO_2) gas is the most widely used shielding gas for FCAW-G since it is cheap and helps in deep penetration. The CO_2 atmosphere can act as either a carburizing or a decarburizing medium, depending on the material being welded. If decarburization occurs (in metals with more than 0.10% carbon) carbon monoxide can be trapped in the weld metal and can cause porosity in the weld.

Gas mixtures are also used to take advantage of the different gas characteristics. Like the GMAW process the most commonly used gas mixture that is used for welding carbon steel is the 75% argon and 25% CO_2 commercial mix.

5.8.1 Attributes of Shielding Gases

Gas-shielded, flux-cored arc welding (FCAW-G) is a very popular and versatile welding process. It is used with mild steel, low-alloy steel and other alloy materials in a variety of applications, such as heavy fabrication, structural, shipbuilding and offshore. The two most common (but not exclusive) shielding gases used with the FCAW-G process are carbon dioxide (CO_2) and a binary blend of 75% argon (Ar.) + 25% CO_2. Other blends, such as 80% Ar. + 20% CO_2, is also be used.

That brings to the question which shielding gas, 100% CO_2 or a blend of Ar. and CO_2, should be selected for the welding?

There is no perfect gas or mix of gases for all types of welding situation. Each type offers some advantages and have some disadvantages. The factors such as quality level of the weld, the parent metal itself, the cost, and desired productivity all play in to the selection process. The selection should be made before the welding procedure is prepared for testing and qualification. The choice of shielding gas affects each of these factors, sometimes they may be complimentary to one another, or they may be in a conflicting in their impact. The merits of the two basic gas options as pointed out earlier for FCAW-G process for welding on steel is the primary focus of this discussion.

5.8.2 How Shielding Gas Works?

To understand the premise of selecting suitable shielding gas or gas blend it is important to know how the shielding works. We have in the earlier paragraphs talked about the subject but here we will elaborate a little more.

The primary function of all shielding gases is to protect the molten weld puddle and electrode from the oxygen, nitrogen and moisture in air. Shielding gases flow through the welding gun and exit the nozzle surrounding the electrode, displacing the air and forming a temporary protective pocket of gas over the weld puddle and around the arc. Both CO_2 and Ar. and CO_2 blends shielding gases accomplish this purpose.

Some shielding gases make it easier to create the arc plasma, providing a current path for the welding arc. The choice of shielding gas also affects the transfer of thermal energy in the arc and forces on the puddle. For these issues, CO_2 and Ar. and CO_2 blends behave in very different ways, that is essential to know for proper selection for the job at hand.

5.8.3 Properties of Shielding Gases

As stated above the carbon dioxide, and argon gases respond in different ways under the heat of the arc. Three basic criteria discussed below, that the shielding gases may have and are important to know to understand the properties of each shielding gas.

1. **Ionization potential** is a measure of the energy required to ionize the gas (i.e. transform to a plasma state in which it is positively charged), enabling the gas to conduct current. The lower the number, the easier it is to initiate the arc and maintain arc stability. The ionization potential for CO_2 is 14.4 eV, versus 15.7 eV for argon. Thus, it is easier to initiate an arc in pure CO_2 than in pure argon.

2. **Thermal conductivity** of a gas is its ability to transfer thermal energy. This affects the mode of transfer (spray versus globular, for example), shape of the arc, weld penetration and temperature distribution within the arc. CO_2 has a higher thermal conductivity level than argon and an Ar. + CO_2 blend.
3. **Reactivity** of a gas is a classification as to whether or not it will chemically react with the molten weld puddle. Gases can be divided into two groups, inert and active. Inert gases, or noble gases, are those that do not react with other elements in the weld puddle. Argon is an inert gas.

Active gases, or reactive gases, are those that chemically react and combine with other elements in the weld puddle to form compounds. At room temperature, CO_2 is inert. However, in the arc plasma, CO_2 will disassociate, forming CO, O_2 and some monotonic O. Therefore, in the heat of the welding arc CO_2 becomes an active gas, allowing the oxygen to react with metals, metals oxidize in the arc. In the Ar. + CO_2 blend the CO_2 portion is an active gas, but less reactive than 100% CO_2. Higher the percentage of the CO_2 gas in the blend more reactive the blend will be.

With all other welding variables being equal the different shielding gases would produce different welding fume at different rates. It is now obvious that the Ar. + CO_2 blend will produce less gas as compared to 100% CO_2. This is due to the oxidizing potential of CO_2. Specific fume generation level would further vary based on the particular application and welding variables, and parameters used in the specific welding procedure.

5.8.4 Limits on the Use of Inert Gases

Although inert gases provide weld puddle shielding, they are not suitable by themselves for FCAW-G welding on ferrous metals and alloys like carbon steel, low-alloy steel, and stainless steel, etc. If, for example, 100% Ar. were used for welding on carbon steel, the resulting weld characteristics would be very poor. The outer steel sheath of the electrode prematurely melts. The arc length is excessive, the arc is wide and uncontrollable, and there is excessive weld build up. Therefore, for ferrous metal FCAW-G welding applications, inert gases are always used in a binary blend with an active gas.

5.8.5 Argon and Carbon Dioxide Gas Blends

The most common blend for carbon steel FCAW-G applications is the 75% Ar. + 25% CO_2 blend. Another blend 80% Ar. + 20% CO_2 is also used for welding carbon steel with FCAW-G process. Yet another blend of 90% argon with balance CO_2 is used with specially designed flux-cored wires for FCAW-G process. However, it may be noted that rarely a gas blend of inert gas less than 75% argon is used for welding carbon steel and alloys. The 75% argon in the blend is the optimal ratio for welding gas blend. This is because as the argon content in the blend is decreased below 75%, the effects of argon on the arc characteristics begin to diminish, yet the costs of having argon in the shielding gas are still incurred. In addition, non-standard percentages of Ar. + CO_2 blended cylinders will typically be more difficult to obtain than standard blended cylinders, like 75% Ar. + 25% CO_2 or 80% Ar. + 20% CO_2 blends.

5.8.6 How the Shielding Gas and Blends Affect the Mechanical Properties of the Weld Metal?

Due to the reactive nature of CO_2, it forms oxides with the alloying elements in the wire, especially with Manganese and silicon, these oxides end up with the flux in the slag, and are lost. A portion of these alloys react or oxidize with the free oxygen from the CO_2, ending up in the slag instead of being recovered in the weld metal.

A higher level of alloy recovery is made from the given electrode in the weld metal when using an Ar. + CO_2 blend shielding gas. With the reduced CO_2 in the blend shielding gas a higher level of Mn and Si is recovered in the weld deposit. The consequences of higher levels of Mn and Si in the weld deposit are an increase in weld strength and a decrease in elongation, as well as changes to the impact properties.

By simply changing from CO_2 to an Argon + CO_2 blend, the increase of 7 to 10 ksi tensile strength can be obtained. The adverse effect of the higher percentage of argon (Inert) gas could lead to higher weld strength and too low ductility. Table 5.6.6.1 below details the impact of shielding gas on the strength and ductility of weld metal.

Table 5.6.6.1 Impact of shielding gases on the mechanical properties of weld metal.

MECHANICAL PROPERTIES[1] – As Required per AWS A5.20/A5.20M: 2005					
	Yield Strength[2] MPa (ksi)	**Tensile Strength MPa (ksi)**	**Elongation %**	Charpy V-Notch J (ft•lbf)	
				@ -18°C (0°F)	**@ -29°C (-20°F)**
Requirements[4] AWS E71T-1C H8, E71T-1M H8	400 (58)	480 - 655	22	27 (20) min.	Not Specified
AWS E71T-9C H8, E71T-9M H8	min.	(70 - 95)	min.	Not Specified	27 (20) min.
Typical Performance[3] As-Welded with:					
100% CO_2	510 - 550 (73 - 79)	570 - 600 (82 - 87)	26 - 28	38 - 95 (28 - 70)	27 - 65 (20 - 48)
75% Ar/25% CO_2	570 - 610 (82 - 88)	620 - 660 (89 - 95)	24 - 26	62 - 111 (46 - 82)	39 - 85 (29 - 63)

MECHANICAL PROPERTIES[1] – As Required per AWS A5.20/A5.20M: 2005						
	%C	%Mn	%Si	%S	%P	**Diffusible Hydrogen (mL/100g weld deposit)**
Requirements[4] AWS E71T-1C H8, E71T-1M H8 AWS E71T-9C H8, E71T-9M H8	0.12 max.	1.75 max.	0.90 max.	0.03 max.	0.03 max.	8 max.
Typical Performance[3] As-Welded with:						
100% CO_2	0.03 - 0.04	1.28 - 1.41	0.42 - 0.49	0.01	0.01 - 0.02	3 - 8
75% Ar/25% CO_2	0.03 - 0.04	1.45 - 1.60	0.54 - 0.62	0.01	001 - 002	4 - 8

Importance of impact on weld properties has been understood and acknowledged for long, and it is reflected in several industrial specification, codes, and regulations. For example, the AWS structural code, D1.1 includes a series of mandatory requirements directing the fabricators and engineers to ensure compliance to the impact of the shielding gas on metal properties. And for that objective in mind the AWS has issued as specification A 5.32 to regulate the shielding requirements. Failing which, the fabricator is required to requalify their WPS to the prove the properties of the gas and filler metal combinations.

The importance of the Filler Metal and gas combination is also included in the electrode (wire) classification, refer Figure 5.5.1 where, the classification is "EXXT-XX", where the last X is the "Shielding Gas Designator." It will either be "C" for 100% CO_2 or "M" for mixed gas of 75 – 80% argon + balance CO_2 as shown in the following two examples.

- E71T-1C, here the C indicates 100% CO_2 gas shielding and
- E71T-1M, here M indicates 75% Argon + 25% CO_2 gas shielding.

For a low alloy steel FCAW-G electrode, the shielding gas designator follows the deposit composition designator.

for example,

- E81T1-Ni1C = here the letter C meaning 100% CO_2 is after the deposit indicator

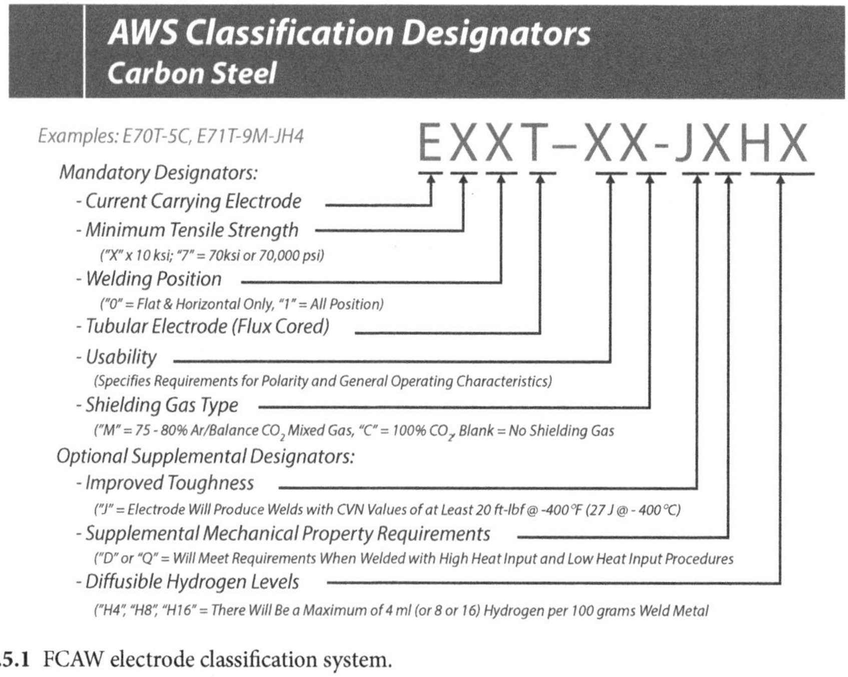

Figure 5.5.1 FCAW electrode classification system.

However, there is no position for the letter indicating the gas in self-shielded flux-cored electrodes (GMAW-S) process. This position is not filled because the process does not require gas shielding.

- Example is, E71T-8

As we now note that some electrodes are designed to be used solely with 100% CO_2. And there are other electrodes that are developed for use only with Argon + CO_2 blend. And the other group of electrodes are developed that can be used either with 100% CO_2 or an Argon + CO_2 blend. It is important to note that this third group of electrodes must comply with requirements of both gas shielding, and meet other requirements of specific classification.

5.8.7 Understanding the Performance of Various FCAW-G Gases

As we know that the choice of shielding gas rests between either 100% CO_2 or an Ar. + CO_2 blend for flux-cored welding. In choosing one from other the following three points need considered.

5.8.7.1 Shielding Gas Cost

For any project, and for any welding contractor the cost control is the most importan tor for their business. So, it is oblivious that the cost of shielding gas assumes impo in what gas to choose for their specific needs. Gas cost as part of the total welding significant, and it factors in their profitability. In general, 80% of total welding cost attributed to labor and overhead expenses and 20% to material costs; with shieldi accounting for as much as a quarter of material costs, that is about 5% of tota costs. If cost of shielding gas is the only deciding factor, then significant cost savi achieved by using CO_2 over an Ar + CO_2 blend. However, this is not a unilater there are several other factors that have over bearing influence the total weldin

It is clear and well known that CO_2 costs less than Ar. + CO_2 blends because it gas to collect and bottle, and the sources are plentiful considering the environ and capture of CO_2 it is even more easily sourced, and widely available all ove the welding industry, a common source is from the processing or crackin Argon, on the other hand, can only be collected from air. With argon con than 1% in the atmosphere, a tremendous amount of air must be process large quantities. Special air separation plants are required to process air. Ai consume large quantities of electricity and are only located in specific are

5.8.7.2 Overall Operator Appeal and Impact on Productivity

When comparing shielding gases for use on the same type and siz softer arc characteristics and lower spatter levels are achieved with resulting in an increased overall weld quality, and deposition rate, quality especially post weld cleaning and reduced deposition rate w

A welding arc in CO_2 shielding gas has a more globular transfe typically larger than the diameter of wire, this results in a har

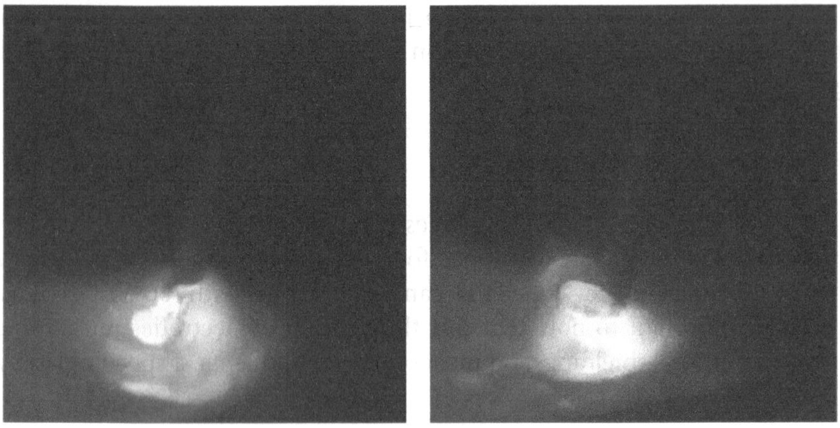

Figure 5.8.7.2.1 Shows the metal transfer through the arc with CO_2 shielding on the left, and 75% Ar. + CO_2 on the right.

higher levels of spatter affecting the deposition rate and post weld cleaning by the welder. On the other hand, the welding arc shielded by an Ar. + CO_2 blend has more of a spray arc type transfer with smaller droplet sizes, typically smaller than the diameter of wire, result-ng in a much smoother and softer arc, this lowers the spatter levels of the weld. Lower ɔatter is less chances of accidental burn for the welder, and less cleaning post weld (Figure 3.7.2.1).

An Ar. + CO_2 blend has much lower thermal conductivity. The lower conductivity means t heat is focused and concentrated in the weld zone, this attribute keeps the weld hotter fluid. This concentration of heat, is not so much for the weld done under CO_2 shielding. ıe focused and low conducting heat provides better wetting of the weld at the toes of 'eld, and it makes it easier to work the puddle. These qualities of the Ar. + CO_2 blend ıproves the out of position like vertical up and overhead positions welding.

limitations of using an Ar. + CO_2 blend primarily comes from the higher argon con-ıd also from the attribute of low conductivity described above.

Due to higher argon content, and inability of heat to dissipate through the metal, the heat radiates up towards the welder, making the welder more ıncomfortable in the heat.

his heat also makes the welding gun much hotter than as compared with ın with CO_2 shielding. In other words, the duty cycle is lower in case of + Co_2 gas shielding.

shielding gas affects the quality of weld. While we have learned that fluidity of d better wetting associated with Ar. + CO_2 gas shielding allows a good fusion of the weld. And that the reduction of spatter level improves the appearance d reduces the post weld cleaning, to allow for inspection activity such as UT ıy.

ity issue associated primarily with the Ar. + CO_2 gas shielding is the suscep-;as marks, on the surface of the weld. The gas marks are not weld defects, face blemish which are small grooves that sometimes appear on the weld caused by dissolved gases in the weld metal that have escaped out of the weld metal, before the puddle freezes, but then are trapped underneath the slag after it has solidified. There are also known as the warm tracks or chicken scratch.

There is a higher susceptibility to gas marks with an Ar. + CO_2 blend than with a CO_2 shielding gas. This is attributed to the fact that the presence of Argon gas in the shielding gas results in smaller droplet size and a greater number of droplets. This increases the total surface area of the molten droplets, resulting in a higher level of dissolved gases in the weld metal.

5.8.7.3 Typical Use of Shielding Gas

Over the years, the type of shielding gas used for FCAW-G welding has been standardized for some main applications and industries. For example, for high deposition applications using flat and horizontal only type wires, CO_2 is preferred, as little benefit is achieved with an Ar. + CO_2 blend in the down hand position.

Heavy fabrication industries like offshore structure fabrications, and shipyards also prefer to use CO_2 for heavy welding because the arc characteristics have proven ability to burn off light scale, rust, and primer on the base material. However, where better quality and appearance is desired as in the offshore structures in fabrication yards, where weld is done in vertical down position, as the final passes of groove welds on T, K and Y configuration of members requiring welds with very smooth weld contour for fatigue sensitive joints, and minimal spatter levels, choice of Ar. plus CO_2 blend as shielding gas is made.

5.9 Welding Various Metals

FCAW process is very useful process for heavy construction and fabrication, the process has good deposition rate and cost effective.

The application of the FCAW process is strongly linked to the available welding wires, and the flux system for the specific type of material being welded.

Its limitations are the availability of wires and flux core, for specific material sought to be welded.

The above discussion is primarily tailored for welding carbon steel, and alloy steel welding. The following is a set of guidelines, that can be converted as a welding procedure of carbon and alloy steel, around AWS structural specification D1.1, and sheet metal code AWS D1.3. The guideline can be used to develop FCAW welding skills, or to develop a simple FCAW-G welding procedure.

The welding is done with semi-automatic Flux Cored Arc Welding (FCAW-G) process.

If fully-automatic process is used an increase in amperage of about 25% should be done, and worked from there to adjust the optimal required amperage.

Semi-automatic welding:
Welding with equipment that supplies continuous wire feed with or without means for mechanical travel. Manual manipulation by the welder of one or more of the variables of speed of travel, guidance, and direction of wire is involved during the welding operation. Joints shall be made following the procedural stipulations indicated in Applicable Codes, and may consist of single or multiple passes as specified in the qualified WPS.

5.9.1 Applicable Base Metals

The base metals shall conform to any of the following groups:

Steels in Groups I, II, III, IV of Table 5.3 of AWS D1.1 Code

Steels in Table 6.9 of AWS D1.1 Code

Steels in Groups I, II, III, IV of Table 1.2 of AWS D1.3 Code

Materials confirming other specifications may be welded providing WPS are prepared, approved, and approved by the engineering.

Base Metal Thicknesses

AWS D1.1: Base metal thicknesses from 3 mm (1/8 in) to unlimited thickness.

AWS D1.3: Structural sheet/ strip steels, including cold formed members which are equal to or less than 5 mm (3/16 in) in nominal thickness

Storage and Conditioning of Wires

Wires shall be dry and free from surface rust and foreign material.

Shielding Gas

The shielding gas shall be of welding grade, that has the dew point of -40 °C (-40 °F) or lower.

Flux cored arc welding with external gas shielding shall not be done in a draught or wind unless the weld is protected by a shelter. This shelter shall be of material and shape appropriate to reduce wind velocity in the vicinity of the weld to 8 km/hr. (5 mph) or less.

Suggested Spec. for FCAW-G

Position(s) of Welding

Preferably, the welding expected to be done in the flat (1G) position, but other positions such as horizontal, vertical and overhead are permissible as specified on WPS, note that variations in position may demand change in welding variables discussed above this includes the amperage, WFS, and CTWD, etc.

Electrical Characteristics

DCEP current and polarity is recommended. The machine should be capable of giving constant voltage (CV) output.

The range of parameters, to be qualified with WPS qualification. And it may start with the guidelines of wire manufacturer's recommendations.

Minimum Preheat and Interpass Temperature

The minimum preheat before welding will comply with Table 5.8 of AWS D1.1 for Prequalified WPS or Notes of Table 6.9 of AWS D1.1 for WPS that need qualifications.

For sheet metal welding refer or as per Clause 7.4 and Annex A of AWS D1.3.

In the event that the welding is interrupted for the temperature of the base metal to falls below the minimum preheat temperature, then the weld area must be preheated again to bring the metal back to the pre hat temperature. Even heating all along the work are ais required, prior to recommencing welding.

The completed weldment is allowed to cool to the ambient temperature, without external quench. Avoid quick cooling.

Heat Treatment and Stress Relieving

Unless required by the governing code or job specification post weld heat treatment of any kind id not specified in tis guideline.

5.9.2 Types of Welding Procedure Specifications (WPS)

There are two types of WPS, Prequalified or that need to be qualified by mechanical and NDT testing.

Prequalified WPS

These use weld joints with fixed welding parameters, that cannot be changed, and weld must be completed in strict adherence to those specified in the Pre-qualified WPSs. Several codes include these pre-qualified WPS to relive the burden of qualification for welding materials that are routine, and follow routine weld parameters, but those parameters are not open to any alterations and changes during welding.

It is assumed that if the specified parameters and conditions are adhered to the mechanical properties of the weld will not change Clause 5 of AWS D1.1 includes these pre-qualified WPSs. Similarly, Clause 5 of AWS D1.3 includes pre-qualified WPS for sheet metal welding. ASME section IX also includes some prequalified WPS.

If a job requires special tests such as Charpy test at certain temperature or any other then use of prequalified procedure is not recommended, and the new WPS should be developed and tested by welding and actual mechanical testing, as required by the work specification.

5.9.3 FCAW Welding Austenitic, Ferritic Stainless Steels and Duplex Steels

Flux-cored welding wires are developed for welding austenitic steels, Ferritic steels and Duplex steels with FCAW-G process, and they are listed in the AWS specification A 5.22. These welding wires are for use with specific shielding gas compositions. However, since these corrosion resistant alloys (CRAs) metals often serve in very challenging environment, a carful metallurgical consideration must be made based on the actual job requirements to select the specific wire, and shielding combination for welding.

Welding wires

Welding wire for common stainless steels grades are very commonly available, however when it comes to less used and very specific type of welding wire the availability decreases and special orders may be required.

As stated above the welding wires for the FCAW of stainless and Duplex steel are described in AWS 5.22 and it should be referred in detail for proper selection of welding wire. Following is the brief description of welding wires that are available for FCAW process.

5.9.3.1 Stainless Steel

E308LTX-X and **ER308LSi** TX-X is predominately used on austenitic stainless steels, such as types 301, 302, 304, 305 and cast alloys CF-8 and CF-3. For high temperature applications such as in the electrical power industry, the high carbon E308H TX-X electrode provides better creep resistance than does 308L.

E316L TX-X and **E316LSi** TX-X filler metal should be used with E316LTX-X and E316 base metals. For welding purpose only, the cast alloys CF-8M and CF-3M can be treated as equivalents of 316 and 316L, respectively.

E309LTX-X and E309LSiTX-X for joining mild steel or low alloy steel to stainless steels, for joining dissimilar stainless steels such as Type 409 (ferritic) to itself or to type 304 stainless, as well as for joining type 309 base metal. For welding purpose only, the cast alloys CG-12 can be considered as the equivalent of type 309 and welded using matching wire. Some ER308L applications may be substituted with E309LTX-X filler metal, but E316LTX-X or E316TX-X applications generally require molybdenum, and 309L contains no molybdenum.

Type **E347TX-X** stainless steel filler metal is ideal for joining stabilized base materials, type 347 and type 321, because the wire matches these stabilized grades. For welding purpose only, the cast alloys CF-8C can be treated as the cast equivalent of Type 347. Type 347 filler metal is also suitable most E308LTX-X filler metal applications.

5.9.3.2 Duplex Steels

The FCAW wire for welding Duplex steel are difficult to find, some basic welding can be done using E 410 TX -X or E 410Ni Mo TX-X welding wires developed for ferritic stainless steels. But welding UNS 2205 and 2304 grade with FCAW is difficult, there are some FCAW wires developed under AWS specification A5.22 like, E 2205T0-1 that is used to weld Duplex steels. But these alloys and even higher alloys are preferably welded with GTAW process.

5.9.3.3 Welding Ferritic Stainless Steels

These alloys are welded in a limited way, using E 410Ni MO TX -X or E 409 TX-X electrode with suitable shielding gases.

5.9.3.4 Choice of Shielding Gases

For austenitic stainless steels, and the Duplex steel the choice of shielding gas is either 100 percent CO_2 or 75 percent argon + 25 percent CO_2. The slag covering on the weld limits carbon absorption, so shielding gases with high CO_2 content can be used.

The chosen gas blend typically depends on the welding position and operating conditions. An argon + CO_2 blend generally provides the widest range of operation and the best operator appeal. With FCAW-G process, argon + 25 percent CO_2 blend offers good control for out-of-position welding and a reduced distortion rate compared with 100 percent CO_2 shielding.

5.9.4 FCAW Welding of Aluminum

Normally, aluminum is not welded with FCAW process. In fact, it is difficult to weld aluminum with FCAW process. There are several obstacles that prevent the use of FCAW aluminum wire. These include usability and quality issues.

Even solid aluminum wire is soft and tricky to feed. Just imagine trying to feed a hollowed-out flux core wire without crushing and jamming. Basic wire drives just wouldn't be up to the task. It would take a specialized drive system.

Because of aluminum's unique characteristics, fluxes designed for carbon or stainless steels do not work. And fluxes used in soldering and brazing aren't effective at arc welding's higher temperatures.

Also, aluminum fluxes tested for welding are extremely corrosive. This is a concern for both users and the environment. Plus, the fluxes are hygroscopic and absorb water from the air. These traits contribute to poor weld quality and excessive spatter.

5.9.5 Welding Nickel and Nickel Alloys by FCAW Process

FCAW process has very limited application for day-to-day welding of nickel or nickel alloys.

Nickel alloys are best welded with GTAW, SMAW and to some extent by the GMAW process. Wires and electrodes are abundantly available for these processes.

The AWS specification A5.34 details requirements and regulation of nickel some of the welding wires for welding Nickle-chromium alloys and Nickle-Iron-Chromium alloys. The welding wire for welding these groups of alloys are E NiCr-3, is available for welding of alloys 600, 601 and 800, and 800HT and joining dissimilar alloy combinations.

Another Nickel chromium alloy welding wire E NiCrMO-3 T1-4 is specially developed to weld alloy 625 and other heat resistant alloys and also for cladding.

If absolute necessary to weld with FCAW process, other groups of nickel alloys like, nickel-iron may be welded using these wires. But a test-weld must be carried out to ensure that the job specifications are fully met to the satisfaction of the project objectives. In case of uncertainty, it is best to use either SMAW, GTAW, or GMAW process to weld some of these alloys.

Limited availability of suitably matched flux-cored wires is the primary reason for low application of FCAW process to weld nickel and its alloys.

There are number of classified by AWS 5.34, as well as some non-classified FCAW welding wires in the market that are used for surface build-up, cladding, and/or hard-facing operations. They contain high percentage of chromium with other metals. These flux cored wires are generally used with FCAW-S process.

5.10 Tips for Good Welding by FCAW Process

It is very important to get a good, solid work connection. This means that the contact surface is thoroughly cleaned to establish good electrical contact.

Work clamp should be tightly attached to the work.

Connect the welding machine through an independent circuit breaker, with suitable fuse rating to protect the machine from surge of high current.

The weld joint should be properly fitted for a god weld. Avoid gaps whenever possible to minimize burn-through problems. This is especially critical on thin sheet metal.

Ensure proper up keep of wire feed and gun cable, any kinks in wire feed cable would be detrimental to good feeding of wirc to wcld.

Refresh the wire contact tip, to look they are without elongated or melted ends.

Cutting the wire contact tip to an angle helps in quick initiation of arc.

Follow suggested electrode stick-out and maintain it for good welds.

Follow established welding procedures.

Ensure wire drive rolls are maintained and that they are capable of feeding the wire smoothly with proper tension.

Relax and try to hold the gun as steady and smooth as possible.

Observe and follow all welding safety precautions as specified in the machine's Operators Manual, and the work place safety rules relating to the potential for electric shock. Heat from the Arc rays can burn skin and eyes, cause fire and explosion, and proper ventilation. For more details, consult ANSI Z 49.1.

5.11 Test Your Knowledge

The following questions are set on the basis of this chapter as well as some information contained in the previous chapters that relate to FCAW process. It is expected that the reader of this book and this chapter will be able to answer all questions.

To be able to master the subject, it is advised that the reader take the test and answer all that they can, then get back in the chapter to verify their responses and that way they revise essential points once again. After a couple of hours take the test again, a noticeable improvement will be noted. If required, repeat this method till you answered all questions correctly.

1. What are two different types of FCAW processes?
2. What is the significance of CV for FCAW process?
3. What are different types of DC welding machines, suitable for FCAW process?
4. What is the role of the flux in the FCAW wire?
5. Can a cored wire be used, for FCAW-S and FCAW-G process? Explain, how the two types of wire different.
6. Explain the effect on impact properties of steel, with the argon blend gas, vs. 100% CO_2 gas shielding on carbon steel welding.
7. How does the drooping characteristics of a welding machine maintain constant voltage?
8. Explain the DCEN and DCEP, in each circuit what direction the electrons flow?
9. What are the dangers of arc flash?
10. From which shielding gas you get better (lower) hydrogen by welding with Ar. + CO_2 or 100% CO_2 shielding?
11. How is welding wire for FCAW classified, give AWS classification example of a carbon steel or a low alloy steel wire?
12. What are the key characteristics of welding austenitic steel?
13. State the importance of wire/electrode extension.
14. Give the effect of argon +CO_2, Vs Argon + O_2 blend of gases on welding carbon and low alloy material.
15. Why it is easier to initiate arc with CO_2 gas shielding, as compared to argon gas, and why?
16. Can you weld Aluminum with FCAW process?
17. What type of gas is primarily used for FCAW welding of stainless steel?

18. What are gas marks in FCAW process and how it can be eliminated?
19. For E71T-14 welding wire what type of current and polarity is used, in horizontal position? Can you use this wire for multiple pass welding? What kind of shielding gas will be your choice?
20. You are tasked to weld a nickel alloy 625, what will be your electrode and shielding gas selection for the FCAW welding?
21. What is the difference between pre-heat and Interpass temperatures?
22. What do you understand by Post weld heat treatment?
23. Using E 71T-1 wire in vertical down progression, and over-head position, with DCEP current, what shielding gas is your choice?

6
Submerged Arc Welding (SAW)

6.1 Synopsis

This chapter about the SAW process details the basic principles of the process and how it works, and shows it with number of pictures and tables. It introduces other variables of the process and why these variants are useful for the industry. The chapter describes the advantages and the limitations of the SAW process.

The chapter include details on the various types of fluxes, and some information on the choice and selection of the welding wire flux combinations.

6.2 Keywords

AC and DC power sources, flux, agglomerated flux, bonded flux, fused fluxes, neutral, basic flux, acid flux.

6.3 Introduction to Submerged Arc Welding (SAW) Process

The Submerged Arc Welding process is generally not available in most of the welders that re starting their career. Welders may encounter the process after their initial works with other process most likely with SMAW or GMAW or both.

Fabrication shops that use this process use it for heavy welding for both its ability to weld thick sections and faster production rate. Often, with the time welders slowly graduate into the process as they grow their efficiency.

Submerged arc welding (SAW) process is an arc welding process which produces coalescence of metal by heating them with an arc or multiple arcs (Where such system is used), between the bare metal electrode or electrodes and the work.

The arc is shielded by a mound of granular flux, fusible material that continuously deposited on the work. These fluxes may be utilized to supplement the alloying elements to the weld that is made using the filler wire which is also an electrode.

Submerged arc welding process is one of the most used process in heavy fabrication industry, but it is also limited by its inability to be potable. The process requires the work to be accessible closer to the machine set up, often to a relatively fixed work station. The process's limited mobility is inherent in its very setup, that includes following.

- Need for the welding head Gun, nozzle etc., to travers along the weld line, that weld line may be longitudinal or radial. But lay in flat or horizonal position.
- Use and delivery of flux from a fixed location.

Figure 6.3.1 below is the schematic view, and descriptive explanation of the process, in this figure a single arc is shown, the figure also shows the cut-out view of the completed weld. From the top to the weld, at the top is the residual flux, flux that is not consumed during the welding operation but worked only to provide blanket coverage to the weld beneath. Below that blanket is the fused layer of the flux in the form of the solidified slag. And finally, below the slag is the weld itself.

Note the positions of the weld head, and the flux feed hopper, and the direction of the weld. The power supply and electrode wire are fed through the welding head, and the flux is through the hopper.

Similar to GMAW process of welding, the SAW process involves formation of an arc between a continuously-fed bare wire electrode through the nozzle and the workpiece. But unlike GMAW the SAW process uses granulated flux to generate protective gases and slag, and to add alloying elements to the weld pool.

The effect of gravity on the flux feeding through the hopper to fully blanket the weld area, this blanket method is a versatile approach to produce welds with minimal contaminations. Figures 6.3.2, 6.3.3 and 6.3.4 show the SAW welding process show that flux coverage in both pipe and plate welds. However, this advantage of protecting the molten weld pool also limits the versatility of Submerged Arc Welding, as shown in the Figure 6.3.1 above, it shows that the process must be performed in the flat, and horizontal fillet positions only. So that the flux is fed vertically down over the area of the weld, except in very special cases. These

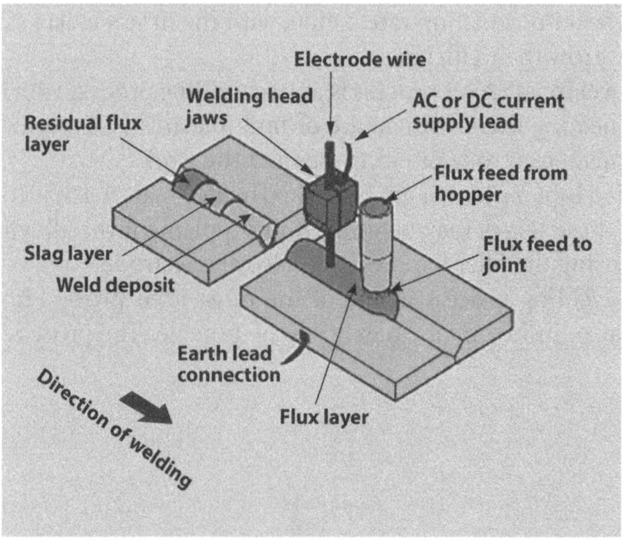

Figure 6.3.1 Schematic display of the SAW process.

Figure 6.3.2 Shows the submerged arc welding of a plate.

Figure 6.3.3 Shows the SAW of a pipe in a fabrication shop – note the arc and flux position as the pipe rotates.

very special cases include vertical and horizontal welds using specially designed equipment, such as belts or shoes that would hold the flux in position.

Much like the GMAW-S, but unlike GMAW-G, both these processes have been described in preceding chapter of this book. The SAW process does not use a shielding gas to protect weld pool and surrounding area from atmospheric contaminations. In the SAW process the

Figure 6.3.4 Shows the completed pipe weld.

shielding is provided by the flux itself. Prior to welding, a thin layer of flux powder is released on the workpiece surface, this activity serves several important functions above is indicated.

1. It ensures proper flow of flux to the weld,
2. It leaves a fine dust of the flux that helps initiate arc in the protective environment,
3. It starts both essential functions of the flux, (i) the protection of weld pool from the atmospheric contamination, and (ii) addition of alloying elements in the weld.

The arc moves along the joint line and as it does so, excess flux is released on the arc through a hopper, see Figure 6.3.1 to 6.3.3 how that flux is fed through the hooper in each case.

As the weld is completed, part of the flux that is closest to the arc heat and the weld-pool is consumed and forms a hard slag covering, remaining flux is recovered, as shown in the figures above, note the suction head trailing the weld, and in some cases, they are recycled for further welds but not all fluxes can be recycled. Type of fluxes are discussed further in the chapter.

Remaining fused slag layers can be easily removed after welding. As the arc is completely covered by the flux layer, heat loss is extremely low. This produces a thermal efficiency as high as 60%, compare this with the 25% heat efficiency of SMAW process. The deposition

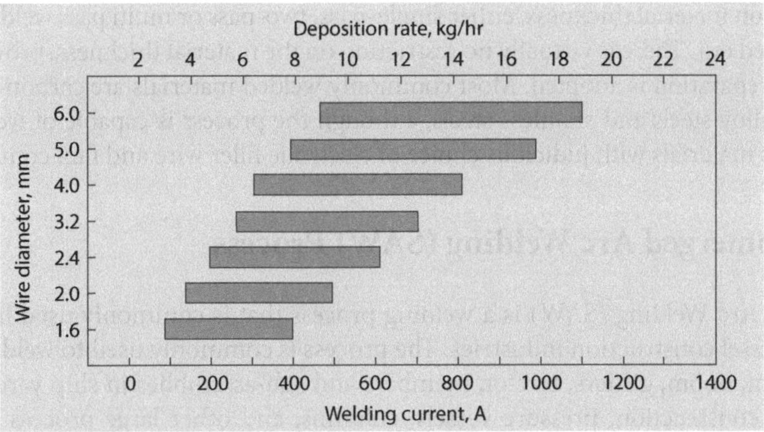

Figure 6.3.5 Higher deposition rate of SAW process.

rate of SAW process is also very high as compared to other arc welding processes; the Figure 6.3.5 above shows the result of one such study about the deposition rate rising as the diameter of the wire is increased and the corresponding current is raised.

Another advantage of the SAW process is that there is no visible arc light, welding is spatter-free, and there is no need for fume extraction.

6.4 Operating Characteristics

SAW is usually operated as a fully-mechanized or automatic process, but it can be semi-automatic.

Welding parameters, like current, arc voltage and travel speed all affect bead shape, depth of penetration and chemical composition of the deposited weld metal, and the rate of deposition. The graph shown in the Figure 6.3.5 above illustrates the higher deposition rate of the SAW process.

Because the operator cannot see the weld pool, greater reliance is on parameter settings.

Similar to the GMAW process of welding the SAW process involves formation of an arc between a continuously-fed bare wire electrode and the work piece. However, the SAW process uses flux to generate protective slag, and to add desired additional alloying elements to the weld metal.

The effect of gravity on the flux feeding into the weld area and the molten weld pool limits the versatility of the process. Due to this limitation, this process must be performed in the flat and horizontal position only, except in some very special cases where specific adaptations are made to deliver and hold the flux in position to blanket the weld area. These special cases include vertical and horizontal welds using special equipment, like special shoes, or belts to support the flux in place.

SAW processes is ideally suited for longitudinal and circumferential butt and fillet welds. However, because of high fluidity of the weld pool, molten slag and loose flux layer, welding is generally carried out on butt joints in the flat position or 1G position for pipes, and fillet joints in both the flat and horizontal-vertical positions. For circumferential joints, the workpiece is rotated under a fixed welding head with welding taking place in the flat position.

Depending on material thickness, either single-pass, two-pass or multi pass weld procedures can be carried out. There is virtually no restriction on the material thickness, provided a suitable joint preparation is adopted. Most commonly welded materials are carbon-manganese steels, low alloy steels and stainless steels, although the process is capable of welding some non-ferrous materials with judicious choice of electrode filler wire and flux combinations.

6.5 Submerged Arc Welding (SAW) Process

Submerged Arc Welding (SAW) is a welding process that is commonly used in the structural and vessel construction industries. The process is commonly used to weld and manufacture beam, boom, girders, tractor, members and sub-assemblies in ship yards, offshore, and bridge construction, pressure vessels, columns, and other large process equipment. Locally, the process is also referred as Sub Arc or SAW, this process uses a blanket of granular fusible flux beneath which both the weld and the arc zone are protected or "submerged." This flux blanket offers the following advantages:

1. Guards against atmospheric contamination
2. Stabilizes the arc during welding
3. Prevents spatter and sparks from flying about
4. Has higher deposition rate and efficiency,
5. Suppresses radiation and fumes that are typical of the other arc welding process like GMAW, or SMAW.

6.5.1 Advantages and Limitations of Submerged Arc Welding

The process had several very important advantages useful for heavy fabrication projects. Some of the advantages of submerged arc welding include:

- Strong, sound welds are readily made
- Minimal welding fume is emitted
- Minimal arc light is emitted
- SAW is suitable for both indoor and outdoor works
- Less distortion
- Deep weld penetration
- Minimal edge preparation
- High deposition rates are possible
- Thick materials may be welded
- At least half or more of the flux is recoverable

Limitations of the Submerged Arc Welding

There are a few limitations with submerged arc welding. One issue is that welding can normally be performed only in the flat position. The use of a granular flux and the fluidity of the molten weld pool mean that welding is limited to 1G, 1F, and 2F positions.

Another disadvantage of SAW is that welding is normally limited to long, straight seams or rotated vessels or pipes. Flux handling systems can be quite elaborate and requires regular monitoring.

Another limitation of the SAW process is that it can't be used to deposit root pass without some support material to deposit the root weld.

6.6 How the SAW Process Works

SAW process requires a continuously-fed tubular or consumable solid electrode and the process may be fully automatic or semi-automatic.

The arc is flat and is maintained between the end of a bare wire electrode and the weld. The electrode is constantly fed into the arc and as it is melted, a layer of granular flux provides a protective cover beneath which the welding occurs. The blanket is created as some of the flux becomes molten. This fusible flux may consist of lime, silica, manganese oxide, calcium fluoride, and other compounds. In a molten or melted state, the flux becomes conductive. This allows the flow of constant current between the electrode and the welding work. After the weld had solidified, the remainder of the flux is recovered and reused, unless it has become contaminated. Refer the flux details on the details about the reuse of fluxes.

In the automatic version of SAW, the process is performed with a set of rollers driven by a controlled motor to ensure that the wire is fed into the arc at the rate that is matched with the rate at which the electrode is melted. As a result, the arc length remains constant. The SAW process is usually automated however, there are semi-automated systems available.

Properly performed Sub Arc welding should consistently result in mechanical properties that are at least equal to that of the base metal. Ductility and impact resistance should be good, and bead appearance should be uniform. The "properly" in this respect means selection of specific welding electrode, and flux combination and maintenance of all other variables listed below for welding.

Figures 6.6.1 and 6.6.2 below show the SAW in progress, and the collection of flux after the welding. This flux is then shifted to remove the fines and pieces of slag and returned to the hopper to be reused.

6.6.1 Depositing a Root Pass with SAW Process

Because the SAW arc is very strong it is not used for depositing the root pass, without providing some support material. That support material could be either a pre deposited root pass with any other process often SMAW or GTAW is used for root pass before the rest of the weld is deposited by the SAW process.

Other option is to either use the backing strip to deposit the root pass, and rest of the weld and then remove the backing strip if that is the requirement of the design specification. In this type of weld, the protection of the back of the weld (root pass) from the atmospheric contamination is also a subject to consider, if that is determined a necessary step for the quality of weld desired, then a channel or trove is specifically designed to hold the flux as backing material for the initial weld pass(es).

6.6.2 Travel Mechanism

The process is most suited for mechanization, and simple to complex mechanized systems are designed and available. The movement along the joint can be either manual, or

Figure 6.6.1 Showing SAW process in progress on a pipe weld.

mechanized. Hand-held travel is a welder operation, and the welder manipulates the at a study rate, alternatively for smaller work a small motor mounted gun is synchronized with the flux delivery system.

The mechanical travel is either radial or lateral, as can be seen in the Figures 6.3.2 and in Figure 6.3.3, flat or horizontal position, along the line of the joint. Often this is set so that the gun position is fixed at a spot and the work travels under the nozzle at a constant pre-set speed. Tractor mounted welding stations are also used. All these variations in mechanizations are dependent on the type of work at the fabrication shop, more routine type weld could have less repositioning opportunity of the machine, as compared to the constantly varying type of work where frequent repositioning of the machine may be required, as in a job-shop fabrication shop. Same concept also applies to the size of the job, a larger work as in case, 40ft long pipe in a pipe-mill will work well with a fixed station unidirectional travers, where as a welding shop with smaller products take for example a valve manufacturer would not require such a long travers.

6.6.3 Variables of the SAW Process

There are some key variables of the submerged arc welding process. These variables include:

The arc voltage

Figure 6.6.2 Shows the collected flus for cleaning and reusing.

The wire feed speed
Travel speed
Contact tip to work (CTTW) or electrode stick-out (ESO)
Polarity and current type (may be either AC or DC), as well as variable balance AC current.

While developing a welding procedure the welding engineer should refer to the applicable welding code and follow the specific directions.

6.7 SAW Process Variants

The submerged arc welding process may be varied in number of ways, the process has several variations but the fundamental concept remains same. The variations are primarily based on the number of welding wire that is fed in the weld pool and aimed at increasing the production rate. Increasing the capacity of the power source is also used but the primary focus remains on the increasing the production rate and that is mostly achieved by adding wire to the weld pool.

Other methods are also sued but they are specifically aimed at a particular type of job, for example, adding iron powder to the flux, and using a strip electrode for surfacing operations. A brief introduction to these variants is given below.

According to material thickness, joint type and size of component, varying the following can increase deposition rate and improve bead shape.

6.7.1 Variants Based on Use of Welding Wire

SAW is normally operated with a single wire on either AC or DC current. Common variants are:

- twin wire
- multiple wire (tandem or triple)
- single wire with hot or cold wire addition
- metal powder addition
- tubular wire

All the above contribute to improved productivity through a marked increase in weld metal deposition rates and/or travel speeds.

A narrow gap process variant is also established, which utilizes a two or three bead per layer deposition technique.

6.7.1.1 Multi-Wire Systems

As the name suggests this system uses more than one wire to initiate arc and feed into the weld pool. The advantage of this method comes from the use of more electrodes (wire) this allows the welding to improve deposition rates and increase the travel speeds of the arc.

To introduce more than one wire for arc the system may either use two or more power units for each wire, or use one large output source that can support the current demand to run more than one wire.

The advantage of using multiple power source for running two or more electrodes allows for the utilization of different polarities on the electrodes. Also, with separate power sources for two electrodes, alternating current may be used on one, while direct current is used on the other electrode. Typically, when three wires are used in the tandem position, where one electrode is placed in front of the other two, the alternating current is used. The electrodes are connected to three-phase power systems, which are used for making high-speed longitudinal seams on large pipes in pipe-mill type of fabrication situation, and the fabricated beams in the fabrication yards and shops.

The Figure 6.7.1 shows the multi wire SAW process used in a pipe mill note the three welding heads in line. Also note the entire arrangement of welding flux delivery from the hopper followed by three welding heads, and then the suction nozzle (Black in color) for the unused flux recovery.

6.7.1.2 Use of Hot-Wire

Some SAW process variants are developed where the filler wire is pre-heated to a predetermined temperature, before they are introduced in the weld arena. The challenge of this hot wire approach is to maintain that pre-heat temperature, and then protection of that hot wire during its travel to the weld zone and receive protection from the welding flux.

Number of a bit complicated approaches have been observed where either an inert gas chamber is used in which the pre heat temperature is applied, and then promptly moved in to the flux envelop in the weld zone, or in some cases a chamber of flux itself is used at the point where the wire is heated, and at the point where the wire gets hot enough to oxidize or exposed to the atmosphere. Most of these methods are very much specific developments for very specific need and it can't be labeled as a regular process variant.

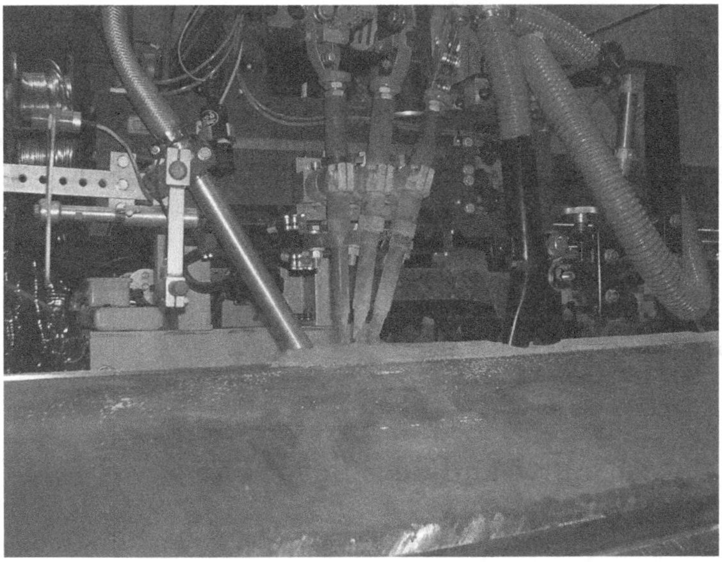

Figure 6.7.1 Multi-wire SAW system.

6.7.2 Adding Iron Powder to the Flux

Adding iron powder to the flux is not a very new method, recall that in SMAW process there are number of electrodes that have iron-powder added to the coating on the electrode, E7014 or E7024 are some of the iron powder added electrodes, the addition of iron-powder to the SAW flux is based on the same basic concept.

This approach of adding iron-powder is both a metallurgical as well as a production enhancement method. The added iron-powder increases rate of deposition. But by itself it does not decrease the properties of the weld metal, unless other alloying compounds and elements are added to the flux.

Figure 6.7.3 Tandem head strip wire SAW process for cladding.

6.7.3 The Utilization of a Strip Electrode for Surfacing

The strip type electrodes are preferable used to cover relatively larger surface area, hence very useful for the surface overlay jobs. This particular welding system uses the strip electrode and flux to make a corrosion-resistant overlay on a less expensive base material such as carbon steel. During this procedure, a wide, uniform bead is produced that has minimum penetration. The uniform bead is necessary to provide a smooth overall surface. The strip electrode welding system is often used for overlaying the inside of vessels.

The Figure 6.7.3 below shows the tandem head SAW cladding process, that is using the strip type electrode (not visible), the flux that is used in strip surfacing is made specifically for that purpose.

6.8 SAW Power Source and Equipment

The constant voltage output machines are used for the SAW processes. The process uses machines that are similar in characteristics with GMAW process except that the SAW machine uses much higher current output than most of the standard GMAW machines. Arc formation between the wire electrode and workpiece is similar to the GMAW process.

Compared to the GMAW the SAW process has an additional advantage of shielding by the granular flux making the SAW a spatter free process, low fumes emission, and free from UV light emission. The equipment has the following in its inventory.

Submerged arc welding machines are rated to work at 100% duty cycle. The SAW welding process is continuous, and the length of one weld may go up to 10 minutes. General power sources with a 60% duty cycle may get derated according to the duty cycle curve of 100%. The voltage sensing wire feeder must be used when a constant current of AC/DC machines are used. The fixed speed wire feeder uses a constant voltage, while the CV system drives with direct current.

Both the process DC generator, and AC transformer may be used but rectifier machines are more popular. The SAW machines are available in the range from 300 amperes to 1500 amperes output. Submerged arc welding power source could be either DC or AC.

The direct current equipment suits semi-automatic applications while alternating current power source is more suited to the automation only.

Typically, the total package set up for SAW system includes the following.

1. Power source – as discussed above,
2. Welding torch/gun and cable assembly
3. Flux hopper and its feeding
4. Travel mechanism for automatic welding

6.9 Welding Heads (Gun)

In automatic submerged arc welding, there are three types of guns that are generally used. These include the side flux delivery gun, the deep groove gun, and the concentrated flux delivery gun.

The concentrated flux delivery gun deposits the flux around the wire. With both the side flux delivery gun and the deep groove gun, the flux is fed from an overhead gravity hopper to the gun's flux shut-off assembly.

The type of gun chosen for a certain job may be dependent upon the joint design and/or the welding operator's preference.

6.10 Fluxes

Without any argument, the SAW process is distinguished by the sue of flux to make the weld. It is flus and the way they are used in the process that identifies the process.

Fluxes used in SAW are granular fusible minerals containing oxides of manganese, silicon, titanium, aluminum, calcium, zirconium, magnesium and other compounds such as calcium fluoride. The flux is specially formulated to be compatible with a given electrode wire type so that the combination of flux and wire yields desired mechanical properties. All fluxes react with the weld pool to produce the weld metal chemical composition and mechanical properties. It is common practice to refer to fluxes as 'active' if they add manganese and silicon to the weld, the amount of manganese and silicon added is influenced by the arc voltage and the welding current level.

The granular flux used in SAW process and introduced above, serve several functions. In addition to providing a protective cover over the weld, the flux shields and cleans the molten puddle. The flux also affects the chemical composition of the weld metal, the weld bead shape, and the mechanical properties of the weld.

Another function of granular flux is to act as a barrier that holds the heat in and concentrates the heat into the weld area to promote deep penetration.

Fluxes may be further categorized in two ways,

- by the method of manufacture (fused or agglomerated) or,
- by its activity (neutral, active or alloying).

Within these broad groupings the fluxes may be classified further by their constituents, silica, manganese oxide, calcium fluoride etc.

6.10.1 Types of Granular Fluxes

The methods used to manufacture fluxes determine the flux types. There are fused fluxes, bonded fluxes, agglomerated fluxes, and mechanically mixed fluxes.

When manufacturing *fused fluxes*, raw materials are melted into a liquid state with a high temperature electric furnace. The material is then cooled and crushed or ground into the desired particle size.

When making *bonded fluxes*, the ingredients are dry mixed, then glued together with a liquid binder. This binder may be a liquid such as sodium silicate. After the particles are bonded, they are baked and then sifted through a sieve to attain flux particles of the desired size.

Agglomerated fluxes are manufactured much the same way that bonded fluxes are made. However, instead of a liquid binder, a ceramic binder is used. A higher drying temperature is used, too. (The higher drying temperature limits the use of deoxidizers and alloy elements.)

Fluxes that are *mechanically mixed* are combinations of two or more bonded or agglomerated fluxes. Although mechanically mixed fluxes make it possible to create special mixtures for more sensitive welds, these fluxes may separate during storage, use, and recovery of flux.

6.10.2 Fused Fluxes versus Bonded Fluxes

Among the various types of fluxes use in Submerged Arc Welding are the fused flux and the bonded flux. Each of these fluxes offers some advantages and some disadvantages.

6.10.3 Fused Fluxes

Fused fluxes are produced by mixing the ingredients, then melting them in an electric furnace to form a chemically homogeneous product, cooled and ground to the required particle size. Smooth stable arcs, with welding currents up to 2000A and consistent weld metal properties, are the main attraction of these fluxes.

When making fused fluxes, the raw materials are dry mixed together, and then they are fused or melted into a liquid state by using a high temperature furnace. After fusion is complete, the fluxes are cooled. This may be accomplished by using a stream of water or with big chill blocks.

Once the fluxes are cooled, they are crushed or ground into particles. A variety of particle sizes are made to ensure optimal performance for different applications.

Advantages of fused fluxes include:

- The non-hygroscopic flux particles do not absorb moisture and, therefore, any surface moisture can be eliminated merely by drying the particles at a low temperature oven setting of 300 degrees F.
- Low temperature drying of condensation on the fused flux particles provides better protection against hydrogen cracking.
- Flux particles create welds that are chemically consistent.
- Recycling of fused flux particles through the flux recovery systems can be achieved without losing sizing or composition.

A disadvantage of fused fluxes is that the high temperature used during the manufacture process makes it difficult to add alloys and deoxidizers.

6.10.4 Bonded Fluxes

Bonded fluxes are produced by drying the ingredients, then bonding them with a low melting point compound such as a sodium silicate. Most bonded fluxes contain metallic deoxidizers which help to prevent weld porosity. These fluxes are effective over rust and mill scale.

The manufacture of bonded fluxes involves combining the dry ingredients, then using a liquid binder such as sodium silicate or potassium silicate to glue the ingredients together. After the bonded mix is made into pellets, the pellets are baked at a low oven temperature. Once the drying of the pellets is complete, the pellets are broken up by using a sieve to attain the desired particle size. The particles are then packaged for shipping.

Advantages of bonded fluxes include:

- Deoxidizers are present in bonded fluxes, protecting against rust and mill scale. These deoxidizers also help to prevent welds from becoming porous.
- Alloys can be added to bonded fluxes. Alloy elements may improve chemical and mechanical properties of the flux.
- Bonded fluxes allow for a thicker flux layer when welding.
- Bonded fluxes can be identified by color.
- Bonded fluxes typically provide better peeling properties than fused fluxes.

There are at least two disadvantages of using bonded fluxes. These are:

- They absorb moisture.
- They can change in composition due to segregation or loss of fine particle size.

6.10.5 Neutral Fluxes

Neutral fluxes are designed to have little or no effect on the chemical analysis of the weld metal and therefore on the mechanical properties. They contain low silica, calcium silicate and alumina and do not add significant amounts of silicon and manganese to the weld.

6.10.6 Acid Fluxes

The acid fluxes contain substantial amounts of silica, silicates in the form of calcium and/or manganese silicate and manganese oxide. These fluxes react with the weld pool and will raise both silicon and manganese content of the weld together with a high oxygen content. The result of this is that the toughness of the weld is poor but the fluxes will tolerate rusty surfaces, will detach easily and give a good weld appearance. They are especially useful for single pass high speed welding such as fillet welding of web to flange girder joints.

6.10.7 Basic Fluxes

The basic fluxes fill much the same role in submerged arc welding as basic coatings do in manual metal arc welding. They have a low silica content and are composed of varying amounts of calcium carbonate and/or fluoride, alumina, calcium, manganese and magnesium oxides and rutile.

This combination of compounds gives a clean, low sulphur, low oxygen weld metal with good to excellent notch toughness. As a general rule, the higher the basicity, the higher the toughness. The transfer of silicon and manganese into the weld metal is also limited. Such fluxes are preferred for the welding of high-quality structural steels, pressure vessels, pipework and offshore structures where either good high or low temperature properties are required.

Basicity is commonly used to describe the metallurgical behavior of a welding flux. The basicity index is a ratio between basic and acid compounds (oxides and fluorides) of which the flux is composed. In this context, perhaps the most convenient method of classifying SAW fluxes, is by reference to the 'basicity index' (BI) of the flux. The index is calculated by dividing the sum of the percentages of the basic constituents in the flux by the sum of the acid constituents.

Basic constituents of flux include the following elements and their compounds.

Calcium, Magnesium, Sodium, Potassium and Manganese oxides, Calcium carbonate and Calcium fluoride.

The acid constituents of flux include the following

Silica and Alumina.

Acid fluxes have a basicity index of 0.5 to 0.8;

neutral fluxes > 0.8 to 1.2;

basic fluxes 1.2 to 2.5 and highly basic fluxes 2.5 to 4.0.

The basicity of a flux has a major effect on the weld metal properties, most importantly the notch toughness. As a general rule the higher the basicity the higher the notch toughness.

There are several ways of calculating basicity and in welding Boniszewski's formula has become the predominant way of calculating basicity.

As stated above, based on the type of chemical compositions, SAW fluxes can be divided into three groups as the table below indicates.

Basicity has great influence on impact toughness of the weld metal. Increasing basicity brings down the oxygen content and hence the inclusion level in the weld metal. Consequently, the impact toughness will increase and also, to a limited extent, the ductility of the weld metal.

The relation between basicity and impact toughness is particularly important for high alloyed grades, such as duplex steels.

The fused fluxes can be acid, neutral or slightly basic. They are manufactured by mixing the constituents together, then melting them in an electric furnace, and crushing the solidified slag that is produced. This gives the flux a glassy appearance.

Fluxes produced by the fused method are homogeneous, resistant to moisture pick-up, and their particle structure is mechanically strong, so that they do not break down and hold the intended particle size. The high temperatures required to melt the constituents, results in the loss of some constituents, particularly the de-oxidants present in the highly basic fluxes. This limits the range of applications of these fluxes to general structural work. They are not suitable for sub-zero service temperature.

The agglomerated fluxes may be neutral, basic or highly basic. They are made from a wet mix that is corned, dried and baked to achieve a low moisture content. This low temperature process means that strong deoxidants and ferro-alloys can be effectively incorporated, they would not be lost during the low heat process. The binders used in the corning process, however, are hygroscopic so that moisture pick-up can be a problem on the shop floor. Baking of the flux prior to use may be necessary and if the flux is not used within a specified timeframe, which is normally a very short. The flux hoppers on the welding equipment

Table 6.10.7 Indicates the basicity of various fluxes.

Type of welding flux	Basicity
Acid fluxes	<0.9
Neutral fluxes	0.9 - 1.2
Basic fluxes	>1.2

should also be heated, to limit moisture pick-up during storage. The flux may also suffer mechanical damage during recirculation, breaking down to form a dust. Although a small particle size is capable of carrying a higher current, too many fines in the flux will give rise to gas being trapped between the slag and the weld pool. This will result in unsightly gas flats or pockmarking on the weld surface. To avoid this, the recirculating system should be equipped with filters to remove both large particles of detached slag and the fine dust.

Commercial availability of fluxes varies, but normally the fluxes are packed in plastic bags of 25 kg to 40kg for the retail use, but they are also supplied in bulk packaging in plastic drums of 250kg. The use of plastic drum or plastic bag is primarily to prevent moisture pick-up and other contaminates. Some suppliers also package the fluxes in the hermetically sealed bags, also known as the vacuum-packed electrodes. These hermetically sealed fluxes can be used straight from the bag with guaranteed low hydrogen levels and would not require pre use baking, but any other heating as in hopper may still be necessary to control the moisture levels during the welding operation, or if the flux is left unused for longer time in the hopper.

6.10.8 Selection of Specific Flux

For carbon and low alloy steels the welding electrode is matched with the flux type and is essential part of the wire classification. This requirement is not adhered to specialized materials and high alloy steel wires. This is due to the varying metallurgical complexities of these alloys during and after the welding.

For this reason, for most of the alloy steels and specialized materials the choice and selection of flux is very supplier specific, and the manufacturers' experts should be consulted.

6.11 Submerged Arc Welding Various Metals

SAW material applications include carbon steels, low alloy steels, stainless steels, nickel-based alloys, and cladding, surfacing building and hard-facing applications. SAW is frequently used in heavy structural construction. It is also used in the pressure vessel industry, chemical plants, and shipbuilding.

The limitation of SAW process to weld any metal is the development and availability of proper flux. Flux for some materials is very well developed and number of options are easily available. For others that may be limited and yet another there do not exit any flux options.

For carbon steel and low alloy steel electrode wire and flux combinations like AWS A 5.17 classification F48A2-EM12K, F48A2-EM13K, F7A2-EM13K, F7A0-EM12K, F7A2-EM12K, F7A4-EM12K.

Similarly, for the SAW process welding of Nickel and its alloys, stainless-steel and other alloys, the wire and flux combination meeting the AWS A 5.14, 5.9 and 5.23 classification are listed in the table below. Note that the specific flux combination is to be selected as per the guide of the wire manufacturer and the applicable specification. Various factors that need to be considered in selection is the storage and reconditioning, removal of slag, usability of specific type of flux for the service and basicity index etc. The Table 6.11 below list some common welding electrode for various class of steels, and also nickel alloys. Note that this is just and indicative list, for more detail the AWS Class listed in the column one, may be more appropriate document to refer.

Table 6.11 Common welding electrodes for SAW process.

AWS Class*	Flux wire combination	Material top be welded	Industrial application
A5.9	ER308L	304L	Reactors, tanks and other chemical equipment
	ER316L	316L	
	ER317L	317L, or modified 317 for PREN > 33 and stabilized austenitic steels type 321 and 347	
	ER 309 L	For dissimilar metals (SS to CS)	
	ER19-10H	304H, 321H, 347H	For Creep resistant, pressure vessels, and High temperature applications.
A 5.14	ERNiCrMo-3	Alloy 625,	General process equipment in chemical and petrochemical industries
	ERNiCr-3	Nickel base alloys, Alloy 600	
	ERNiCrMo-4	Nickel base alloys, Alloy 600, 9% Ni alloys etc.	
A5.9	ER410NiMo(mod.)	Joining UNS 410 12% Cr. steel, Cladding Steels with 13% Cr such as X3CrNiMo13-4	Steam power Hydro turbines
	ER410NiMo(mod.)		Crack resistant, Hydro turbines, Steam power
A 5.9	ER430(mod.)	For hard surfacing cupro-nickel alloy 430 and similar	Gas, water and steam fittings, Crack resistant Cladding. Hardness 320 – 420 HB with PWHT hardness can be 200 HB
A5.17	F7AZ-EM12	Carbon steel	Power generation Membrane walls High speed fillet welds Non-alloyed no requirements
A5.17	F7AZ-EM12K		Power generation Membrane walls High speed fillet welds Non-alloyed no requirements
A5.17	F7A2-EM12K	API 5L Pipe X 60	Pipe manufacturing mills
A5.23	F6TA0G-EM12K		

(*Continued*)

Table 6.11 Common welding electrodes for SAW process. (*Continued*)

AWS Class*	Flux wire combination	Material top be welded	Industrial application
A5.17	F7A8-EC1/ F7P8-EC1	Cr. Ni and Cr-Ni-Mo alloys, Martensitic alloys, S355, S420, S 460	Offshore and other heavy construction fabrication, requiring high productivity and high toughness values.
A5.23	F8TA6G-EG	API 5L X 70 and X 60 pipes	Pipe manufacturing mills
A5.23	F9TA6G-EA2TiB	API 5L X 65 to X 80 pipes	
A5.23	F8A10-ENi2-Ni2/ F7P10-ENi2-Ni2	Alloys P460NL1, 12Ni14	Pressure vessels Offshore constructions
A5.23	F7A15-ENi3-Ni3/ F7P15-ENi3-Ni3	3.5 % Ni steel, 10Ni14	Pressure vessels Offshore constructions
A5.23	F8A10-ENi1-Ni1/ F8P10-ENi1-Ni1	S460 - S500	Offshore fabrication or other heavy constructions

*Refer the AWS specification listed for full information.

For most of the alloy steels and specialized materials the choice and selection of flux is very supplier specific, and their experts should be consulted.

Properly performed Sub-Arc welding should consistently result in mechanical properties that are at least equal to that of the base metal. Ductility and impact resistance should be good, and bead appearance should be uniform.

6.12 Test Your Knowledge

1. How is SAW process compared with SMAW and GMAW processes.
2. Can a good quality root pass be deposited using the SAW process, support your response with details of how?
3. What type of power source is used for SAW process, what is the rated duty cycle for these machines?
4. Describe if there any similarity between the GMAW-S and SAW process.
5. Electrode coating in SMAW process have iron powder to increase deposition rate, what are the methods used by SAW process to enhance its already high production rate?
6. What Gas is used to shield SAW welds from atmospheric contamination?
7. What number of shade a welder should use while welding with the SAW process?
8. How is the hot weld pool and surrounding weld area protected from atmospheric contamination in SAW process?

7

Useful Data and Information Related to Welding and Fabrication

7.1 Common Weld Symbols and Their Meanings

When welds are specified on engineering and fabrication drawings, a cryptic set of symbols is used as a sort of shorthand for describing the type of weld, its size and other processing and finishing information. Here we will introduce you to the common symbols and their meaning. The complete set of symbols is given in a standard published by the American National Standards Institute (ANSI) and the American Welding Society (AWS): ANSI/AWS A2.4, Symbols for Welding and Nondestructive Testing.

The horizontal line — called the reference line — is the anchor to which all the other welding symbols are tied. The instructions for making the weld are strung along the reference line. An arrow connects the reference line to the joint that is to be welded. In the example above, the arrow is shown growing out of the right end of the reference line and heading down and to the right, but many other combinations are allowed.

Quite often, there are two sides to the joint to which the arrow points, and therefore two potential places for a weld. For example, when two steel plates are joined together into a T shape, welding may be done on either side of the stem of the T.

The weld symbol distinguishes between the two sides of a joint by using the arrow and the spaces above and below the reference line. The side of the joint to which the arrow points is known (rather prosaically) as the arrow side, and its weld is made according to the instructions given below the reference line. The other side of the joint is known (even more prosaically) as the other side, and its weld is made according to the instructions given above the reference line. The rule that below the line equals the arrow side and above the line equals the other side applies regardless of the arrow's direction. The flag growing out of the junction of the reference line and the arrow is present if the weld is to be made in the field during erection of the structure. A weld symbol without a flag indicates that the weld is to be made in the shop. In older drawings, a field weld may be denoted by a filled black circle at the junction between the arrow and the reference line.

The open circle at the arrow/reference line junction is present if the weld is to go all around the joint, as in the example below.

The tail of the weld symbol is the place for supplementary information on the weld. It may contain a reference to the welding process, the electrode, a detail drawing or any information that aids in the making of the weld that does not have its own special place on the symbol.

Ramesh Singh. Arc Welding Processes Handbook (349–368) © 2021 Scrivener Publishing LLC

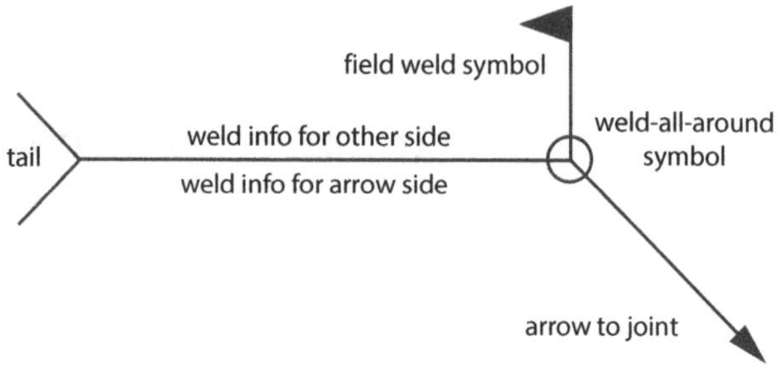

Figure 7.1 Structure of the welding symbol.

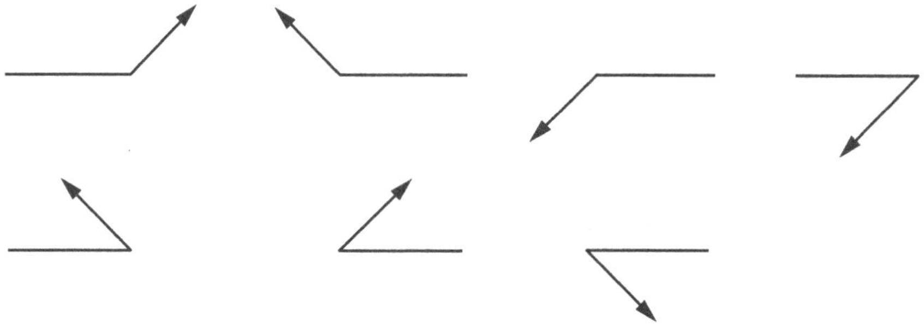

Figure 7.2 Welding symbol arrows.

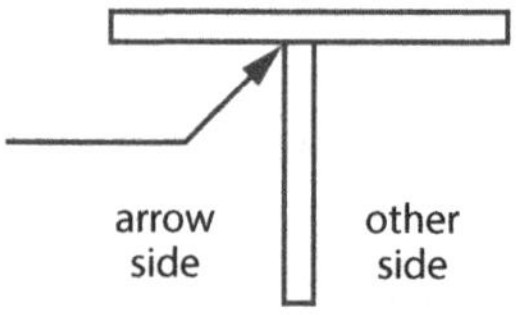

Figure 7.3 Welding symbol position of the arrows.

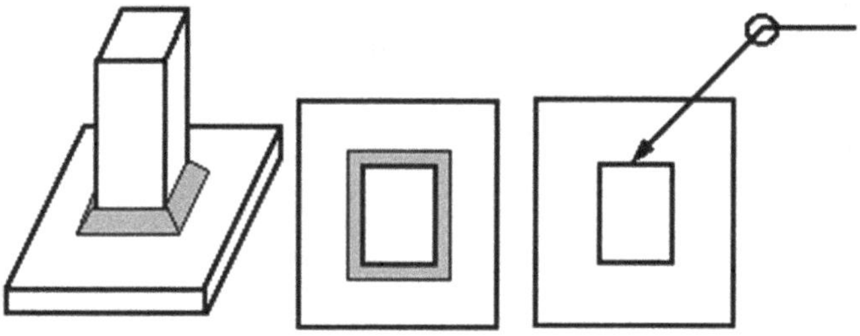

Figure 7.4 Significance of the circle on the arrows.

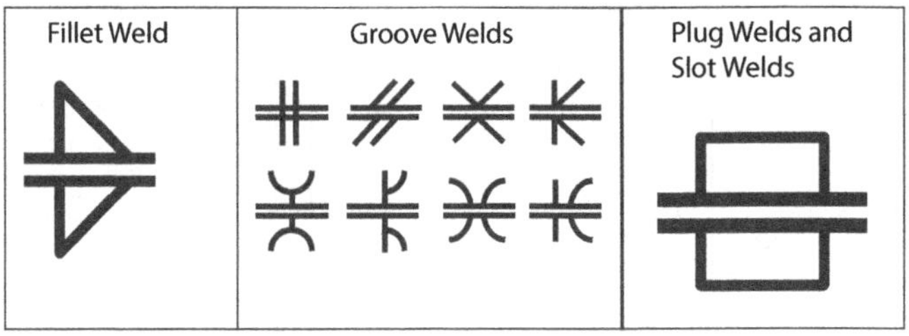

Figure 7.5 Symbols for type of welds.

Types of welds and their symbols
Each welding type has its own basic symbol, which is typically placed near the center of the reference line (and above or below it, depending on which side of the joint it›s on). The symbol is a small drawing that can usually be interpreted as a simplified cross-section of the weld. In the descriptions below, the symbol is shown in both its arrow-side and other-side positions.

7.2 Fillet Welds

The **fillet weld** is used to make lap joints, corner joints and T joints. As its symbol suggests, the fillet weld is roughly triangular in cross-section, although its shape is not always a right triangle or an isosceles triangle. Weld metal is deposited in a corner formed by the fit-up of the two members and penetrates and fuses with the base metal to form the joint. (Note: for the sake of graphical clarity, the drawings below do not show the penetration of the weld metal. Recognize, however, that the degree of penetration is important in determining the quality of the weld.)

The perpendicular leg of the triangle is always drawn on the left side of the symbol, regardless of the orientation of the weld itself. The leg size is written to the left of the weld

Figure 7.6 Symbol of fillet weld.

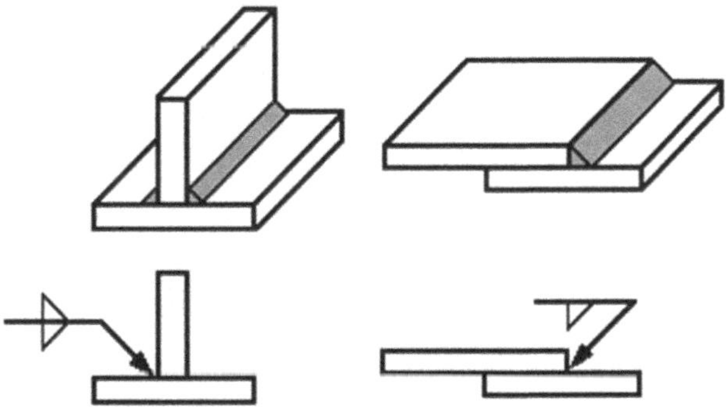

Figure 7.7 Shows the side of the metal where the fillet weld is required to be made.

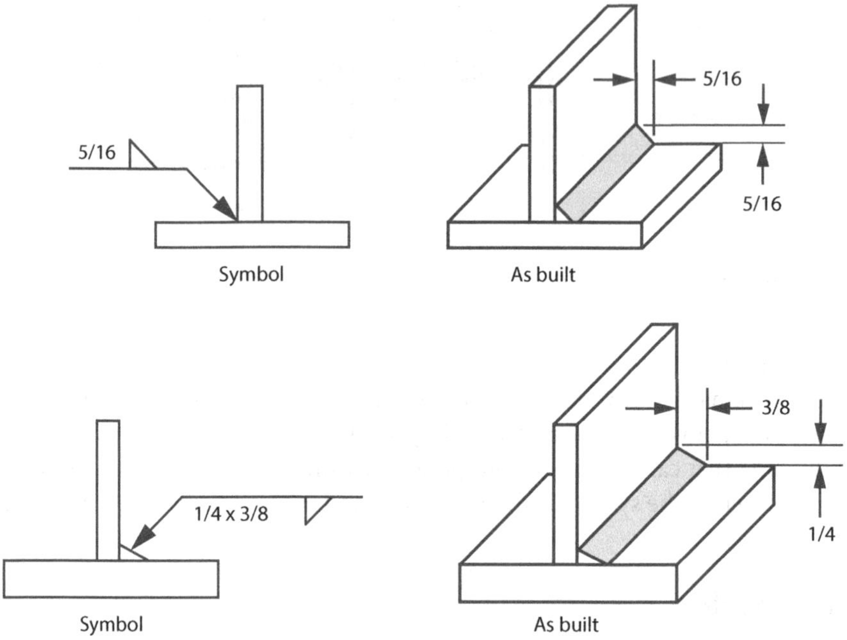

Figure 7.8 Graphic and as built depiction of welds – note the weld sizes shown in the symbol on left and its corresponding annotation on the actual weld.

symbol. If the two legs of the weld are to be the same size, only one dimension is given; if the weld is to have unequal legs (much less common than the equal-legged weld), both dimensions are given and there is an indication on the drawing as to which leg is longer.

The length of the weld is given to the right of the symbol.

If no length is given, then the weld is to be placed between specified dimension lines (if given) or between those points where an abrupt change in the weld direction would occur (like at the end of the plates in the example above).

For intermittent welds, the length of each portion of the weld and the spacing of the welds are separated by a dash (length first, spacing second) and placed to the right of the fillet weld symbol.

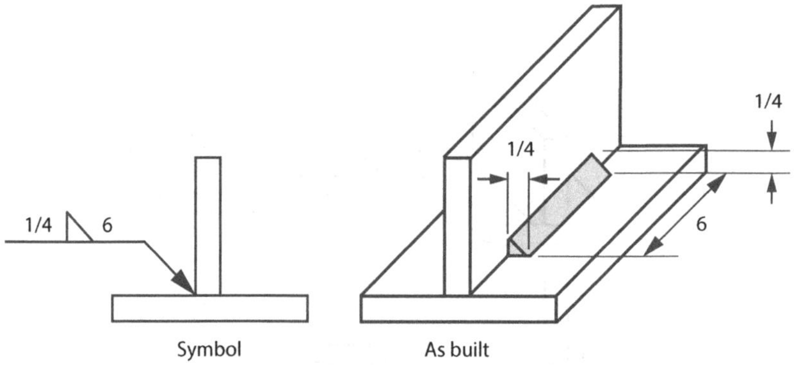

Figure 7.9 Shows the addition of the length of the weld to the symbol at the left, and what it means is shown in the as built figure on the right.

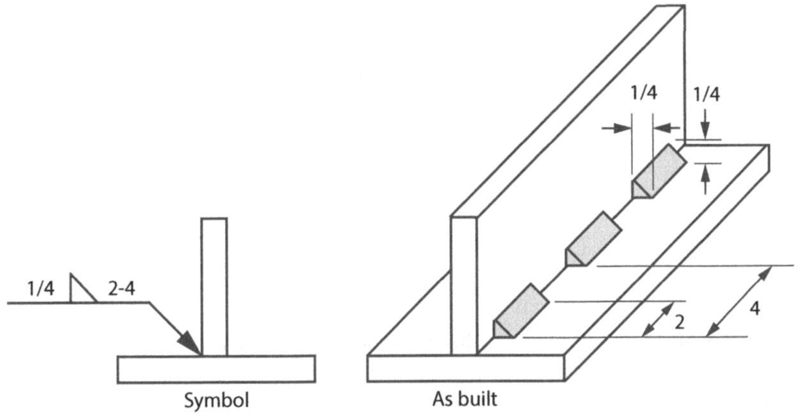

Figure 7.10 Adding pitch of the weld.

Notice that the spacing, or **pitch**, is not the clear space between the welds, but the center-to-center (or end-to-end) distance.

7.3 Groove Welds

The **groove weld** is commonly used to make edge-to-edge joints, although it is also often used in corner joints, T joints, and joints between curved and flat pieces. As suggested by the variety of groove weld symbols, there are many ways to make a groove weld, the differences depending primarily on the geometry of the parts to be joined and the preparation of their edges. Weld metal is deposited within the groove and penetrates and fuses with the base metal to form the joint. (Note: for the sake of graphical clarity, the drawings below generally do not show the penetration of the weld metal. Recognize, however, that the degree of penetration is important in determining the quality of the weld.)

The various types of groove weld are:

Square groove welds
The groove is created by either a tight fit or a slight separation of the edges. The amount of separation, if any, is given on the weld symbol.

Figure 7.11 Symbols of various types of Groove Welds.

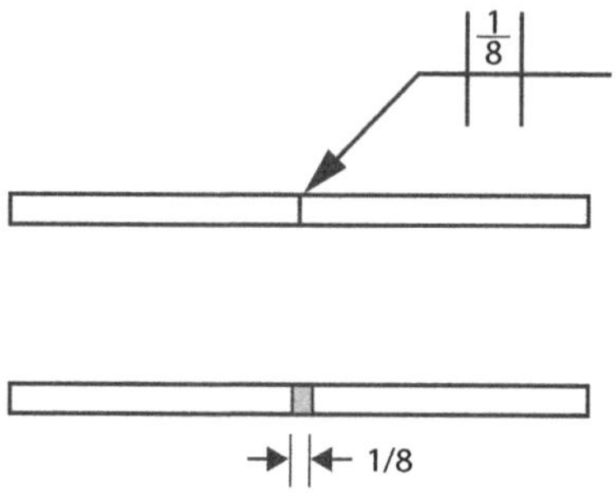

Figure 7.12 Symbol of Sq. groove weld – note the annotation of root opening.

V-groove welds

The edges of both pieces are chamfered, either singly or doubly, to create the groove. The angle of the V is given on the weld symbol, as is the separation at the root (if any).

If the depth of the V is not the full thickness — or half the thickness in the case of a double V — the depth is given to the left of the weld symbol.

If the penetration of the weld is to be greater than the depth of the groove, the depth of the **effective throat** is given in parentheses after the depth of the V.

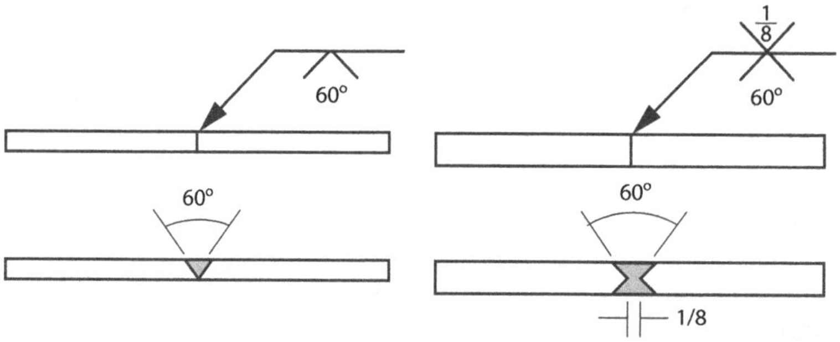

Figure 7.13 Symbol and as built of V-groove welds, note how the root gap (opening) is shown.

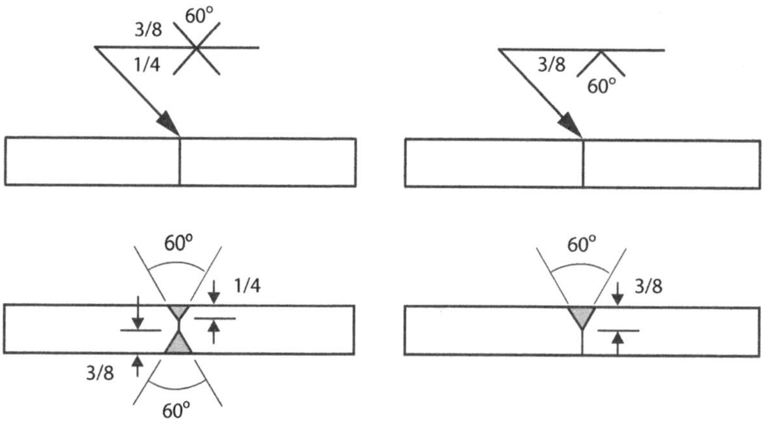

Figure 7.14 Shows the (1) depth of V groove on both sides of the weld, (2) shows the depth of the penetration desired of the weld.

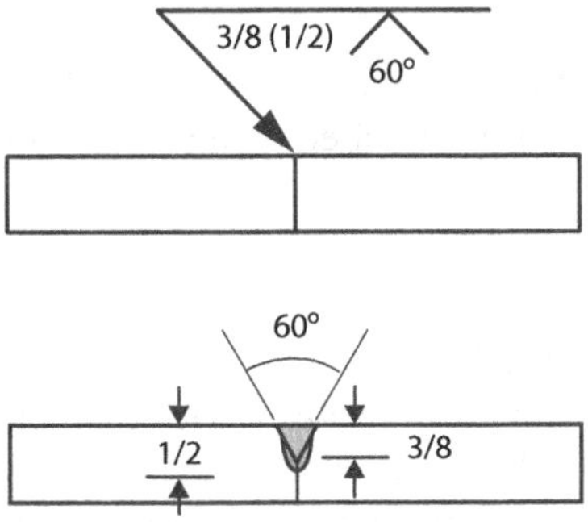

Figure 7.15 Shows the specific depth of the groove weld (effective throat) desired.

Bevel groove welds

The edge of one of the pieces is chamfered and the other is left square. The bevel symbol's perpendicular line is always drawn on the left side, regardless of the orientation of the weld itself. The arrow points toward the piece that is to be chamfered. This extra significance is emphasized by a break in the arrow line. (The break is not necessary if the designer has no preference as to which piece gets the edge treatment or if the piece to receive the treatment should be obvious to a qualified welder.) Angle and depth of edge treatment, effective throat and separation at the root are described using the methods discussed in the V-groove section.

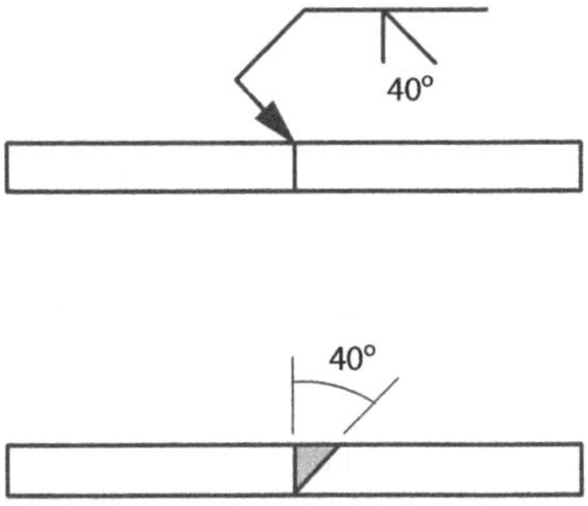

Figure 7.16 Symbol of a bevel groove note which side of the plate is to be beveled and to what degree.

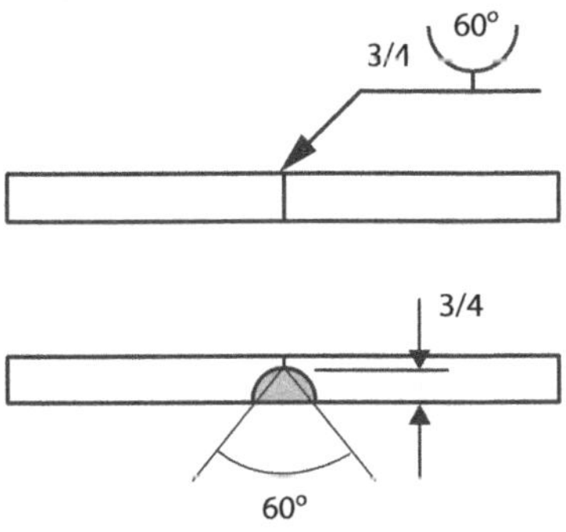

Figure 7.17 Shows U-groove symbol.

U-groove welds
The edges of both pieces are given a concave treatment. Depth of edge treatment, effective throat and separation at the root are described using the methods discussed in the V-groove section.

J-groove welds
The edge of one of the pieces is given a concave treatment and the other is left square. It is to the U-groove weld what the bevel groove weld is to the V-groove weld. As with the bevel, the perpendicular line is always drawn on the left side and the arrow (with a break, if necessary) points to the piece that receives the edge treatment. Depth of edge treatment, effective throat and separation at the root are described using the methods discussed in the V-groove section.

Flare-V groove welds
Commonly used to join two rounded or curved parts. The intended depth of the weld itself is given to the left of the symbol, with the weld depth shown in parentheses.

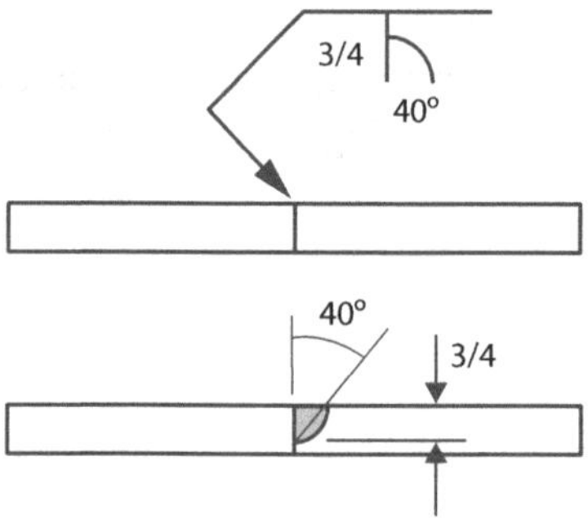

Figure 7.18 Shows the J-groove symbol and the weld. Note the indicated depth of the weld.

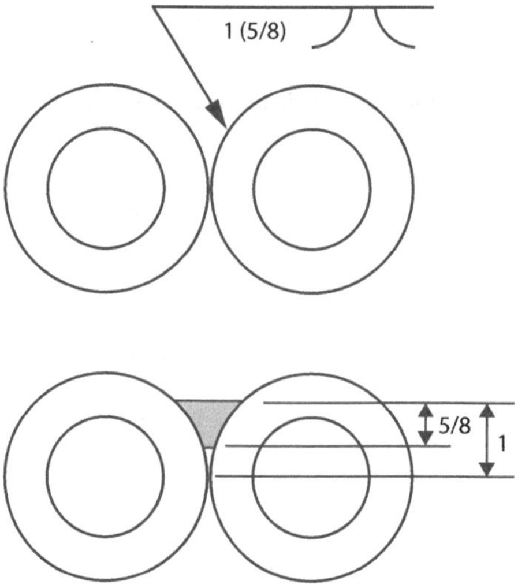

Figure 7.19 Symbol of Flare-V groove weld and as built weld.

Flare bevel groove weld
Commonly used to join a round or curved piece to a flat piece. As with the flare-V, the depth of the groove formed by the two curved surfaces and the intended depth of the weld itself are given to the left of the symbol, with the weld depth shown in parentheses. The symbol's perpendicular line is always drawn on the left side, regardless of the orientation of the weld itself.

Common supplementary symbols used with groove welds are the **melt-thru** and **backing bar** symbols. Both symbols indicate that complete joint penetration is to be made with a single-sided groove weld. In the case of melt-thru, the root is to be reinforced with weld metal on the back side of the joint. The height of the reinforcement, if critical, is indicated to the left of the melt-thru symbol, which is placed across the reference line from the basic weld symbol.

When a backing bar is used to achieve complete joint penetration, its symbol is placed across the reference line from the basic weld symbol. If the bar is to be removed after the weld is complete, an "R" is placed within the backing bar symbol. The backing bar symbol has the same shape as the plug or slot weld symbol, but context should always make the symbol's intention clear.

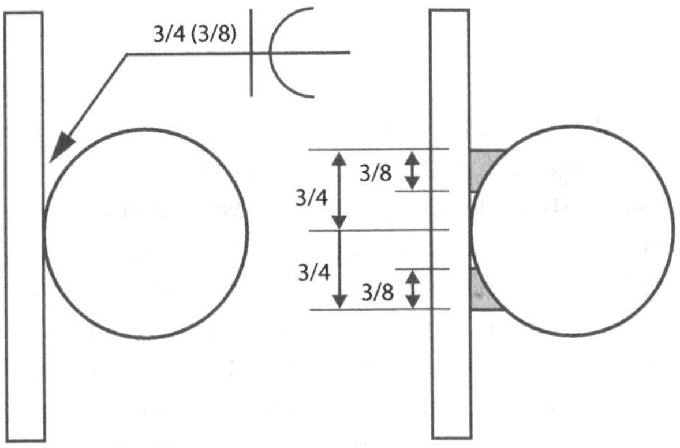

Figure 7.20 Symbol of and as built flare bevel and the weld.

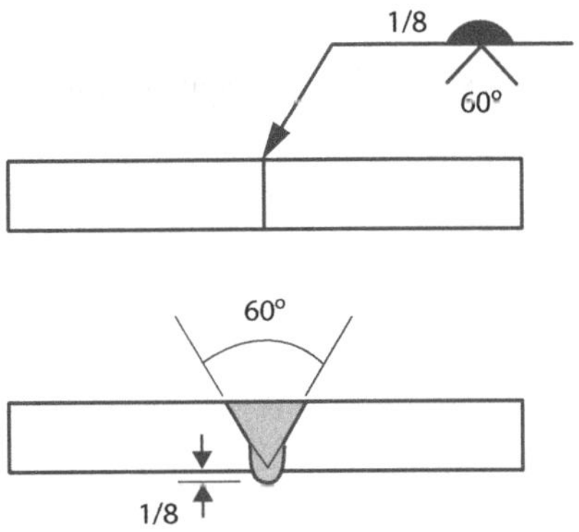

Figure 7.21 Shows the melt-thru weld.

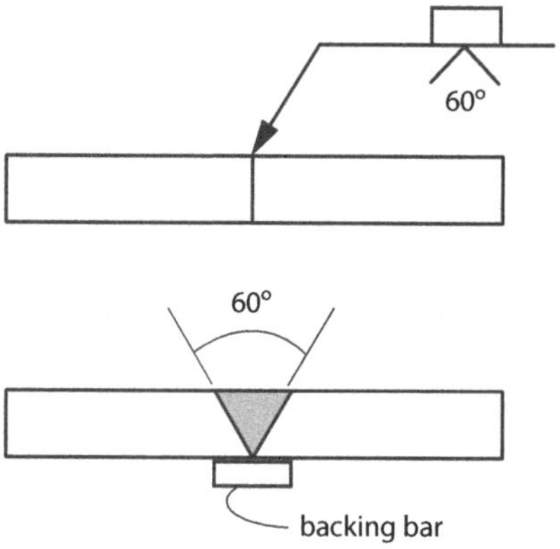

Figure 7.22 Shows the supplementary symbol of backing bar for the weld.

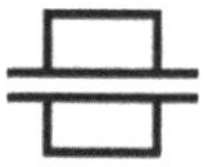

Figure 7.23 Symbol of a plug weld.

Plug and slot welds

Plug welds and **slot welds** are used to join overlapping members, one of which has holes (round for plug welds, elongated for slot welds) in it. Weld metal is deposited in the holes and penetrates and fuses with the base metal of the two members to form the joint. (Note: for the sake of graphical clarity, the drawings below do not show the penetration of the weld metal. Recognize, however, that the degree of penetration is important in determining the quality of the weld.) For plug welds, the diameter of each plug is given to the left of the symbol and the plug-to-plug spacing (pitch) is given to the right. For slot welds, the width of each slot is given to the left of the symbol, the length and pitch (separated by a dash) are given to the right of the symbol, and a detail drawing is referenced in the tail. The number of plugs or slots is given in parentheses above or below the weld symbol. The arrow-side and other-side designations indicate which piece contains the hole(s). If the hole is not to be completely filled with weld metal, the depth to which it is to be filled is given within the weld symbol.

For more information, see ANSI/AWS A2.4, Symbols for Welding and Nondestructive Testing.

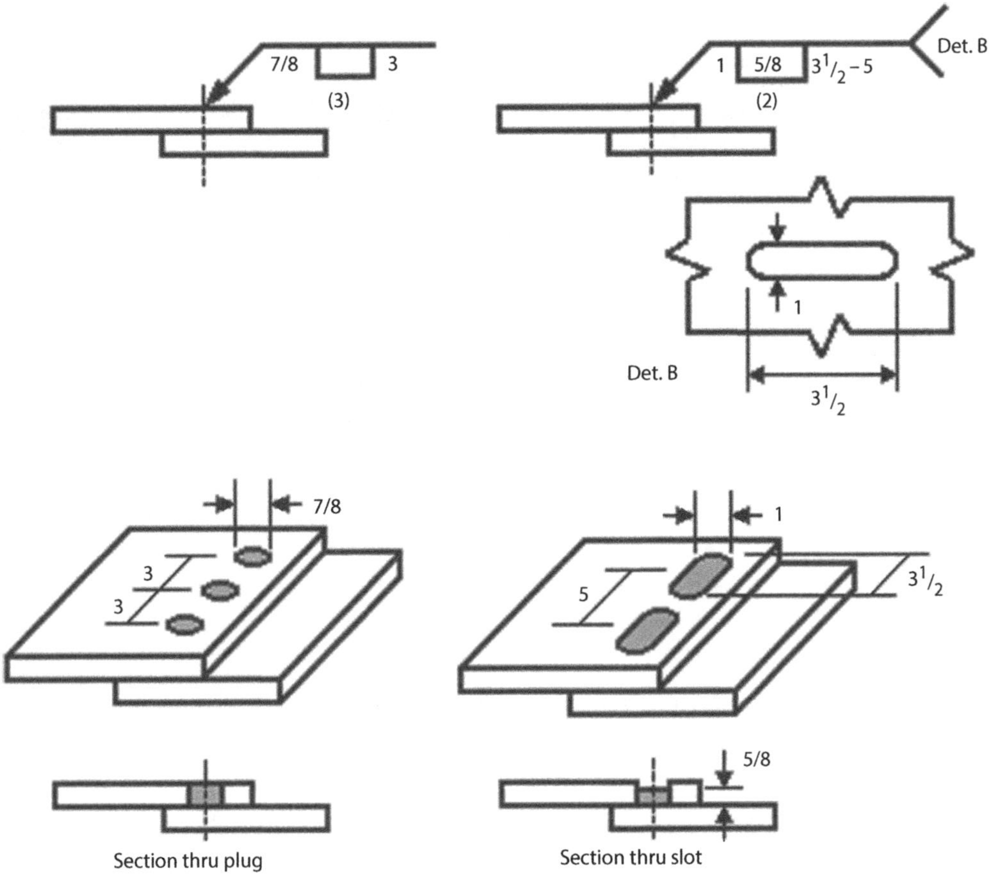

Figure 7.24 Shows symbols of plug and slot welds, with weld sizes, spacing and depth of the weld.

7.4 Pipe Schedule

The following table of pipe schedule as copied from ANSI/ASME B36.10 and also ANSI/ASME 36.19 gives the nominal size of pipes from 1/8 inch to 48 inch with actual outside diameter, and the wall thickness. The table also includes the metric (DN) sizes equivalent to the US customary units as in column 2.

The Table 7.1 below, is useful for selecting suitable pipe sizes.

Table 7.1 Pipe schedule.

Steel Pipe Dimensions Chart ANSI B36.10 & 36.19

Nominal Pipe Size		Outside Diameter (mm)	Nominal Wall Thickness Schedule																
NPS	DN	OD	SCH 5s	SCH 10s	SCH 10	SCH 20	SCH 30	SCH 40s	SCH STD	SCH 40	SCH 60	SCH 80s	SCH XS	SCH 80	SCH 100	SCH 120	SCH 140	SCH 160	SCH XXS
1/8	6	10.3		1.24				1.73	1.73	1.73		2.41	2.41	2.41					
1/4	8	13.7		1.65				2.24	2.24	2.24		3.02	3.02	3.02					
3/8	10	17.1		1.65				2.31	2,31	2.31		3.20	3.20	3.20					
1/2	15	21.3	1.65	2.11				2.77	2.77	2.77		3.73	3.73	3.73				4.78	7.47
3/4	20	26.7	1.65	2.11				2.87	2.87	2.87		3.91	3.91	3.91				5.56	7.82
1	25	33.4	1.65	2.77				3.38	3.38	3.38		4.55	4.55	4.55				6.35	9.09
1 1/4	32	42.2	1.65	2.77				3.56	3.56	3,56		4.85	4.85	4.85				6.35	9.70
1 1/2	40	48.3	1.65	2.77				3.68	3.68	3.68		5.08	5.08	5.08				7.14	10.15
2	50	60.3	1.65	2.77				3.91	3.91	3.91		5.54	5.54	5.54				8.74	11.07
2 1/2	65	73	2.11	3.05				5.16	5.16	5.16		7.01	7.01	7.01				9.53	14.02
3	80	88.9	2.11	3.05				5.49	5.49	5.49		7.62	7.62	7.62				11.13	15.24
3 1/2	90	101.6	2.11	3.05				5.74	5.74	5.74		8.08	8.08	8.08					
4	100	114.3	2.11	3.05				6.02	6.02	6.02		8.56	8.56	8.56		11.13		13.49	17.12
5	125	141.3	2.77	3.40				6.55	6.55	6.55		9.53	9,53	9.53		12.70		15.88	19.05
6	150	168.3	2.77	3.40				7.11	7.11	7.11		10.97	10.97	10.97		14.27		18.26	21.95
8	200	219.1	2.77	3.76		6.35	7.04	8.18	8.18	8.18	10.31	12.70	12.70	12.70	15.09	18.26	20.62	23.01	22.23
10	250	273.1	3.40	4.19		6.35	7.80	9.27	9.27	9.27	12.70	12.70	12.70	15.09	18.26	21.44	25.40	28.58	25.40
12	300	323.9	3.96	4.57		6.35	8.38	9.53	9.53	10.31	14.27	12.70	12.70	17.48	21.44	25.40	28.58	33.32	25.40
14	350	355.6	3.96	4.78	6.35	7.92	9.53		9.53	11.13	15.09		12.70	19.05	23.83	27.79	31.75	35.71	
16	400	406.4	4.19	4.78	6.35	7.92	9.53		9.53	12.70	i6.66		12.70	21.44	26.19	30.96	36.53	40.49	
18	450	457.2	4.19	4.78	6.35	7.92	11.13		9.53	14.27	19.05		12.70	23.83	29.36	34.93	39,67	45.24	
20	500	508	4.78	5.54	6.35	9.53	12.70		9.53	15.09	20.62		12.70	26.19	32.54	38.10	44.45	50.01	
22		559	4.78	5.54	6.35	9.53	12.70		9.53		22.23		12.70	28.58	34.93	41.28	47.63	53.98	
24	600	610	5.54	6.35	6.35	9.53	14.27		9.53	17.48	24.61		12.70	30.96	38.89	46.02	52.37	59.54	
26		660			7.92	12.70			9.53				12.70						
28	700	711			7.92	12.70	15.88		9.53				12.70						
30		762	6.35	7.92	7.92	12.70	15.88		9.53				12.70						
32	800	813			7.92	12.70	15.88		9.53	17.48			12.70						
34		884			7.92	12.70	15.88		9.53	17.48			12.70						
36	900	914			7.92	12.70	15.88		9.53	19.05			12.70						
38		965							9.53				12.70						
40	1000	1016						9.53		12.70									
42		1067				12.70	15.88		9.53	19.05			12.70						
44	1100	1118							9.53				12.70						
46		1168							9.53				12.70						
48	1200	1219							9.53			12.70							

7.5 Terms and Abbreviations

The Table 7.2 below lists some of the welding related terms and their explanations.

The table below lists some welding and construction related abbreviations that are commonly used in high demand welds in oil and gas industry.

Table 7.2 Terms and abbreviations relating to welding and construction.

Welding related abbreviations used	Expanded version of abbreviations	Explanation of abbreviations (If required) or street names
LEL	Low Emission Level	
UEL	Upper Emission Level	
H_2S	Hydrogen Sulfide gas	
GTAW	Gas Tungsten Arc Welding	Noun: Normally referred in various names as, TIG welding, or Argon arc welding etc.
GMAW	Gas Metallic Arc Welding	Noun: Often referred as MIG welding.
FCAW	Flux Core Arc Welding	Noun: No known street name for this process
*p*GMAW	Pulsed Gas Metallic Arc welding	Noun: A variant of GMAW described above, in that pulsed arc is used (the prefix p is for pulsed)
STT	Surface Tension Transfer	Noun: A variant of GMAW described above, in that the metal globule is transferred to weld pool by metal's surface tension. (this is a short circuit arc process)
CMT	Cold Metal Transfer	Noun: A variant of GMAW described above, in that the metal globule is transferred to weld pool at the point where the arc cycle is at its lowest point. (this is a short circuit arc process)
PAW	Plasm Arc Welding	Noun:
1G, 5G etc.	Are various positions of welding	European terminology is also included.
*p*WPS	Propped Welding Procedure specification	Noun: A proposal of welding procedure that can be changed and modified before actual testing is conducted.
WPS	Welding Procedure specification	Noun: A document prepared by the welding contractor to say that this is the way "I am going to weld to meet all specified properties of the required weld". The proof of the pudding is in eating it, so the contractor welds test coupons and it is tested to the required standards and results are analyzed to see if weld holds up to the required standards.

(Continued)

Table 7.2 Terms and abbreviations relating to welding and construction. (*Continued*)

Welding related abbreviations used	Expanded version of abbreviations	Explanation of abbreviations (If required) or street names
PQR	Procedure qualification Records	Records of the data collected during the welding and post welding inspection and testing. Note that a WPS is just a proposal without a supporting PQR, or number of PQRs.
WPQT	Welders' qualification Test	It is a term often used to describe Welder's qualification test.
IPW	Internal Pipe Welder	A machine that can be used to weld a pipe from inside. This is a nonstandard term used by some welding contractors.
HAZ	Heat Affected Zone	A narrow band of heated section of the parent metal along the weld, that has different metallurgical structures than either the weld or the parent metal, often an issue of possible hardness and failure.
NDT or NDE	Non-Destructive testing, or Nondestructive Examination	
UT	Ultra-sonic testing	
AUT	Automatic Ultrasonic testing	
RT	Radiographic testing	This is a generic term that comprises both the X-Ray and Gamma-Ray versions.
FSH	Full screen height	Term is sued to describe the level of sonic amplitude in UT testing.
PWHT	Post Weld heat treatment	A metallurgical process of controlled heating and cooling to take advantage of allotropic properties of material, in this case steel, this is done to achieve the desired properties from material. In case of welding, it is either stress relief (Normalizing) or Tempering or both.
RWC	Responsible Welding Coordinator	A term used in EN ISO 3834 for a person who is responsible for quality of all welding products in a fabrication set up.
QA	Quality Assurance	
WQAS	Welding Quality Assurance System	As described in EN ISO 3834

7.5.1 ASME Section IX QW 432 - F Number Table for Carbon and Alloy Steel

Other F number for Austenitic and Duplex steels, Aluminum, Nickel and alloys, Copper and alloys can be found in the ASME section IX QW 432. Table 7.3 below shows the F-Number and corresponding ASME specification and AWS Classification.

Table 7.3 F-Number, ASME specification and AWS classification.

Grouping of electrode and welding rods (wires) for qualification		
F number	**ASME specification**	**AWS classification**
1	SFA 5.1	E XX20
1	SFA 5.1	E XX22
1	SFA 5.1	E XX24
1	SFA 5.1	E XX27
1	SFA 5.1	E XX28
1	SFA 5.4	E XXX(X)-25
1	SFA 5.4	E XXX(X)-26
1	SFA 5.5	E XX20-X
1	SFA 5.5	E XX27-X
2	SFA 5.1	E XX12
2	SFA 5.1	E XX13
2	SFA 5.1	E XX14
2	SFA 5.1	E XX19
2	SFA 5.5	E (X) XXX11-X
3	SFA 5.1	EXX10
3	SFA 5.1	EXX11
3	SFA 5.5	E(X)XX10-X
3	SFA 5.5	E(X)XX11-X
4	SFA -5.1	EXX15
4	SFA -5.1	EXX16
4	SFA -5.1	EXX17
4	SFA -5.1	EXX18M
4	SFA -5.1	EXX48
4	SFA -5.4 Other than Austenitic and Duplex	EXXX(X)-15
4	SFA -5.4 Other than Austenitic and Duplex	EXXX(X)-16
4	SFA -5.4 Other than Austenitic and Duplex	EXXX(X)-17

7.6 Procedure Qualification Range as Per the Material Group

The Table 7.4 below lists P-numbers for some very common materials.

Table 7.4 P-number, group number, and type of material.

P-number	Group number	Type of material
1	1	Carbon Steel
1	2	C-Mn Steel
1	3	Carbon Steel Castings
3	1	Cr-Mo (½Cr –½Mo)
4	1	Cr-Mo (1 Cr – ½ Mo)
5	1	Cr-Mo (2-1/4 Cr – 1 Mo)
3	3	Mn – Mo Steel
5	2	Cr – Mo (5Cr – ½ Mo)
6	1	13 Cr. Alloy Forgings (Flanges)
6	2	Wrought 15 Cr. Alloy steel
6	3	13 Cr. Forgings (Check details in the QW 422)
6	4	Cr. Ni Alloy castings (13 Cr. – 4 Ni)
7	1	Alloy steel tubes (12 Cr. – 1 Al)
7	2	Alloy Steel plates (17 Cr. - Ti)
7	3	St. Steel Bars (Type XM-30 Annealed)
8	1	Alloy flanges (18 Cr. – 8 Ni)
8	2	Alloy Pipe and flanges (25Cr. – 20 Ni)

P number from 9A, 9B, 10A, 10B that goes up to P number 61, they are assigned to numerous other materials.

7.7 Material Qualification Rage for Procedure Qualification Based on P-Numbers

Base metal used for procedure qualification and that qualification covers for several other materials in the same P number grouping. The following table is from ASME Section IX QW 424. The Table 7.5 below lists the qualification of metal according to the weld coupon welded for the procedure qualification.

Table 7.5 Qualification of metals based on the procedure qualification.

Base metal used for the procedure qualification	Base metal qualified
One metal from a P-number to any metal from that same P-number	Any metal assigned that P-number
One metal from a P-number to any metal from any other P-number	Any metal assigned to the first P-number to any metal assigned the second P-number.
One metal from P-number 3 to any metal from P-number 3	Any P-number 3 metal to any P-number 3 or P-number 1 metal.
One metal from P-number 4 to any metal from P-number 4	Any P-number 4 metal to any other P-number 4, 3, or 1 metal
One metal from P number 5A to any metal from P-number 5A	Any P 5A metal to any metal from P 5A, 4, 3, or 1 metal
One metal from P 5A to a metal from P 4, or P 3, or P-number 1	Any P 5A metal to any metal from P-4, P-3 or P1 metal
One metal from P4, to a metal from P3 or P1	Any P4, metal to any metal assigned P3 or P1 number.
Any unassigned metal to the same unassigned metal	The unassigned metal to itself.
Any unassigned metal to any P number	The unassigned metal to any metal assigned the same P numbers in the qualification.
Any unassigned metal to any other unassigned metal	The first unassigned metal to the second unassigned metal.

7.8 Temperature Conversion

Following table gives the conversion of temperatures in degrees, from Fahrenheit to Celsius and Kelvin. The Table 7.6 is in two columns.

Table 7.6 Temperature conversion.

Temperature conversion Fahrenheit to Celsius and Kelvin					
F°	C°	K°	F°	C°	K°
0	32	269.2	510	266	538.7
10	-12	224.9	520	271	544.2
20	-7	230.5	530	277	549.8
30	-1	236.0	540	282	555.3

(Continued)

Table 7.6 Temperature conversion. (*Continued*)

Temperature conversion Fahrenheit to Celsius and Kelvin					
F°	C°	K°	F°	C°	K°
40	4	241.6	550	288	560.9
50	10	247.1	560	293	566.5
60	16	252.7	570	299	572.0
70	21	258.3	580	304	577.6
80	27	263.8	590	310	583.1
90	32	269.4	600	316	588.7
100	38	274.9	610	321	594.2
110	43	280.5	620	327	599.8
120	49	286.0	630	332	605.3
130	54	291.6	640	338	610.9
140	60	297.1	650	343	616.4
150	66	302.7	660	349	622.0
160	71	308.3	670	354	627.6
170	77	313.8	680	360	633.1
180	82	319.4	690	366	638.7
190	88	324.9	700	371	644.2
200	93	330.5	710	377	649.8
210	99	336.0	720	382	655.3
220	104	341.6	730	388	660.9
230	110	347.1	740	393	666.4
240	116	352.7	750	399	672.0
250	121	358.2	760	404	677.6
260	127	363.8	770	410	683.1
270	132	369.4	780	416	688.7
280	138	374.9	790	421	694.2
290	143	380.5	800	427	699.8
300	149	386.0	810	432	705.3

(*Continued*)

Table 7.6 Temperature conversion. (*Continued*)

Temperature conversion Fahrenheit to Celsius and Kelvin					
F°	C°	K°	F°	C°	K°
310	154	391.6	820	438	710.9
320	160	397.1	830	443	716.4
330	166	402.7	840	449	722.0
340	171	408.2	850	454	727.5
350	177	413.8	860	460	733.1
360	182	419.4	870	466	738.7
370	188	424.9	880	471	744.2
380	193	430.5	890	477	749.8
390	199	436.0	900	482	755.3
400	204	441.6	910	488	760.9
410	210	447.1	920	493	766.4
420	216	452.7	930	499	772.0
430	221	458.2	940	504	777.5
440	227	463.8	950	510	783.1
450	232	469.3	960	516	788.7
460	238	474.9	970	521	794.2
470	243	480.5	980	527	799.8
480	249	486.0	990	532	805.3
490	254	491.6	1000	538	810.9
500	260	497.1	1010	543	816.4

7.9 Useful Calculations

Carbon Equivalent

$$C_{eq} \text{ (or } C_E) = C + Mn + \{(Cr + Mo - V)/5\} + (Ni + Cu)/15$$

Percent Ferrite

$$\% F = 3 \; (C_{eq} - 0.93 \; Ni_{eq} - 6.7)$$

Determine Carbon equivalent (Ceq)

$$Ceq = C + Mn/6 + (Ni + Cu)/15 + (Cr + Mo + V)/5$$

Determine Pre heating temperature

$$C_t = Ceq * 0.005 * t_{min}$$

$$C_E = Ceq + C_t$$

Now the pre heat temperature °C = 350 $\sqrt{}$ (C_E -0.25)

7.10 Effect of Temperature on Gas Cylinder Pressure

The pressure within the cylinder varies with the variation in temperature. In colder temperature the pressure drops and in hotter temperature the pressure rises. For example, a full cylinder at 21°C (about 71°F) at full 6.9 m^3 (244-Ft^3) capacity, would have 15,169 kPa (2200 psi). The Table 7.7 below gives incremental pressure with changes with temperature of oxygen gas cylinder of 6.9 m^3 (244-Ft^3) capacity.

Table 7.7 Temperature and pressure.

Temperature O_2		**Pressure gauge reading**	
Celsius °C	**Fahrenheit °F**	**kPa**	**Psi(g)**
38	100	16,030	2325
32	90	15,741	2283
27	80	15,458	2242
21	70	15,169	2200
16	60	14,879	2158
10	50	14,596	2117
5	41	14,307	2075
-1	30	14,024	2023
-7	20	13, 734	1992
-12	10	13,452	1951
-17.8	0	13,162	1909

Index

Printed in the USA/Agawam, MA
September 17, 2026

910661.045